The Endeavour Journals

From what I have said of the Natives of New-Holland they may appear to some to be the most wretched people upon Earth, but in reality they are far more happier than we Europeans; being wholly unacquainted not only with the superfluous but the necessary conveniences so much sought after in Europe, they are happy in not knowing the use of them. They live in a Tranquillity which is not disturbed by the Inequality of Condition: The Earth and sea of their own accord furnishes them with all things necessary for life; they covet not Magnificent Houses, Household-stuff &c. They live in a warm and fine Climate and enjoy a very wholesome Air, so that they have very little need of Clothing and this they seem to be fully sensible of, for many to whom we gave Cloth &c. to, left it carelessly upon the Sea beach and in the woods as a thing they had no manner of use for. In short they seemed to set no Value upon any thing we gave them, nor would they ever part with anything of their own for any one article we could offer them; this, in my opinion argues that they think themselves provided with all the necessaries of Life and that they have no Superfluities.[1]

James Cook, 23rd August 1770

The Endeavour Journals

Captain Cook in Australia

John & Clare MacDonald

First published in Great Britain in 2024 by
Pen & Sword History
An imprint of Pen & Sword Books Limited
Yorkshire – Philadelphia

Copyright © John and Clare MacDonald 2024

ISBN 978 1 39906 409 5

The right of John and Clare MacDonald to be identified as Authors of this Work has been asserted by them in accordance with the Copyright, Designs and Patents Act 1988.

A CIP catalogue record for this book is available from the British Library

All rights reserved. No part of this book may be reproduced or transmitted in any form or by any means, electronic or mechanical including photocopying, recording or by any information storage and retrieval system, without permission from the Publisher in writing.

Typeset by Mac Style
Printed in the UK by CPI Group (UK) Ltd, Croydon, CR0 4YY.

Pen & Sword Books Limited incorporates the imprints of After the Battle, Atlas, Archaeology, Aviation, Discovery, Family History, Fiction, History, Maritime, Military, Military Classics, Politics, Select, Transport, True Crime, Air World, Frontline Publishing, Leo Cooper, Remember When, Seaforth Publishing, The Praetorian Press, Wharncliffe Local History, Wharncliffe Transport, Wharncliffe True Crime and White Owl.

For a complete list of Pen & Sword titles please contact

PEN & SWORD BOOKS LIMITED
47 Church Street, Barnsley, South Yorkshire, S70 2AS, England
E-mail: enquiries@pen-and-sword.co.uk
Website: www.pen-and-sword.co.uk
or
PEN AND SWORD BOOKS
1950 Lawrence Rd, Havertown, PA 19083, USA
E-mail: Uspen-and-sword@casematepublishers.com
Website: www.penandswordbooks.com

For Clare

A Girl for all Seasons

Sea-Fever

I must go down to the seas again,
to the lonely sea and the sky,
And all I ask is a tall ship
and a star to steer her by;
And the wheel's kick and the wind's song
and the white sail's shaking,
And a grey mist on the sea's face
and a grey dawn breaking.

I must go down to the seas again,
for the call of the running tide
Is a wild call and a clear call
that may not be denied;
And all I ask is a windy day
with the white clouds flying,
And the flung spray and the blown spume,
and the sea-gulls crying.

I must go down to the seas again,
to the vagrant gypsy life,
To the gull's way and the whale's way
where the wind's like a whetted knife;
And all I ask is a merry yarn
from a laughing fellow-rover,
And quiet sleep and a sweet dream
when the long trick's over.

- by John Masefield

Contents

Contents by Geographical Division	x
Colour Plates	xi
The Endeavour Journals	xvii
The Authors	xviii
Prologue	xix

Continent of Smoke   1

Chapter 1	A Certain Sign the Country is Inhabited	2
Chapter 2	Resolved to Dispute our Landing to the Utmost	27
Chapter 3	All They Seemed to Want Was For Us To Be Gone	41
Chapter 4	To Try To Form Some Connections With The Natives	51
Chapter 5	A Great Quantity of Smoke	69
Chapter 6	Mister Orton He Is A Man Not Without Faults	86
Chapter 7	Not One Drop of Which We Could Find	105
Chapter 8	Fear of Death Now Stared Us in the Face	127
Chapter 9	Nor Have We Seen One Since We Have Been in Port	150
Chapter 10	Their Unaccountable Timidity	164
Chapter 11	Two of Them Embarked and Came Towards the Ship	176
Chapter 12	Our Very Good Friends	188
Chapter 13	A Countenance Full of Disdain	193
Chapter 14	Lumber Not Worth Carriage	203

Chapter 15	Cape Flattery	214
Chapter 16	The Indians Had Been Here	226
Chapter 17	Great Dangers Swallow Up Lesser Ones	231
Chapter 18	More Happier Than We Europeans	245

Epilogue	262
Glossary of Terms	279
Ship's Muster on Endeavour	291
Notes	301
Acknowledgments	311

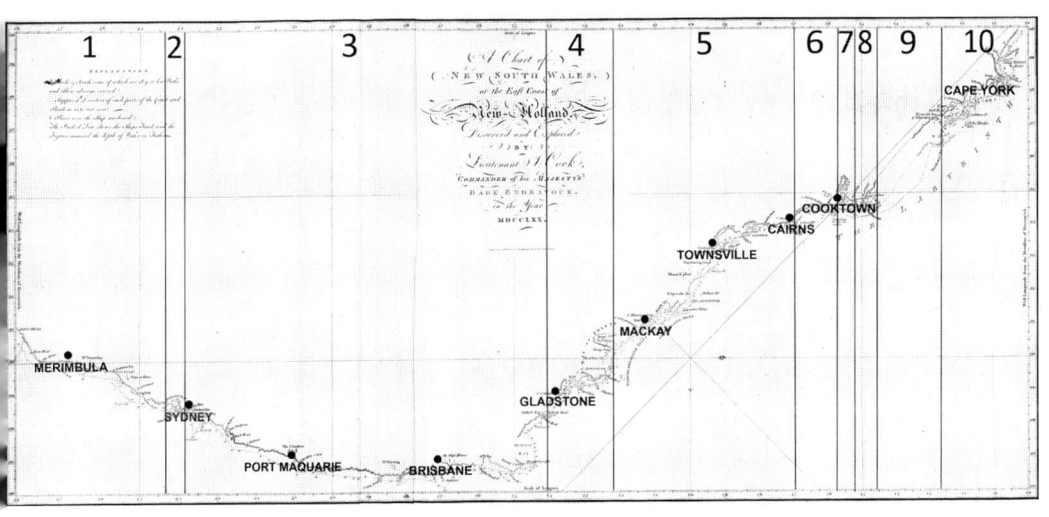

James Cook's chart of New South Wales.[1]

Contents by Geographical Division

1.	Point Hickes To Botany Bay	2–26
2.	Botany Bay	27–68
3.	Botany Bay To Bustard Bay	69–88
4.	Bustard Bay To Thirsty Sound	89–104
5.	Thirsty Sound To Cape Grafton	105–126
6.	Cape Grafton To Endeavour River	127–149
7.	Endeavour River	150–213
8.	Endeavour River To Lizard Island	214–231
9.	Lizard Island To Providential Channel	232–244
10.	Providential Channel To Cape York	245–261

Colour Plates

1. (a) *Holothuria obtusata* – now *Physalia physalis*. Pencil with some watercolour (Sydney Parkinson 1770). Catalogue of Natural History drawings commissioned by Joseph Banks on the *Endeavour* Voyage 1768–1771. Vol 3. 261. (3:41). Courtesy of Trustee of the Natural History Museum, London.
    (b) *Mimus volutator* – now *Glacus atlantius* – Blue sea slug. Watercolour (Sydney Parkinson 1770). Catalogue of Natural History drawings commissioned by Joseph Banks on the *Endeavour* Voyage 1768–1771. Vol 3. 240.(3:23). Courtesy of Trustee of the Natural History Museum, London.
    (c) *Medusa pelagica* – now *Pelagia noctiluca*. Watercolour (Sydney Parkinson 1770). Catalogue of Natural History drawings commissioned by Joseph Banks on the *Endeavour* Voyage 1768–1771. Vol 3. 274.(3:54). Courtesy of Trustee of the Natural History Museum, London.
2. Giant cuttlefish – *Sepia apama*. Courtesy of Diveplanit Travel.
3. Cabbage tree – *Livistona australs*. Courtesy of Schomynv. From Wikimedia Commons, the free media repository. Public Domain in UK. Also, PD-US.
4. *Epacris longiflora*. Outline drawing (Sydney Parkinson 1770) and finished drawing (James Miller). Catalogue of Natural History drawings commissioned by Joseph Banks on the *Endeavour* Voyage 1768–1771. Vol 1. No A5/225. Courtesy of Trustee of the Natural History Museum, London.
5. *Lambertia Formosa*. Outline drawing (Sydney Parkinson 1770) and finished drawing (John Frederick Miller 1773). Catalogue of Natural History drawings commissioned by Joseph Banks on the *Endeavour* Voyage 1768–1771. Vol 1. No A7/316. Courtesy of Trustee of the Natural History Museum, London.
6. Rainbow lorikeet (Moses Griffith 1772.) Courtesy of National Library of Australia.
7. Magenta Lilly Pilly – *Syzygium paniculatum*. Courtesy of John Tan. From Wikimedia Commons, the free media repository. Public Domain in UK. Also, PD-US.
8. Brown quail – *Synoicus ypsilophorus*. Courtesy of J.J. Harrison. From Wikimedia Commons, the free media repository. Public Domain in UK. Also, PD-US.

9. Warrigal greens – Now *Tetragonia tetragonioides*. Courtesy of Ixitixel. From Wikimedia Commons, the free media repository. Public Domain in UK. Also, PD-US.
10. Wedge-tailed shearwater – *Puffinus pacificus*. Courtesy of Mike Prince. From Wikimedia Commons, the free media repository. Public Domain in UK. Also, PD-US.
11. Brown booby – Sula *leucogaster plotus*. Courtesy of Aviceda. From Wikimedia Commons, the free media repository. Public Domain in UK. Also, PD-US.
12. Pandanus – *Pandanus tectorius*. Courtesy of Eric Guinther. From Wikimedia Commons, the free media repository. Public Domain in UK. Also, PD-US.
13. *Dendrobium discolor*. Outline drawing (Sydney Parkinson 1770) and finished drawing (Fredrick Polydore Nodder 1780). Catalogue of Natural History drawings commissioned by Joseph Banks on the *Endeavour* Voyage 1768–1771. Vol 1. No A8/366. Courtesy of Trustee of the Natural History Museum, London.
14. Large-leafed orange mangrove – *Bruguiera gymnorhiza*. Finished drawing (Frederick Polydore Nodder 1777) based on the outline drawing (Sydney Parkinson 1770). From Wikimedia Commons, the free media repository. Public Domain in UK. Also, PD-US.
15. Weaver ant – *Oecophylla smaragdina*. Courtesy of PHGCOM. From Wikimedia Commons, the free media repository. Public Domain in UK. Also, PD-US.
16. Cup moth caterpillars – *Doratifera stenora* (*Limacodidae*). Courtesy of Jeff Wright, Queensland Museum.
17. Australian pelicans – *Pelecanus conspicillatus temminck*. Courtesy of Jinesh PS. From Wikimedia Commons, the free media repository. Public Domain in UK. Also, PD-US.
18. White-headed shelduck – *Tadorna radjah rufitergum*. Courtesy of Ben's Waterfowl.
19. Australian bustard – *Ardeotis australis* (*Otididae*). Courtesy of J.J. Harrison. From Wikimedia Commons, the free media repository. Public Domain in UK. Also, PD-US.
20. Sucker shark – *Echeneis neucrates*. Unfinished watercolour (Sydney Parkinson 1770). Catalogue of Natural History drawings commissioned by Joseph Banks on the *Endeavour* Voyage 1768–1771. Vol 3. 79.(2:). Courtesy of Trustee of the Natural History Museum, London.
21. 'Egg bird' – Crested tern – *Sterna bergii*. Courtesy of Gopala Krishna A. From Wikimedia Commons, the free media repository. Public Domain in UK. Also, PD-US.

Colour Plates xiii

22. Spear grass – *Heteropogon contortus*. Courtesy of Bernard Dupont. From Wikimedia Commons, the free media repository. Public Domain in UK. Also, PD-US.
23. Nest of Tree termite – *Nasutitermes walker*. Courtesy of Robert Webster. From Wikimedia Commons, the free media repository. Public Domain in UK. Also, PD US
24. Blue tiger butterfly – *Tirumala hamata hamata*. Courtesy of Daniela. From Wikimedia Commons, the free media repository. Public Domain in UK. Also, PD-US.
25. Silver pupa and butterfly of the two-brand crow – *Euploea sylvester sylvester*. Courtesy of Museums Victoria (askus@museum.vic.gov.au) and Atlas of Living Australia – CC-BY – Queensland Museum.
26. Marine snail – now *Architectonica perspectiva*. Courtesy of Nick Hobgood. From Wikimedia Commons, the free media repository. Public Domain in UK. Also, PD-US.
27. Silver lined mud-skipper – *Gobiidae, Periophthalmus argentilineatus*. Courtesy of Bernard Dupont. From Wikimedia Commons, the free media repository. Public Domain in UK. Also, PD-US.
28. Spurred mangrove – *Ceriops tagal*. Outline drawing (Sydney Parkinson 1770) and finished drawing (unknown artist). Catalogue of Natural History drawings commissioned by Joseph Banks on the *Endeavour* Voyage 1768–1771. Vol 1. No A3/117. Courtesy of Trustee of the Natural History Museum, London.
29. Audubon's shearwater – now *Puffinus l'herminieri*. Courtesy of Dominic Sherony from Wikimedia Commons, the free media repository. Public Domain in UK. Also, PD-US.
30. *Hibiscus meraukensis*. Outline drawing (Sydney Parkinson 1770) and finished drawing (Frederick Polydore Nodder 1778). Catalogue of Natural History drawings commissioned by Joseph Banks on the *Endeavour* Voyage 1768–1771. Vol 1. No A1/24. Courtesy of Trustee of the Natural History Museum, London.
31. *Planchonia careya* – Cocky apple. Outline drawing (Sydney Parkinson 1770) and finished drawing (Frederick Polydore Nodder 1777). Catalogue of Natural History drawings commissioned by Joseph Banks on the *Endeavour* Voyage 1768–1771. Vol 1. No A3/146. Courtesy of Trustee of the Natural History Museum, London.
32. Endeavour River from Grassy Hill. Courtesy of John MacDonald.
33. Topknot Pigeon – *Lopholaimus antarcticus*. Courtesy of Julia Burgher.
34. Flying fox – *Pteropus sp*. By Andrew Mercer. From Wikimedia Commons, the free media repository. Public Domain in UK. Also, PD-US.

35. Kale/taro – *Colocasia esculenta*. Courtesy of Wildfeuer. From Wikimedia Commons, the free media repository. Public Domain in UK. Also, PD-US.
36. Wild plantain – *Musa acuminata subsp. Banksia*. Courtesy of Miya.m. From Wikimedia Commons, the free media repository. Public Domain in UK. Also, PD-US.
37. Burdekin plum – *Pleiogymium cerasiferum*. Courtesy of Tatiana Gerus. From Wikimedia Commons, the free media repository. Public Domain in UK. Also, PD-US.
38. Threadfin Salmon – *Eleutheronema tetradactylum*. Unfinished watercolour (Sydney Parkinson 1770) Endeavour River. Catalogue of the Natural History drawings commissioned by Joseph Banks on the *Endeavour* Voyage 1768–1771. Part 3: Zoology. Catalogue number 203. (2:111). Courtesy of Trustee of the Natural History Museum, London.
39. Giant clam – *Tridacna gigas*. Courtesy of John MacDonald.
40. Blue-black urchin – *Echinothrix diadema*. Courtesy of Florence Trentin. From Wikimedia Commons, the free media repository. Public Domain in UK. Also, PD-US.
41. 'Garfish' – Barred Longtom – *Ablennes hains*. Watercolour (Sydney Parkinson 1770) Courtesy of National Library of Australia Call Number –PIC MSR 14/1/6 PIC Solander Box Small Items #PIC/20894.
42. Coconut-opening crab – *Birgus latro*. Courtesy of fearlessRich. From Wikimedia Commons, the free media repository. Public Domain in UK. Also, PD-US.
43. Encrusted coconut on north shore beach Endeavour River. Courtesy of John MacDonald.
44. Endeavour River. Courtesy of John MacDonald.
45. Sea hibiscus – *Hibiscus tiliaceus*. Courtesy of Dr. Avishai Teicher. From Wikimedia Commons, the free media repository. Public Domain in UK. Also, PD-US.
46. Dingo – *A portrait of a large dog from New Holland* (George Stubbs, 1772). Commissioned by Joseph Banks and based on his observations of dingoes on the east coast of New Holland in 1770. From Wikimedia Commons, the free media repository. Public Domain in UK. Also, PD-US.
47. Fresh water mussels – *Velesunio wilsoni*. Courtesy of John MacDonald.
48. Whistling tree duck – *Dendrocygna arcuate*. Courtesy of B.J. Hensen.
49. Estuarine crocodile – *Crocodilus porosus*. Courtesy of Djambalawa. From Wikimedia Commons, the free media repository. Public Domain in UK. Also, PD-US.
50. *Cochlospermum gillivraei* – Kapok. Outline drawing (Sydney Parkinson 1770) and finished drawing (Frederick Polydore Nodder 1778). Catalogue of

Natural History drawings commissioned by Joseph Banks on the *Endeavour* Voyage 1768–1771. Vol 1. No A1/13. Courtesy of Trustee of the Natural History Museum, London.

51. Common baler shell – *Melo amphora*. Courtesy of Marcus Stigwan. From Wikimedia Commons, the free media repository. Public Domain in UK. Also, PD-US32.
52. *Xanthorrhoea resinosa* – Grass tree. Outline drawing (Sydney Parkinson 1770) and finished drawing (James Miller 1775). Catalogue of Natural History drawings commissioned by Joseph Banks on the *Endeavour* Voyage 1768–1771. Vol 1. No A8/389. Courtesy of Trustee of the Natural History Museum, London.
53. *Macropus robustus* – Eastern wallaroo. Watercolour of the skull and lower jaw of specimen shot on 27th July at Endeavour River in 1770 (Nathanial Dance). Catalogue of the Natural History drawings commissioned by Joseph Banks on the *Endeavour* Voyage 1768–1771. Part 3: Zoology. Catalogue number 5.(1:5). Courtesy of Trustee of the Natural History Museum, London.
54. Green turtle – *Chelonia mydas*. Courtesy of Bernard Dupont. From Wikimedia Commons, the free media repository. Public Domain in UK. Also, PD-US.
55. Turtle Grass – *Thalassia hemprichii*. Courtesy of Mudasir Zainuddin. From Wikimedia Commons, the free media repository. Public Domain in UK. Also PD-US.
56. Loggerhead turtle – *Caretta caretta*. Courtesy of Strobilomyces. From Wikimedia Commons, the free media repository. Public Domain in UK. Also, PD-US.
57. Native Yam – *Dioscorea transversa*. Courtesy of Mark Marathon. From Wikimedia Commons, the free media repository. Public Domain in UK. Also, PD-US.
58. Typical Endeavour River bush land. Courtesy of John MacDonald.
59. Native cashew – now *Semecarpus australiensis*. Courtesy of Steve Fitzgerald. From Wikimedia Commons, the free media repository. Public Domain in UK. Also, PD-US.
60. *The Kongouro of New Holland*. Oil painting (George Stubbs 1772). Commissioned by Joseph Banks and based on the inflated skin of an animal shot and collected on the 29th July 1770 at Endeavour River. From Wikimedia Commons, the free media repository. Public Domain in UK. Also, PD-US.
61. *Dillenia alata*. Outline drawing (Sydney Parkinson 1770) and finished drawing (Frederick Polydore Nodder 1778). Catalogue of Natural History drawings commissioned by Joseph Banks on the *Endeavour* Voyage

1768–1771. Vol 1. No A1/1. Courtesy of Trustee of the Natural History Museum, London.
62. *Josephinia imperatricis*. Outline drawing (Sydney Parkinson 1770) and finished drawing (Frederick Polydore Nodder 1778). Catalogue of Natural History drawings commissioned by Joseph Banks on the *Endeavour* Voyage 1768–1771. Vol 1. No A6/276. Courtesy of Trustee of the Natural History Museum, London.
63. Yellow-spotted monitor – *Varanus panoptes*. Courtesy of Nate Lawrence, Lizard Island Research Station.
64. Osprey – *Pandion haliaetus*. Courtesy of J J Harrison. From Wikimedia Commons, the free media repository. Public Domain in UK. Also, PD-US.
65. Orange-footed scrub fowl – *Megapodius reinwardt*. Courtesy of Toby Hudson. From Wikimedia Commons, the free media repository. Public Domain in UK. Also, PD-US.
66. Sea snake – now *Aipysurus duboisi*. Courtesy of Andrew Green, Reef Life Survey.
67. Organ pipe coral – *Tubipora musica*. Courtesy of Chaloklum Diving. From Wikimedia Commons, the free media repository. Public Domain in UK. Also, PD-US.
68. Common Noddy – *Anous stolidus pileatus*. Courtesy of J J Harrison. From Wikimedia Commons, the free media repository. Public Domain in UK. Also, PD-US.
69. King George III (Johan Zoffany 1771). From Wikimedia Commons, the free media repository. Public Domain in UK. Also, PD-US.

The Endeavour Journals: Captain Cook in Australia

The European description of New Holland and its people in 1770.

A single narrative account of the European discovery of the east coast of New Holland in 1770, taken from the combined journals of the four principal journal keepers on board His Majesty's Bark *Endeavour* – Lieutenant James Cook, Joseph Banks (gentleman naturalist), Sydney Parkinson (natural history artist) and James Magra (midshipman).

Focusing particularly on the contact with the Indigenous, and illustrated with numerous original images executed during the voyage including details of charts, plans of harbours, coastal profiles, botanical, zoological, marine, landscape and figurative drawings, and contemporary portraits of the mariners, *The Endeavour Journals* is an adaption of the discovery story made easily accessible to the general reader, without compromising the integrity of the original document.

The Authors

James Cook   Joseph Banks   Sydney Parkinson   James Magra

The Authors.[i] [ii] [iii]

Theirs was a voyage of science when man sought understanding through the strength of measurement and record. Be it measurement of the Earth's distance from the Sun, or measurement from cape to cape, measurement was all. Of the depth of water, of the speed of travel, of the shape of landforms, of the strength of currents, of the Earth's magnetism and the variation of compass. And the measurement of the hour itself, where the ship's bell counted the division of the day. So too, they recorded the new flora and fauna, and took the measure of man himself, of his differing beliefs, his practices, and his cultures.

There is a maxim in writing of others. We only see what qualities in them we can imagine in ourselves. We are best to let those others speak for themselves.

JRM

Their story contains a miscellany of all that interests me. Adventure, danger, bravery, excellence, discovery, risk, intellect, skill, stoicism, talent, perfection, humanity, both artistic and scientific endeavour, all driven by astonishing imagination. In conclusion: a summation of all that is great about being human.

CBM

Prologue

Dates are as Cook gives them for whom the new day begins at noon. Underlined text refers to locations marked on associated maps. Some expressions and spellings are left as original, and supported by a Glossary of Terms.

Earl of Pembroke.[1]

The Vessel

The *Endeavour* began life as the *Earl of Pembroke* in the shipbuilding yards of Whitby, a seaside town in Yorkshire, northern England, famous for the sturdy colliers built there to service the coal trade between Newcastle and London. With her large storage capacity and shallow draft, she was ideal for the long voyage of exploration the Royal Navy had in mind.

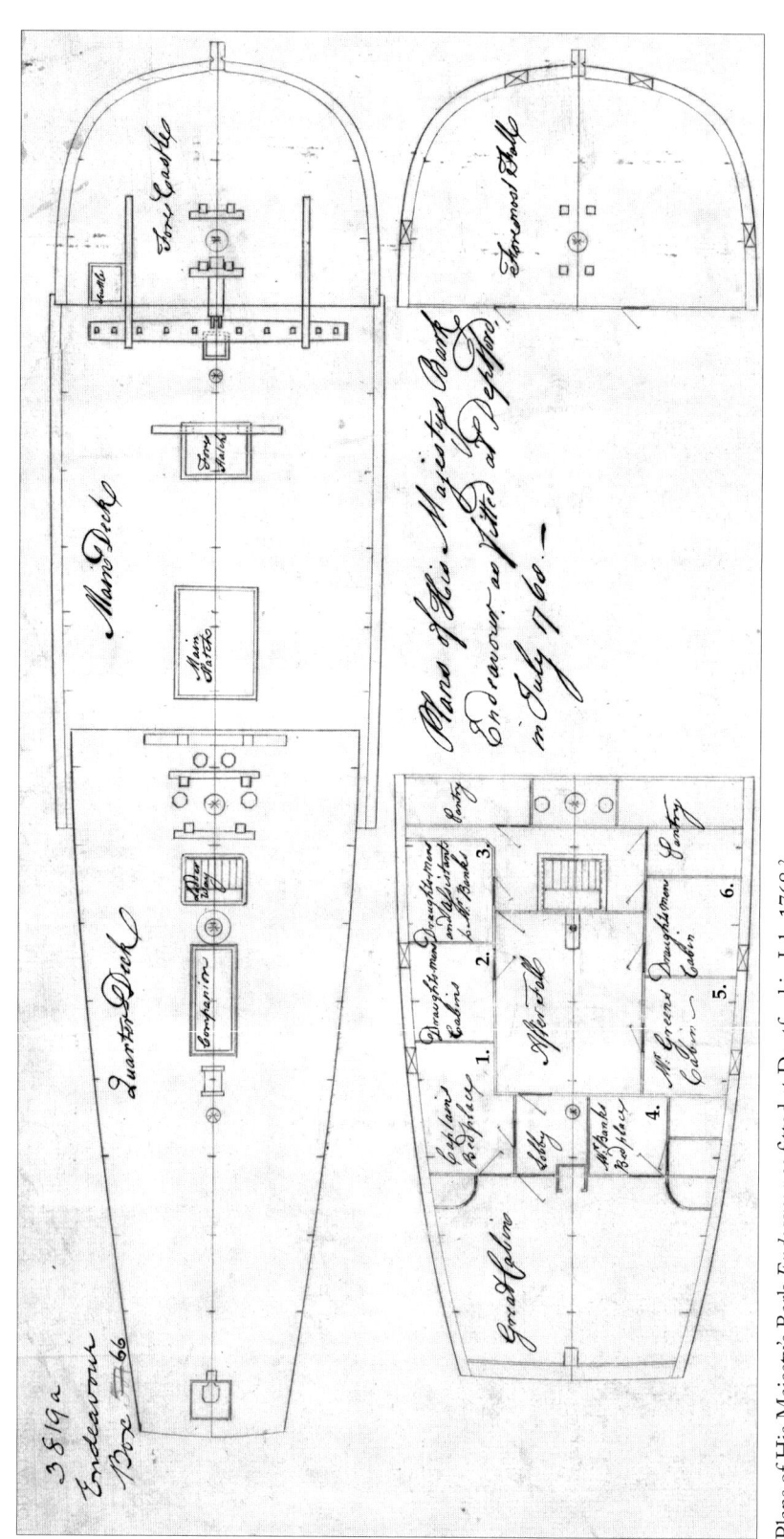

Plans of His Majesty's Bark *Endeavour* as fitted at Deptford in July 1768.[2]

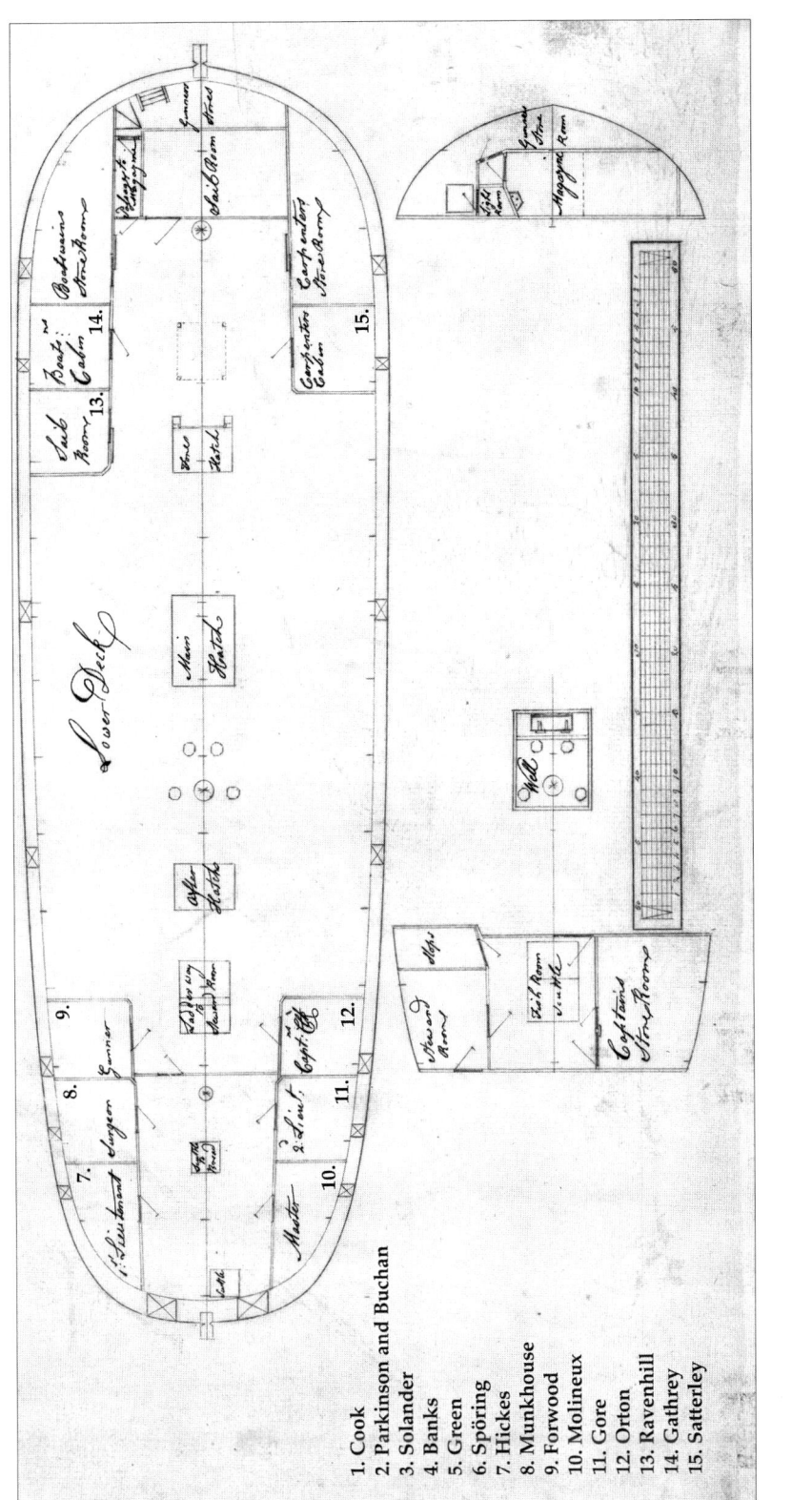

In March of 1768 she was purchased for £2,840 and sent up the Thames to Deptford Yard to be refitted for her new role. There she underwent an elaborate conversion to ready her for a scientific expedition deep into the little-known waters of the South Seas.

The Ship's Company and their Voyage to New Holland (Australia)[3]

His Majestys Bark *Endeavour* by Samuel Atkins (1794).[4]

On 25th August 1768, now recommissioned as His Majesty's Bark *Endeavour*, the ship sailed from Plymouth with ninety-four souls on board. A Royal Navy contingent of officers, midshipmen, carpenters, sailmakers, their mates and servants, and threescore of ordinary seamen, along with twelve marines, 'and near 18 Month' Provisions, 10 Carriage Guns, 12 Swivels, with good Store of Ammunition and Stores of all kinds'. **(Cook)**

In command of all was **First Lieutenant James Cook**, born in Marton, Yorkshire, now 39, with twenty years of seagoing service behind him; much of it in colliers. Cook had served thirteen years in the Royal Navy before receiving his *Endeavour* command. Those years had seen him in active service surveying the St Lawrence River (present-day Canada) in the weeks before the capture of Quebec from the French, and afterward the labyrinthine coast of Newfoundland and southern Labrador. His remarkable skills in surveying had been noted by his superiors. And the Royal Society had invited him to submit his observations of the eclipse of the sun to their elite assembly.

The Royal Society had set the ball rolling on the *Endeavour* enterprise. A transit of Venus was shortly to occur. This very rare astronomical event would not be seen again until 1874. It promised to advance man's understanding of the universe and was not to be missed. King George III had given his support and ordered a ship to be prepared. Cook was to command the *Endeavour* and be one of the Society's official observers.

This was a most unusual expedition. Aboard were a number of civilians. A team of naturalists and artists under the leadership of a young and wealthy landowner, plant enthusiast, 25-year-old **Joseph Banks**.

Banks's interest in the natural sciences had driven him since boyhood. In 1766, at 23, he was elected to the Royal Society, and in that same year, when Cook was in Newfoundland and Labrador, Banks too was there, studying the local natural history. Did they meet? We can only guess.

Banks was a born networker and had friends in high places. The 4th Earl of Sandwich was one of them. With his lordship's help, Banks secured a passage on the *Endeavour*, for himself and his entourage of assistants. He hired a small staff of scientists and draftsmen to complete his team, Sydney Parkinson was among them.

Sydney Parkinson was a Quaker and hailed from Edinburgh. He was 23, with no formal training in art, but he had proved his ability painting the collection of birds and insects Banks had brought back from Newfoundland, and Banks hired him for the voyage. Parkinson was indefatigable. During the voyage he made 955 drawings of flora – 675 sketches and 280 finished drawings. Of these, 233 were drawings of Australian plants and 377 drawings of Australian fauna.

James Mario Magra was 22. He joined the Royal Navy in New York in 1761. He was a member of a prominent Corsican family. Magra was educated in England, and there he found himself enlisted as a midshipman on the *Endeavour*. On the voyage, Magra became acquainted with Joseph Banks, and their friendship lasted until his death.

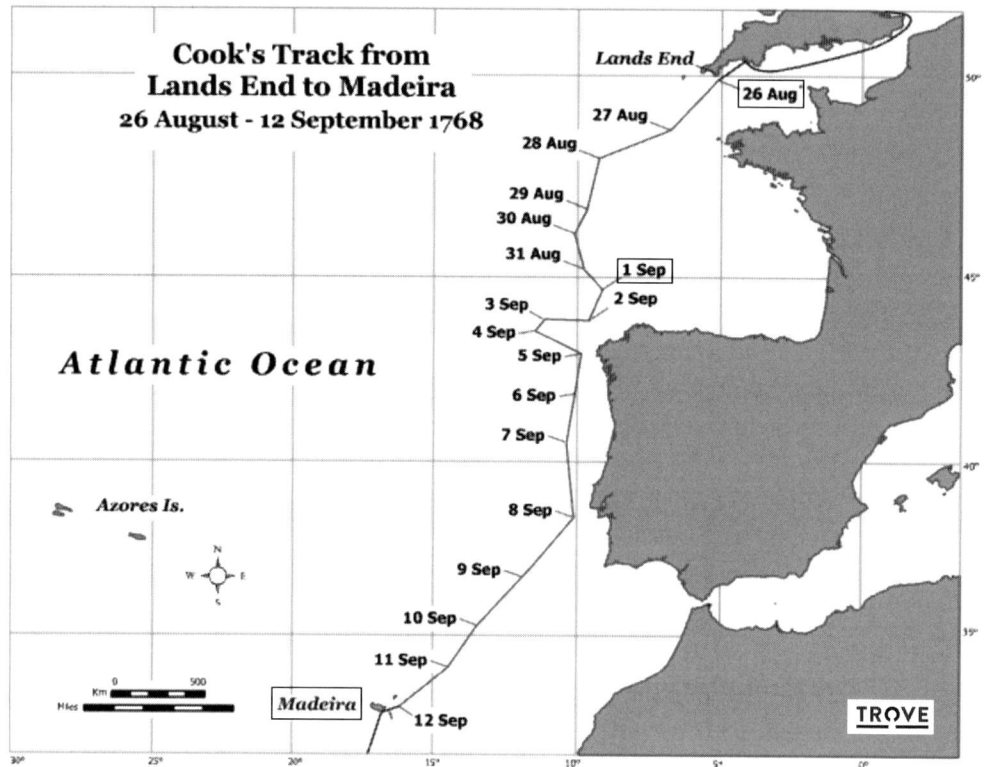

Cook's Track from Lands End to Madeira.[5]

<u>26th August</u> – The *Endeavour* set sail on a course for Madeira. Immediately, Banks became profoundly seasick. Despite this he and his team began collecting specimens and Parkinson began to draw them.

<u>1st September</u> – Soon the weather grew foul, a hard gale blew up, washing overboard a small boat belonging to the boatswain, John Gathrey, and drowning 3 and 4 dozen of their poultry.

John Gathrey was a warrant officer responsible for the masts, yards, sails, rigging, anchors, boats and cordage. The distinctive boatswain's pipe or whistle was used to issue signals to the crew, and the boatswain was assisted by boatswain's mates.

Forby Sutherland was 34 and Able Seaman. He was the ship's poulterer, in charge of looking after their fowl, which included ducks, chickens and geese, and preparing any game birds shot for the table. He was Orcadian, an ethnic group native to the Orkney Islands.

Shortly afterwards, Thomas Richmond, one of Banks's four personal servants, let slip their precious cast-net from his wrist, which sank to the depths, forever gone from the sight of man. 'A misfortune equalling almost the worst which our enemies could have wished.' **(Banks)**

Upon arrival in Madeira, Cook lost one of his faithful. 'Mr Weir, Master's Mate, was carried overboard by the Buoy rope and to the Bottom with the Anchor. Hove up the Anchor by the Ship as soon as possible and found his Body entangled in the Buoy rope.' **(Cook)**

Banks and Dr Solander went ashore.

It was at the British Museum that Banks first met his lifelong friend, the Swede, **Daniel Solander.** Solander was 35 and the principal disciple of Linnaeus. (Carl Linnaeus was famous for his work in taxonomy: the science of identifying, naming and classifying organisms – plants, animals, bacteria, fungi and more.) With Solander's assistance, Banks published the first Linnean descriptions of the plants and animals of Newfoundland and Labrador. Dr Solander agreed to accompany him on the *Endeavour* voyage. It was to prove a bountiful expedition. Dr Solander and Banks brought home more than 30,000 botanical specimens representing more than 3,500 species. Of them, 110 genera and about 1,400 species were new to science.

But their five-day stay in Madeira proved disappointing. The flowering season was over; the country was dry as sticks. The local meat, however, was very good, as was their mutton and pork, and the beef more especially, so Cook received on board 270lb of fresh beef, a live bullock and 3,032 gallons of wine. Henry Stevens Seaman, and Thomas Dunster Marine, refused to take their allowance of fresh beef and were punished with twelve lashes each. (It was a Friday, were they Catholics who abstain from eating meat on Fridays, or just sticklers for Royal Navy regulations that stipulated that Friday was a non-meat day? We can only guess.)

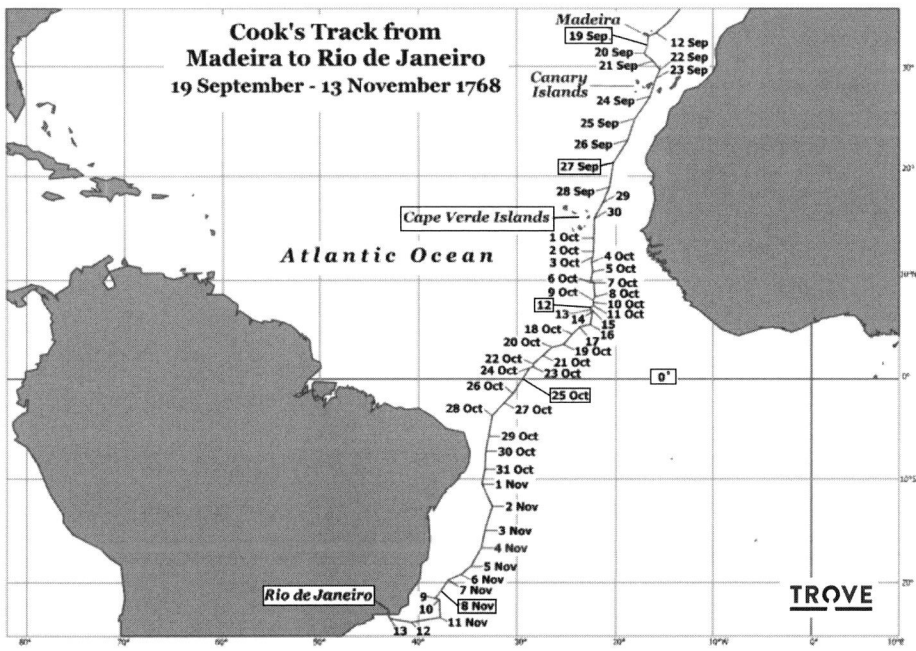

On <u>19th September,</u> at midnight, they sailed from Funchal/Fonchiale, Madeira.

<u>27th September</u> – They were now well in the north-east trade winds and sailing at 7 knots. The nights became intolerably hot as they crossed the Tropic and every cabin window was opened. For the first time they saw plenty of flying fish, their sides shining like burnished silver. One flew into Mr Green's cabin.

Charles Green was 36. In 1760, he became an assistant to James Bradley, the Astronomer Royal at Greenwich. He helped Bradley observe the Transit of Venus at Greenwich in 1761. In 1763, Green, along with the astronomer Nevil Maskelyne, was instructed by the Board of Longitude to make the voyage to Barbados to act as monitors of the test of John Harrison's H4 chronometer, which was in the running for the Longitude Prize. The prize was offered by the British government for the devisor of an accurate method of determining the longitude of a ship at sea. When Nevil Maskelyne was appointed Astronomer Royal in 1765, Green briefly served under him as assistant. Green was appointed by the Royal Society to sail on the *Endeavour* as astronomer to observe the 1769 transit of Venus.

As they passed the <u>Island of Bonavista, one of the Cape Verde Islands</u>, Alexander Buchan was employed in taking views of the land.

Alexander Buchan. His age remains uncertain. Banks took him on as the expedition's second artist to record the scenery and make a general pictorial account of the expedition, thus freeing Parkinson to concentrate on drawing the botanical and zoological specimens collected. Buchan proved to be an epileptic.

Cook and his officers, and their subordinates, attended to their daily tasks of running the ship and navigating their course. The *Endeavour* had been equipped with the very latest scientific instruments and navigational equipment, including accurate sextants.

Robert Molineux was the ship's master. His main duty was to navigate the ship under the direction of the captain. He also acted as surveyor, taking soundings and drawing charts, and he was responsible for 'trimming' the ship – distributing its load so that it sat correctly in the water. He had to ensure the safe anchorage of the *Endeavour* and oversaw the day-to-day running of the ship.

<u>12th October</u> – 'Mates and Midshipmen commanded to scrape and clean between decks, which Mr Pickersgill (only) having the spirit to refuse was ordered before the Mast.' **(Ship's Log)**

Richard Pickersgill was 19 when he joined *Endeavour*. He began the voyage as a master's mate but succeeded Robert Molineux as master after Molineux's death near Cape Town. He drew numerous charts of the lands they visited.

After the voyage, Cook wrote to the Admiralty Secretary in August 1771, 'Mr Richard Pickersgill, Master – deserving of a Lieutenants Commission.'

Earlier in 1766, Pickersgill had obtained a berth on the *Dolphin* under Samuel Wallis for a voyage to the Pacific when Tahiti was discovered.

He joined the *Resolution* as third lieutenant for Cook's second voyage, during which he played a prominent role and was something of a personality.

On Cook's third voyage, when Cook was preparing to lead an expedition to search for the Northwest Passage from the Pacific, a companion voyage was proposed to search from the Atlantic – to the west of Greenland. Pickersgill was given command of the companion voyage and allocated the armed brig *Lyon*. However, Michael Lane had been the commander of the vessel up till then and was unhappy about being replaced and having to serve as Pickersgill's number two. The expedition was badly equipped and had left far too late in the year. Pickersgill returned unsuccessful. Lane wrote to the Admiralty claiming that Pickersgill had been constantly drunk and had been incapable of carrying out his duties. A court martial on 6th February 1777 found the charges partly proven and Pickersgill was dismissed from the navy for 'drunkenness and other irregularities'.

Pickersgill may have then obtained command of a privateer. However, he drowned in the River Thames in 1779. Johann Reinhold Forster wrote: 'once, going on board his ship late in the evening, his foot slipped, and falling into Thames, he was drowned.'

For the next month they headed south-west for their next port of call, Rio de Janeiro.

On 25th October they crossed the Equator (0° on map), and the traditional ceremony took place. This was an initiation rite to commemorate a person's first crossing of the Equator:

> Every one that could not prove upon a Sea Chart that he had before crossed the Line was either to pay a bottle of Rum or be ducked in the sea. **(Cook)**

> Everybody was then called upon the quarter deck and examined by one of the lieutenants who had crossed, he marked every name either to be ducked or let off according as their qualifications directed. Captain Cooke and Doctor Solander were on the Blacklist, as were myself my servants and dogs, which I was obliged to compound for by giving the Duckers a certain quantity of Brandy for which they willingly excused us the ceremony.
>
> Many of the Men however chose to be ducked rather than give up 4 days allowance of wine which was the price fixed upon, and as for the boys they are always ducked of course; so that about 21 underwent the ceremony which was performed thus:
>
> A block was made fast to the end of the Main Yard and a long line reeved through it, to which three Cross pieces of wood were fastened, one of which

was put between the legs of the man who was to be ducked and to this he was tied very fast, another was for him to hold in his hands and the third was over his head least the rope should be hoisted too near the block and by that means the man be hurt. When he was fastened upon this machine the Boatswain gave the command by his whistle and the man was hoisted up as high as the cross piece over his head would allow, when another signal was made and immediately the rope was let go and his own weight carried him down, he was then immediately hoisted up again and three times served in this manner which was every man's allowance. Thus ended the diversion of the day, for the ducking lasted till almost night, and sufficiently diverting it certainly was to see the different faces that were made on this occasion, some grinning and exulting in their hardiness whilst others were almost suffocated and came up ready enough to have compounded after the first or second duck, had such proceeding been allowable. **(Banks)**

<u>8th November</u> – At daybreak today we made the land which proved to be the Continent of South America. **(Banks)**

Upon arrival at Rio de Janeiro, Cook sent Mr Hickes and a midshipman in the pinnace to the Portuguese Viceroy to obtain a pilot to bring them into a proper anchoring ground.

Zachary Hickes was 32, second lieutenant, and second-in-command on *Endeavour*. He was described as, 'a steady, efficient and competent officer … dependable … the soul of naval discipline'. He was already showing signs of tuberculosis when they sailed from Plymouth.

The pinnace returned to the ship without either of the officers. The viceroy had sent back the pinnace with three of his own officers in it, but no pilot. The cockswain related the news that Hickes and the midshipman were detained until Cook went ashore. They were told that this procedure was the custom of the port.

Cook went ashore in the morning to meet the viceroy and obtain leave to purchase provisions and refreshments for the ship. This was granted provided Cook employ a person to buy them. Surgeon Munkhouse was appointed to the task, and went ashore every day to buy necessaries for their table and to assist the agent in buying things for the ship.

William Brougham Munkhouse was 35. He had been in Newfoundland waters, having sailed as surgeon aboard HMS *Niger*. He remained on the Newfoundland station from 1763 to 1767. In 1767, Joseph Banks visited the island and was taken to Labrador and the northern parts by the *Niger*. Munkhouse is credited with saving the botanist's life when he fell seriously ill at Croque. Banks was reportedly, 'very ill with ague and fever and at one time not expected to recover'. Munkhouse joined *Endeavour* as surgeon; his younger brother, 20-year-old **Jonathan Munkhouse**, was also aboard and was a midshipman.

The viceroy's permission was absolute; a soldier was to accompany the boat that brought anything to or from the ship.

A second and third visit to the viceroy by Cook brought no loosening of the restriction. Instead, a paper war began between himself and the viceroy, with Banks conducting his own written remonstrations.

Things took a further turn for the worse when Cook sent Lieutenant Hickes in the pinnace, whereupon all the boat's crew were taken out of the boat by armed force, struck many times, and hurried off to prison.

Meanwhile, the *Endeavour's* longboat, which was coming on board with four casks of rum, went adrift when the weather worsened and a rope broke. Banks's small boat that was in tow was also gone (Banks had a small boat for his own pursuits), 'the loss of our long boat which we much feared was perhaps the greatest misfortune that could happen to people who were going as we were upon discoveries'. **(Banks)**

The morning saw the imprisoned crew and the pinnace returned to the ship, and on Cook's request for assistance, the viceroy sent help and, just on dark, the two lost boats were recovered.

However, the dispute continued with the viceroy arguing that he doubted the ship was one of the king's and accusing Cook's people of smuggling.

Banks decided to look out for an opportunity to smuggle himself ashore. Parkinson gives an account of one such clandestine adventure:

> Mr Banks and Dr Solander appeared much chagrined at their disappointment: but, notwithstanding all the viceroy's precautions, we determined to gratify our curiosity, in some measure, and having obtained a sufficient knowledge of the river and harbour, by the surveys that we had made of the country, we frequently, unknown to the sentinel, stole out of the cabin window at midnight, letting ourselves down into a boat by a rope; and, driving away with the tide till we were out of hearing, we then rowed to some unfrequented part of the shore, where we landed, and made excursions up into the country.

They managed to explore and were well pleased with the natural history specimens they saw and collected. But suspicion ashore grew, obliging Banks and Solander to remain aboard for the rest of the stay.

During the twenty-five days spent at Rio de Janeiro, Cook was obliged to punish John Thurman with twelve lashes for refusing to assist the sailmaker in repairing the sails. As well, Robert Anderson Seaman and William Judge Marine received twelve lashes each; the former for leaving his duty ashore and attempting to desert from the ship, and the latter for using abusive language to the Officer of the Watch. John Readon, boatswain's mate, was also punished with twelve lashes for not doing his duty in punishing the above two men. Punishments were often a daily fact of life at sea.

Having replenished their supplies with water, fresh beef, yams, fruit and greens for the ship's company (which they were served every day during their stay), on 2nd December they weighed and came to sail and turned down the bay. Unfortunately, 'Mr Flower, an experienced seaman, fell from the main shrouds into the sea, and was drowned before we could reach him.' **(Parkinson)**

Peter Flower had one of the longest naval relationships with Cook. He had served with Cook for five years in Newfoundland, and had been one of Cook's two assistants who helped him survey the islands of St Pierre and Miquelon.

On their departure, to their great surprise the Portuguese fort fired two shots at them, one of which went just over their mast.

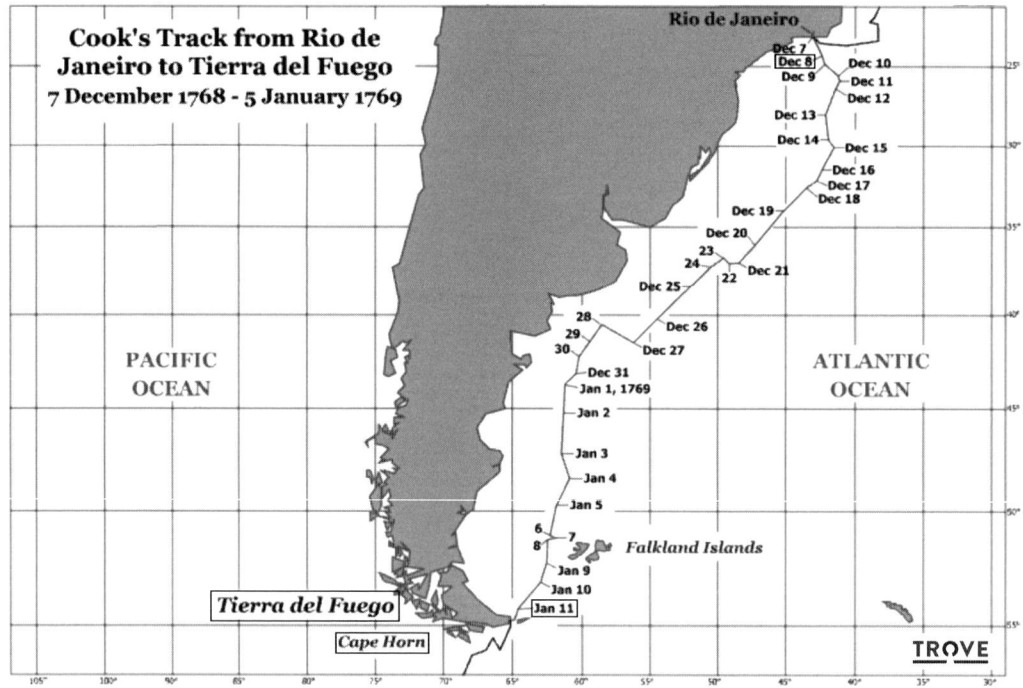

By 8th December they were finally at sea and the ship's routine fell into place again. As they headed south-west down the coast of South America, the weather began to cool. A new suit of sails was bent for Cape Horn, to be ready for the bad weather they were to expect.

Christmas day. All good Christians that is to say all hands get abominably drunk so that at night there was scarce a sober man in the ship, wind thank god very moderate or the lord knows what would have become of us. **(Banks)**

> New Year's Day today made us pass many compliments and talk much of our hopes for success in the year 69. Many whales were about the ship today. **(Banks)**
>
> Large bunches of seaweed floated past the ship, they caught some with hooks, the leaves were four feet long and the stalk about twelve. Lieutenant Gore said he had seen this weed grow quite to the top of the water in 12 fathom [22 metres].

John Gore was American, 38 years of age, and a sailor of great experience. He had twice circumnavigated the world. First as midshipman under Captain John Byron (Lord Bryon's grandfather) on *Dolphin,* and a second time as master's mate under Captain Wallis, when they discovered Tahiti (King George Island), again on *Dolphin.* Gore joined *Endeavour* as third lieutenant, and third in command. It was his third trip to the Pacific. Gore was a crack shot with a musket and signed his journal 'The Master Hunter'. He became friends with Joseph Banks, assisting the naturalists by shooting specimens for their collections.

> The Southeast wind now became very cold, to us at least so lately come from the Torrid Zone. All hands bend their Magellan Jackets (made of a thick woollen stuff called fearnought). **(Banks)**

On 11th January 1769, at daybreak, they saw the land of Tierra del Fuego, and three days later Banks and Solander went ashore and got their first taste of its natural history, collecting 100 plant specimens, every one new and entirely different to any they had seen before.

They anchored in the Bay of Success. Here they met the indigenous Fuegians, three of whom came on board and were shown around the ship. Parkinson and Buchan drew and painted the people and their conical houses. A survey was done of the bay and their water casks filled.

While this was happening, Banks, with his four servants (Peter Briscoe, James Roberts, Thomas Richmond and George Dorlton), Dr Solander, Mr Munkhouse, Mr Green, Mr Buchan and two seamen set off very early in the morning, endeavouring to get to the top of some hills where they could see a place that was not overgrown with trees.

The weather being 'vastly fine much like a sunshiny day in May' **(Banks)** continued fine as they climbed, but by 3 o'clock it began to worsen, becoming intensely cold with frequent blasts of snow. They became stranded with no hope of reaching the ship before dark, so decided to seek shelter in some woods and build a fire. The cold increased 'beyond what I have ever experienced' **(Banks)**. The party began to straggle and disperse. Dr Solander collapsed and lay down in the snow and had to be carried. Banks's two 'negro' servants, Thomas Richmond and George Dorlton, who had lagged behind, were lost. By morning, they were found frozen to death.

These two men being intrusted with great part of the Liquor that was for the whole party had made too free with it and stupefied themselves to that degree that they either could or would not travel but laid themselves down in a place where there was not the least thing to shelter them from the inclemency of the night. **(Cook)**

Four days later, they departed the Bay of Success and headed for Cape Horn. They rounded the cape eight days later:

without ever being brought once under our close reefed Topsails since we left Strait La Maire a circumstance that perhaps never happened before to any Ship in those seas so much dreaded for hard gales of wind insomuch that the doubling of Cape Horn is thought by some to be a mighty thing. **(Cook)**

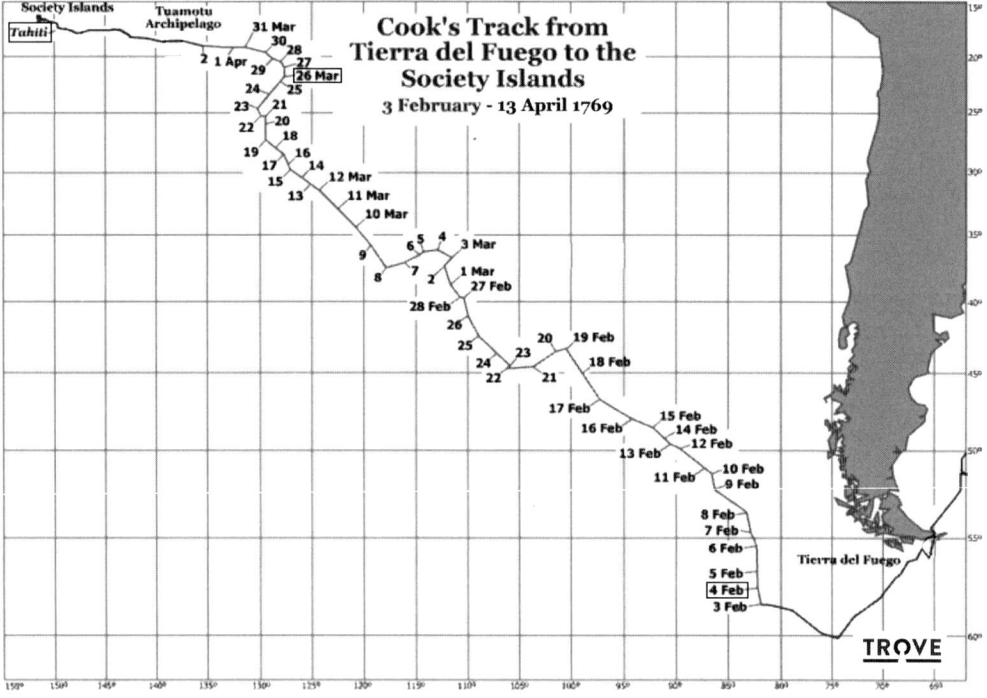

It was 25th January. Ahead lay the vast expanse of the Pacific Ocean. On 4th February they turned north-west and set their course for the long haul to King George Island (Tahiti).

Banks resumed his collecting, Solander his classifying and Parkinson his sketching.

They shot and ate albatrosses:

> which were so good that everybody commended and Eat heartily of them though there was fresh pork upon the table. The way of dressing them is thus: Skin them overnight and soak their carcases in Salt water till morn, then parboil them and throw away the water, then stew them well with very little water and when sufficiently tender serve them up with Savoury sauce. **(Banks)**

The weather warmed and Banks began the new month 'by pulling off an under waistcoat'.

For ten long weeks they held their course to the north-west, out of sight of any land with only the sea for company.

On 26th March, as they recrossed the Tropic, a tragedy occurred. **William Greenslade**, marine, went overboard and was drowned.

> The following circumstances makes it appear as though it was done designedly, he had been sentinel at the steerage door between 12 and 4 o'clock where he had taken part of a seal skin put under his charge which was found upon him. The other Marines thought themselves hurt by one of their party committing a crime of this nature, and he being a raw young fellow, and is very probable made him resolve upon committing this rash action; for the Sergeant, not being willing that it should pass over unknown to me, was about 7 o'clock going to bring him aft to have it inquired into when he gave him the slip between decks and was seen to go upon the Fore Castle, and from that time was seen no more. **(Cook)**

> He was a very young man scarce 21 years of age, remarkably quiet and industrious, and to make his exit the more melancholy was drove to the rash resolution by an accident so trifling that it must appear incredible to everybody who is not well acquainted with the powerful effects that shame can work upon young minds. **(Banks)**

John Edgecumbe was marine sergeant from 48th Company, Plymouth Division. He was the senior marine on board. Cook described Edgecumbe as, 'A very good soldier very much of a gentleman and well deserving of promotion in the Marine Service'.

On arrival at King George Island (Tahiti) they were welcomed by a great number of the natives in their canoes. An elderly man whose name was Owhaa came on board and was recognised by several of those who had already visited Tahiti in the *Dolphin*. These included John Gore, Robert Molineux, Richard Pickersgill, Francis Wilkinson and Samuel Evans.

Francis Wilkinson was 21 and one of the master's mates.

Samuel Evans was 32. He joined *Endeavour* as quartermaster, when boatswain John Gathrey died on the 4th February 1771. Evans was appointed boatswain on the following day.

The politics of the island had changed since the *Dolphin* visit and the old order, where Queen Oboreah ruled, had been overthrown. However, the Tahitian habit of stealing remained, and their visit began tragically when a midshipman ordered the marines to fire on the crowd, a sentry's musket was stolen, and a man was killed. 'When Mr Banks heard of the affair, he was highly displeased, saying, "If we quarrelled with those Indians, we should not agree with angels."' **(Parkinson)**

Soon after their arrival, Alexander Buchan, an epileptic, died, 'an ingenious and good young man'. **(Banks)** He was buried at sea as a precaution in case a land burial offended the Tahitians.

A substantial land fort, complete with trenches and gun emplacements, was erected to house and protect their shore equipment. 'Within, we had an observatory, an oven, forge, and pens for our sheep. Sentinels were also appointed as usual in garrisons, and military discipline observed.' **(Parkinson)**

Banks and Dr Solander stationed themselves near the fort and took charge of trading, receiving hogs, breadfruit, coconuts, bunches of plantains and fruit in exchange for axes, hatchets, linen cloth, beads, looking-glasses, knives and such things as were valued. 'The rates, or terms, on which we trafficked with the natives, were a spike for a small pig; a smaller for a fowl; a hatchet for a hog; and twenty cocoa-nuts, or bread-fruit, for a middling-sized nail.' **(Parkinson)**

On 28th April, a woman, 'a fat, bouncing, good-looking dame', paid them a visit. Mr Molineux, as soon as he saw her, declared her to be the *Dolphin*'s queen, Oboreah:

> Our attention was now entirely diverted from every other object to the examination of a personage we had heard so much spoken of in Europe: She appeared to be about 40, tall and very lusty, her skin white and her eyes full of meaning, she might have been handsome when young but now few or no traces of it were left. **(Banks)**

A chief came aboard and dined in the Great Cabin. While the others ate, he 'sat in the chair like a statue without once attempting to put one morsel to his mouth', **(Cook)** until it was realised he was without his handmaiden to feed food to him. Cook ordered one of the servants to supply the want.

During the three months they were to remain on the island, thefts followed one upon another. Pockets were picked; an earthen vessel was stolen out of Parkinson's cabin; Banks lost his pocket pistols; a rake was cleverly lifted from out of the fort; a barrel went missing, then a horse pistol; the top of the lightening-chain was pilfered; and while they slept some missed their stockings, others their jackets and waistcoats, and Banks lost his white jacket and waistcoat, with silver frogs, in the pockets of which were a pair of pistols. Reprisals were enacted. A large number of canoes were seized. Some items were recovered, but some were never returned.

The most vital of things that were stolen was the quadrant. After a long chase, the quadrant was returned. It was damaged, but Herman Spöring managed to repair it in time for the observation of the transit of Venus.

Herman Spöring was 35. He had worked as a draughtsman and watchmaker before becoming Daniel Solander's personal clerk, and when Dr Solander joined the *Endeavour* expedition he joined as Banks's secretary. Spöring transcribed Solander's notes on the flora and fauna, and after the death of the artist Buchan, he took on a crucial role drawing coastal profiles and natural history specimens.

Peace was restored, and they were entertained with displays of wrestling, and trading resumed.

Cultural differences were arresting:

> This Morning a Man and two young women with some others came to the Fort whom we had not seen before: Mr Banks was as usual at the gate of the Fort trading with the people, when he was told that some Strangers were coming and therefore stood to receive them, the company had with them about a Dozen young Plantains Trees and some other small Plants these they laid down about 20 feet from Mr Banks, the people then made a lane between him and them, when this was done, the Man who appeared to be only a Servant to the 2 Women / brought the young Plantains singly, together with some of the other Plants and gave them to Mr Banks, and at the delivery of each pronounced a Short sentence, which we understood not, after he had thus disposed of all his Plantain trees he took several pieces of Cloth and spread them on the ground, one of the Young Women then stepped upon the Cloth and with as much Innocence as one could possibly conceive, exposed herself entirely naked from the waist downwards, in this manner she turned her Self once or twice round, then stepped off the Cloth and dropped down her clothes, more Cloth was then spread upon the Former and she again performed the same ceremony; the Cloth was then rolled up and given to Mr Banks and the two young women went and embraced him which ended the Ceremony. **(Cook)**
>
> I took her by the hand and led her to the tents accompanied by another woman her friend, to both of them I made presents but could not prevail upon them to stay more than an hour. **(Banks)**

A spike nail was the price expected by the Tahitian women for their favours. Almost every man had a paramour, including 'shy-boots' Parkinson. Banks had his amour:

> In the evening Oboreah, and her favourite attendant Othéothéa payed us a visit, much to my satisfaction as the latter (my flame) has for some days been reported either ill or dead.
>
> Oboreah seems to us to act in the character of a Ninon d'Enclos who satiated with her lover resolves to change him. I am offered if I please to supply his place, but I am at present otherwise engaged; indeed was I free as air her majesties person is not the most desirable. **(Banks)**

Cook remained celibate.

The Tahitians were invited to witness the Europeans' Sunday divine service, in which they took part, kneeling and sitting in imitation of their visitors:

> this day closed with an odd Scene at the Gate of the Fort where a young Fellow above 6 feet high lay with a little Girl about 10 or 12 years of age publicly before several of our people and a number of the Natives. What makes me mention this, is because, it appeared to be done more from Custom than Lewdness, for there were several Women present particularly Oboreah and several others of the better sort and these were so far from shewing the least disapprobation that they instructed the girl how she should act her part, who young as she was, did not seem to want it. **(Cook)**

Worm [Teredo worm – *Teredo navalis*] had eaten the bottom out of the longboat. Her whole bottom needed to be replaced. This was work for carpenter John Satterley and his crew.

John Satterley, 'a man much esteemed by me and every Gentleman on board'. **(Cook)**

On one of their excursions, they witnessed the Tahitians riding the giant waves on makeshift surf skis fashioned out of the stern of an old canoe.

As the day of the transit approached, all eyes turned to the weather. To increase the chance of clear skies for the observation they divided into three teams and headed for three different locations. Gore, Dr Munkhouse and Spöring went to York Island; Hickes, Pickersgill, Clarke (Masters Mate) and Saunders (Midshipman) went in the pinnace to the Eastward; while Cook, Green and Dr Solander observed from the fort on Point Venus. Except for 'an Atmosphere or dusky shade round the body of the Planet which very much disturbed the times of the contacts' **(Cook)**, 'the observation was attended with as much success as Mr Green and the Captain could wish'. **(Banks)**

Gore took part in an archery contest. His rival, Tubourai ('one of their chiefs'), shot his arrow 274 yards (250 metres), but Gore was a more accurate shot.

They met a company of travelling musicians, a band, which consisted of two flutes and three drums, the drummers accompanying their music with their voices. 'They sung many songs generally in praise of us, for these gentlemen like Homer of old must be poets as well as musicians.' **(Banks)** The Europeans replied with songs of their own.

There was some disagreement between Banks and Mr Munkhouse about two of Oboreah's attendants who were very persistent in getting themselves husbands. Surgeon Munkhouse took one, and one of the lieutenants the other. The girls seemed agreeable enough till bedtime, and then they determined to lie in Banks's tent. The surgeon insisted that one of them should not sleep there and thrust her out. 'Mr Monkhouse and Mr Banks came to an éclaircissement sometime after; had very high words, and I expected they would have decided it by a duel, which, however, they prudently avoided.' **(Parkinson)**

They learned to eat dog. 'Captain Cook, Mr Banks, and Dr Solander, commended it highly, saying it was the sweetest meat they had ever tasted; but the rest of our people could not be prevailed on to eat any of it.' **(Parkinson)**

> Here is a species of rats, of which there are great numbers in this island; we caught some of them and had them fried; most of the gentlemen in the bell-tent ate of them and commended them much; and some of the inferior officers ate them in a morning for breakfast. **(Parkinson)**

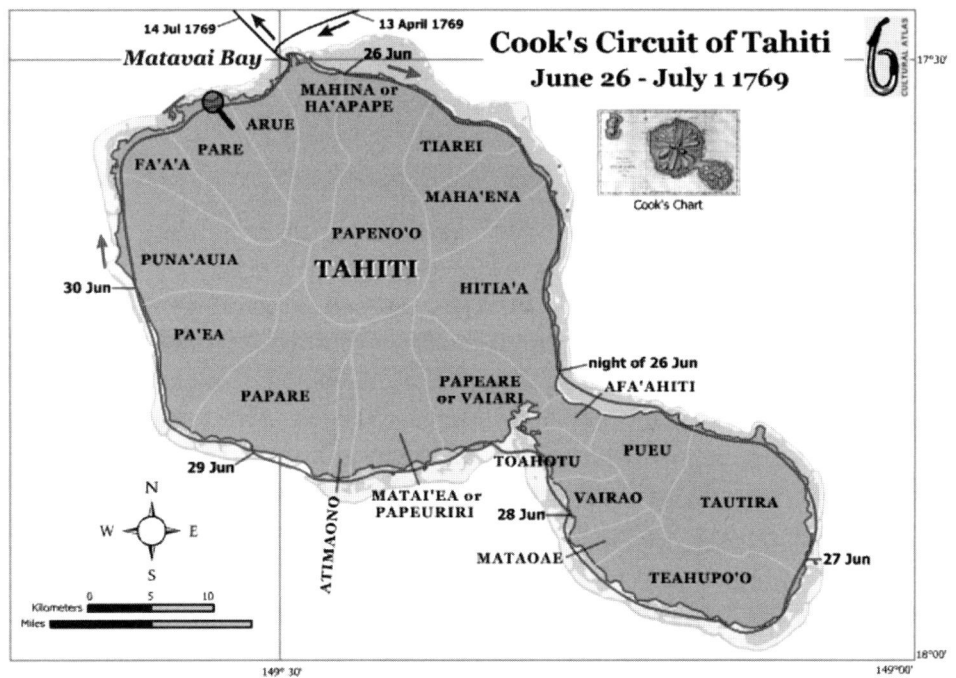

Cook and Banks circumnavigated and charted the island. During their expedition, they discovered a spectacular Mariai. It was pyramidal, stood 44 feet high (13 metres), 207 feet in length (82 metres) and 71 wide (22 metres), and was made of large blocks of carved and polished coral rocks. This was the Mariai of the deposed Queen Oboreah. Numberless human bones of her defeated forces lay scattered all around.

Clement Webb and Samuel Gibson, both marines and young men, attempted to desert, having formed liaisons with two local girls. They were returned to the ship and placed under confinement, later to receive two dozen lashes each.

By 13th July, the tents had been struck, the fort dismantled and, as they prepared to sail, Tupia, and his boy-servant Taiyota, who asked to join the expedition, taken on board.

Tupia [Tupaia] was around 45 years old. He was a Tahitian high priest and skilled navigator. He was Queen Oboreah's right-hand man, who was with her in the *Dolphin*'s time. While in Tahiti, Tupia sometimes acted as Banks's deputy, which brought them together sufficiently for Banks to accept the responsibility of taking him to England. In his role as high priest, Tupia had travelled widely among the islands of the region. He drew up a map of seventy-four of them, which proved an aid to Cook on the next leg of their journey among the Society Islands:

> During our stay here, Mr Banks and Dr Solander were very assiduous in collecting whatever they thought might contribute to the advancement of Natural History; and, by their directions, I made drawings of a great many curious trees, and other plants; fish, birds, and of such natural bodies as could not be conveniently preserved entire, to be brought home. **(Parkinson)**

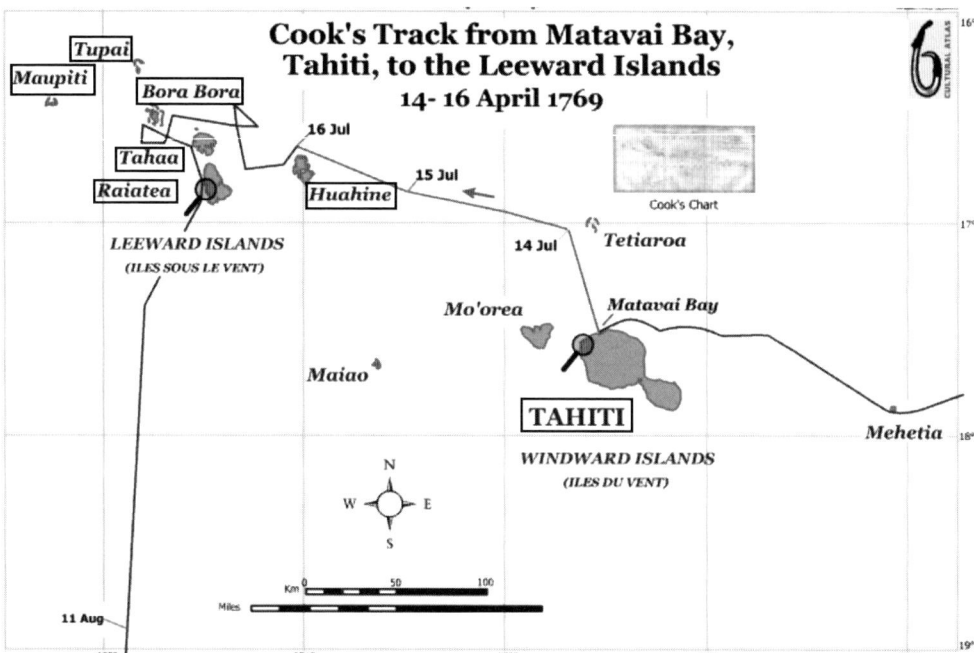

Cook decided to investigate the several islands lying to the north-west of Tahiti before heading south. He landed on Hauhine, and as 'first discoverer' presented their chief Oree with a plate stamped with the ship's name and the date of encounter. They landed too on Tupia's home island of Raiatea where, 'I then hoisted an English Jack and took possession of the Island & those adjacent in the name of His Britannick Majestys.' **(Cook)** The adjacent islands sighted included Tahaa, Bora Bora, Tupai and Maupiti.

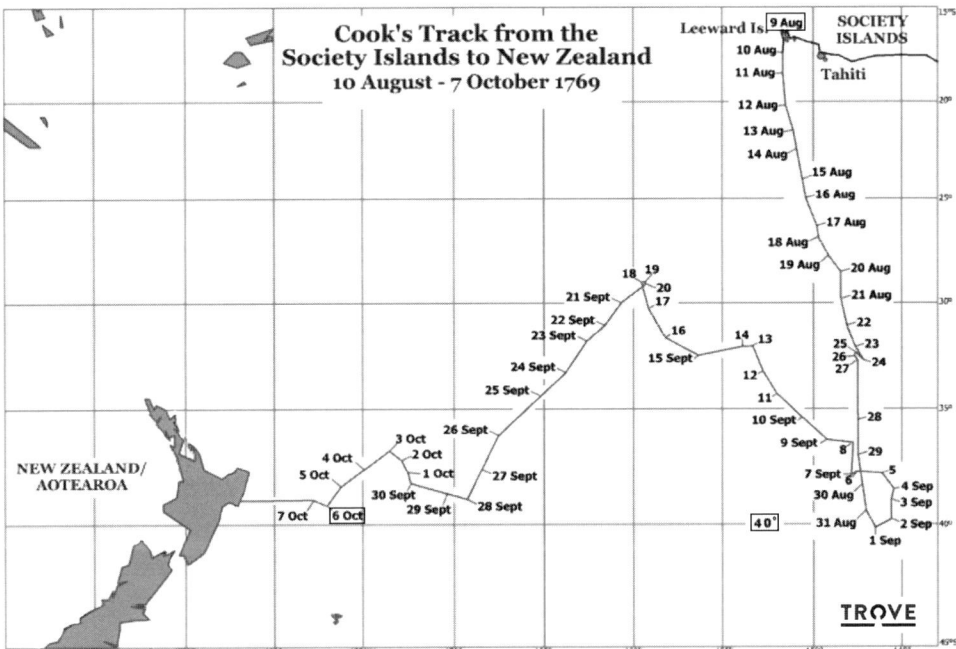

Cook opened his 'secret instructions'. These sealed instructions held the Admiralty's reasons for supporting the voyage: in addition to observing the transit of Venus, Cook was commanded to find the Great South Land, a 'Land of great extent' that was thought to exist in southern latitudes. 'You are to proceed to the Southward in order to make discovery of the Continent above-mentioned until you arrive in the latitude of 40°, unless you sooner fall in with it.'

The second page instructs Cook, 'with the Consent of the Natives to take possession of Convenient Situations in the Country in the Name of the King of Great Britain'.

Before closing, the Admiralty's letter issued a final catch-all directive to Cook. 'You will also observe with accuracy the Situation of such Islands as you may discover in the Course of your Voyage that have not hitherto been discover'd by any Europeans and take Possession for His Majesty.'

On <u>9th August</u> the *Endeavour* turned southward in search of the Great South Continent as Cook's orders instructed.

> On the 28th August at 10. a.m. departed this life Jn Radon [John Reading] Boatswains Mate, his death was occasioned by the Boatswain, out of mere good nature, giving him part of a Bottle of rum last night, which it is supposed he drank all at once, he was found to be very much in Liquor last night, but as this was no more than what was common with him when he could get any, no farther notice was taken of him then to put him to Bed where this morning about 8 o'clock he was found speechless and past recovery. **(Cook)**

Their three-week journey south, to <u>40°</u> latitude, brought no sign of a new continent. Instead, they hauled north-west, then south-west for the scrap of New Zealand shown on a chart created by Francoijs Visscher (Abel Tasman's hydrographer).

Five weeks later Banks wrote:

> Now do I wish that our friends in England could by the assistance of some magical spying glass take a peep at our situation: Dr Solander sits at the cabin table describing, myself at my Bureau Journalizing, between us hangs a large bunch of sea weed, upon the table lays the wood and barnacles; they would see that notwithstanding our different occupations our lips move very often, and without being conjurors might guess that we were talking about what we should see upon the land which there is now no doubt we shall see very soon.

On <u>6th October:</u>

> At half past one a small boy who was at the mast head called out land. I was luckily upon deck and well I was entertained, within a few minutes the cry circulated and up came all hands, this land could not then be seen even from the tops, yet few were there who did not plainly see it from the deck till it appeared that they had looked at least 5 points wrong. **(Banks)**

It was 'Young Nick' who first saw New Zealand. He received his reward of a gallon of rum and had the headland named in his honour.

Nicolas Young. Young's identity remains uncertain. He may have been a stowaway on the *Endeavour*. He is thought to have been around 12 years old. At the time he spotted New Zealand, he was the personal servant of the *Endeavour's* surgeon, William Brougham Munkhouse.

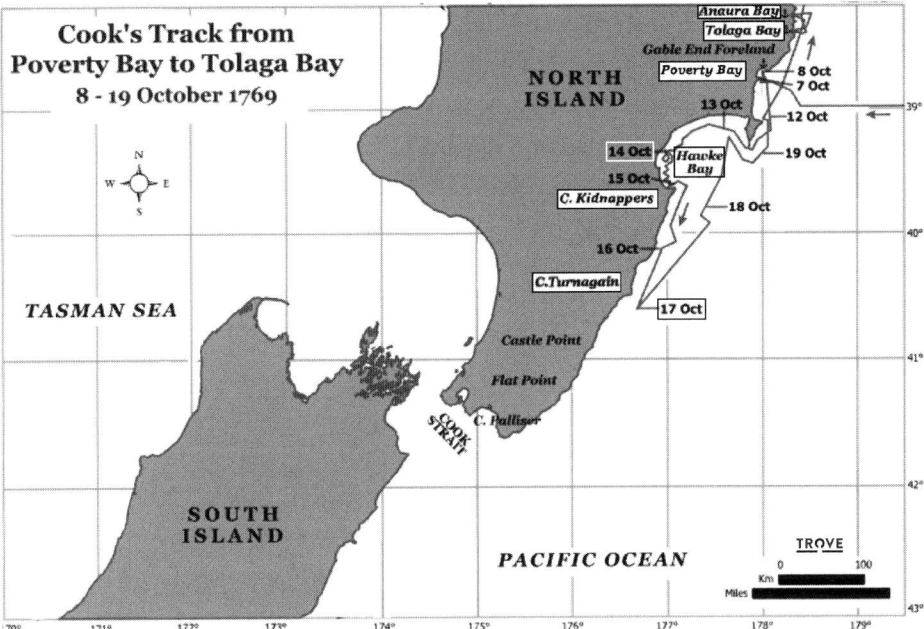

They landed in Poverty Bay. Again, their arrival was marred by a killing. A Māori warrior, preparing to throw a spear at the boat crew, was shot dead. The next day they went ashore again to where fifty warriors were gathered. Another was killed when he snatched Mr Green's short sword and would not give it up. It made no difference that Tupia spoke to them in their own language, explaining that they wanted provisions and water for which they would give them iron in exchange. That afternoon, four more died when Cook tried to kidnap some men in a canoe and he and his men were attacked. Three boys in the canoe leapt overboard and were captured and taken back to the ship.

I am aware that most humane men who have not experienced things of this nature will censure my conduct in firing upon the people in this boat nor do I myself think that the reason I had for seizing upon her will at all justify me. And had I thought that they would have made the least resistance I would not have come near them, but as they did I was not to stand still and suffer either myself or those that were with me to be knocked on the head. **(Cook)**

Thus ended the most disagreeable day my life has yet seen, black be the mark for it and heaven send that such may never return to embitter future reflection. **(Banks)**

These men, while on board, ate an Immoderate quantity of everything that was set before them, taking pieces at one time into their mouths six times larger than we did, and drank a quart of wine and water at one draught. **(Parkinson)**

Salt pork seemed to please them better than anything else. They seemed to have entirely forgot everything that had happened, put on cheerful and lively countenances and asked and answered questions with a great deal of curiosity. We then made them beds upon the lockers and they laid down to sleep with all seeming content imaginable.

In the middle of last night one of our boys seemed to shew more reflection than he had before done sighing often and loud; Tupia who was always upon the watch to comfort them got up and soon made them easy. They then sung a song of their own, it was not without some taste, like a Psalm tune and contained many notes and semitones; they sung it in parts which gives us no indifferent Idea of their taste as well as skill in music. The oldest of them is about 18, the middlemost 15, the youngest 10; the middlemost especially has a most open countenance and agreeable manner; their names are Tahourange, Koikerange, and Maragooete, the two first brothers. In the morning they were all very cheerful and eat an enormous quantity, after that they were dressed and ornamented with bracelets, anklets and necklaces after their own fashion. The boats were then hoisted out and we all got into them: the boys expressed much joy at this till they saw that we were going to land at our old Landing place near the river, they begged very much that they might not be set ashore at that place where they said were Enemies of theirs who would kill and eat them. The Captain resolved to go ashore at that place and if the boys did not choose to go from us, in the evening to send a boat with them to the part of the bay to which they pointed and called their home. Accordingly we went ashore and crossed the river. The boys at first would not leave us. No method was used to persuade them; it was even resolved to return and carry them home when on a sudden they seemed to resolve to go and with tears in their eyes took leave. **(Banks)**

They departed Poverty Bay and set off along the coast to the south, surveying, naming places and drawing up charts. Isaac Smith assisted Cook in this.

Isaac Smith, Able Seaman, was 16. Elizabeth Cook (Cook's wife) and Isaac Smith were first cousins, once removed. Smith became adept at drawing and copying charts for Cook. Later, Cook wrote, 'Mr Isaac Smith and Mr Isaac Manley, both too young for preferment, yet their behaviour merits the best recommendation. The former was of great use to me in assisting to make Surveys, Plans, Drawings etc in which he is very expert.'

By 14th October they were in Hawke Bay:

In the morning, nine canoes came to us, in which were one hundred and sixty of the natives: they behaved in a very irresolute manner. We did all we could to make them peaceable, but to no purpose, for they seemed, at length, resolved to do us some mischief; coming along-side of the ship again, and threatening us, we fired one of our guns, loaded with grapeshot, over their heads. **(Parkinson)**

No sooner had they seen the grape which scattered very far upon the water than they paddled away in great haste. **(Banks)**

Their canoes had from eighteen to twenty-two men in them. Some were between fifty and sixty feet long [18 metres], and rowed with eighteen paddles. They gave us two Heivos [Haka, Māori war dance], in their canoes, which were very diverting. They beat time with their paddles, and ended all at once with the word Epaah; at the same instant striking their paddles on the thwarts: all which afforded a truly comic act.

There were some good-looking people in these canoes, others were disfigured, and had a very savage countenance. The principals amongst them had their hair tied up on the crown of their heads; and some feathers, with a little bundle of perfume, hung about their necks. Most of them were tattooed in the face, and many of them quite naked, who seemed to be servants to the rest. Several of them had pieces of a green stone hung about their necks, which seemed to be pellucid, like an emerald. **(Parkinson)**

Relations with the Māori continued by turns to be friendly and hostile. Trading began:

Several canoes came off with nets and other fishing implements in them. After they had sold all their fish they began to put the stones with which they sink their nets into baskets and sell them but this was soon stopped as we were not in want of such commodities. **(Banks)**

But trouble soon flared up again …

A large armed boat wherein were twenty-two men came alongside. One man wore a black skin something like a bearskin for which Cook offered him a piece of red cloth. He took the cloth but made off without making the exchange.

Then a fishing boat came alongside and offered them some fish. Tupia's boy servant Taiyota, who was over the side to receive the fish, was seized and pulled into their boat. They were fired at, and Taiyota leapt out of their boat and was rescued. 'Two or Three paid for this daring attempt with the loss of their lives, and many more would have suffered had it been for fear of killing the boy.' **(Cook)**

As soon as Taiyota was a little recovered from his fright he brought a fish in to Tupia and told him that he intended it as an offering to his Eatua or god in gratitude for his escape. Tupia approved it and ordered him to throw it into the sea which he did. **(Banks)**

'Cape Kidnappers' took its name from this incident.

16th October

'Mountains covered with snow were in sight again this morn so that there is probably a chain of them runs within the country.' **(Banks)**

On 17th October, seeing no likelihood of meeting with a harbour towards the south, they turned to the north, and named the bluff head 'Cape Turnagain'.

21st October

Several canoes followed them and seemed very peaceably inclined, inviting them to go into a bay (Anaura Bay) where they said that there was plenty of fresh water; the ship followed them in and by 11 a.m. came to an anchor. In the evening, they went ashore and were received with great friendship by the natives:

> The women were plain and made themselves more so by painting their faces with red ocre and oil which generally was fresh and wet upon their cheeks and foreheads, easily transferrable to the noses of any one who should attempt to kiss them; not that they seemed to have any objection to such familiarities as the noses of several of our people evidently showed, but they were as great croquettes as any Europeans could be and the young ones as skittish as unbroke fillies.
>
> One part of their dress I cannot omit to mention: besides their cloth which was very decently rolled round them each wore round the lower part of her waist a string made of the leaves of a highly perfumed grass, to this was fastened a small bunch of the leaves of some fragrant plant which served as the innermost veil of their modesty.
>
> Their food at this time of the year consisted of Fish with which instead of bread they eat the roots of a kind of Fern Pteris crenulate. These were a little roasted on the fire and then beat with a stick which took off the bark and dry outside, what remained had a sweetish clamminess in it not disagreeable to the taste.
>
> They certainly have plenty of excellent vegetables. Their plantations were now hardly finished but so well was the ground tilled that I have seldom seen even in the gardens of curious people land better broke down. In them were planted sweet potatoes, cocos and some one of the cucumber kind. These plantations were from 1 or 2 to 8 or 10 acres each, in the bay might be 150 or 200 acres in cultivation tho we did not see 100 people in all. Each distinct patch was fenced in generally with reeds placed close one by another so that scarce a mouse could creep through.
>
> One piece of cleanliness in these people I cannot omit as I believe it is almost unexampledled among Indians. Every house or small knot of 3 or 4 has

a regular necessary house [lavatory] where everyone repairs and consequently the neighbourhood is kept clean which was by no means the case at Otahite [Tahiti]. They have also a regular dunghill upon which all their offalls of food &c. are heaped up and which probably they use for manure.

In the evening all the boats being employed in carrying on board water we were likely to be left ashore till after dark; the loss of so much time in sorting and putting in order our specimens was what we did not like so we applied to our friends the Indians for a passage in one of their Canoes. They readily launched one for us, but we in number 8 not being used to so ticklish a convenience overset her in the surf and were very well soused; 4 then were obliged to remain and Dr Solander, Tupia, Tayeto and myself embarked again and came without accident to the ship well pleased with the behaviour of our Indian friends who would the second time undertake to carry off such Clumsy fellows. **(Banks)**

'This Bay is called by the Natives Tegadoo' [Anaura Bay]. **(Cook)** But the bay proved unsuitable for watering. The surf was too great, so they weighed anchor and departed. But next morning several canoes of the same people came alongside:

and told us that there was a small bay to leeward of us where we might anchor in safety and land in the boats without a surf where there was fresh water; we followed their directions and they soon brought us into a bay called <u>Tolaga</u> where at 1 we anchored. and found as they had told us a small cove where the boat might land without the least surf, and water near it, so the Captain resolved to wood and water here. **(Banks)**

Banks and Dr Solander found many new plants here at Tolaga Bay.

Tupia who stayed with the waterers had much conversation with one of their priests; they seemed to agree very well in their notions of religion only Tupia was much more learned than the other and all his discourse was heard with much attention. He asked them in the course of his conversation with them many questions, among the rest whether or no they really eat men which he was very loth to believe; they answered in the affirmative saying that they eat the bodies only of those of their enemies who were killed in war.

The people at our desire sung their war song in which both men and women joined, they distorted their faces most hideously rolling their eyes and putting out their tongues but kept very good time often heaving most loud and deep sighs. **(Banks)**

In Tolaga Bay they found celery and scurvy grass in large quantities:

I have caused it to be boiled with portable soup and oatmeal every morning for the Peoples breakfast, and this I design to continue as long as it will last or any is to be got because I look upon it to be very wholesome and a great antiscorbutic [counteracting scurvy]. **(Cook)**

John Thompson, the ship's one-handed cook, was responsible for this and their other meals.

Traffic with the Māori resumed, and fish and sweet potatoes were added to the menu.

Next morning on 29th October at 4 a.m., they unmoored, and at 6 a.m. weighed and put to sea.

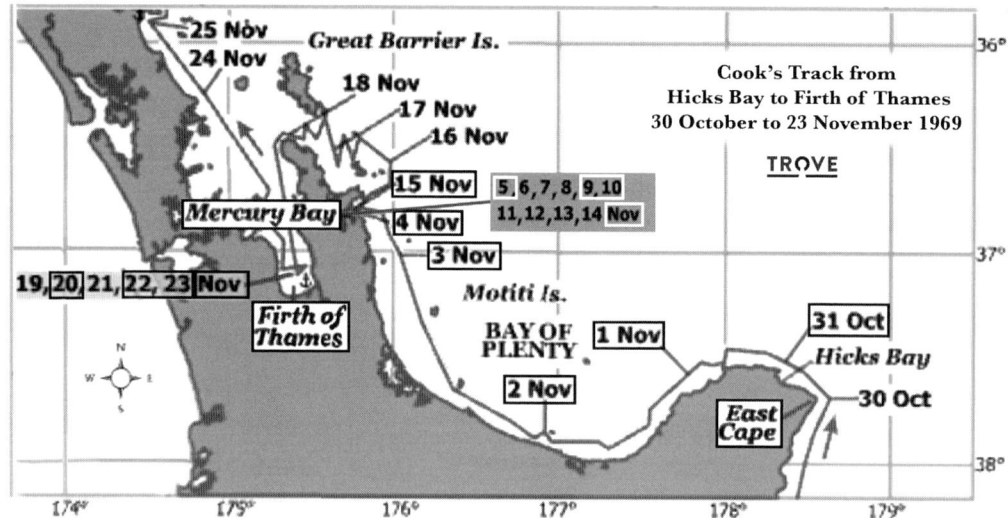

31st October

After we had rounded East Cape we saw as we run along shore a great number of villages and a great deal of cultivated land. **(Cook)**

In the forenoon seven canoes came off to us in a hostile manner, brandishing their lances, and waving their paddles. Some of them gave us an heivo; They kept paddling about us, calling out to us Kaka kee, no Tootwais, harre yoota patta pattoo; that is to say, if we would go on shore they would beat us with their patta pattoos [hand weapons]. **(Parkinson)**

Soon after we saw an immense large canoe coming from the shore crowded full of People, all armed with long lances. We Judged there could not be less than 60 people in her, 16 paddlers of a side, besides some who did not paddle and a long row of people in the middle from stem to stern crowded as close as possible. They pulled briskly up towards the ship as if to attack. A gun loaded with grape was therefore fired ahead of them: they stopped paddling but did not retreat: a round shot was then fired over them: they saw it fall and immediately took to their paddles rowing ashore with more haste than I ever saw men, without so much as stopping to breathe till they got out of sight. **(Banks)**

This occasioned our calling the point of land off which this happened Cape Runaway. **(Cook)**

1st November

When some Māori stole linen that was towing over the side and would not return it a musket ball was fired through their boat, and after that another musket loaded with small shot, neither of which they minded, so a four pounder was fired which sent them off.

2nd November

These villages are built upon eminences near the sea and are fortified on the land side with a bank and a ditch, and palisaded all round, besides this some of them appeared to have outworks. I rather think that they are places of retreat or stronghold where they defend themselves against the attack of an enemy as some of them seemed not ill designed for that purpose. **(Cook)**

3rd November

At breakfast a cluster of Islands and rocks were in sight which made an uncommon appearance from the number of perpendicular rocks or needles (as the seamen call them) which were in sight at once: these we called the Court of Aldermen in respect to that worthy body and entertained ourselves some time with giving names to each of them from their resemblance, thick and squat or lank and tall, to someone or other of those respectable citizens. **(Banks)**

4th November

At half past 7 Anchored in 7 fathom a little within the south entrance of the Bay or inlet [Mercury Bay]. We were Accompanied in here by several Canoes, who stayed about the Ship until dark. And before they went away they were so generous as to tell us that they would come and attack us in the morning, but some of them paid us a Visit in the night, thinking no doubt but what they should find all hands a sleep, but as soon as they found their mistake they went off. Each of these Canoes were made out of one large tree and were without any sort of ornament and the people in them were mostly quite naked — Between 5 and 6 o'clock in the Morning several Canoes came to us from all Parts of the Bay. In them were about 130 or 140 People, to all appearances their first design was to attack us being all completely Armed in their way. **(Cook)**

We all got up to see the event. **(Banks)**

After Parading about the Ship near three hours, sometimes trading with us and at other times tricking of us, they dispersed but not before we had fired a few Musquets and one great gun, not with any design to hurt any of them, but to shew them what sort of Weapons we had and that we could revenge any insult they offered to us. **(Cook)**

5th November

This morn some canoes came off but brought nothing to sell. One old man whose name was Torava came on board; he seemed to be the chief both today and yesterday but in all the transactions of yesterday he was observed to behave sensibly and well, laying in a small canoe always near the ship and at all times speaking civilly to those on board. With some persuasion he ventured down into the cabin and had presents, Cloth, Iron &c. given him; he told us that the Indians were now very much afraid of us, we promised friendship if they would supply us with provision at their own price. **(Banks)**

9th November

At daybreak this morn a vast number of boats were on board almost loaded with mackerel of 2 sorts, one exactly the same as is caught in England. We concluded that they had caught a large shoal and sold us the overplus what they could not consume, as they set very little value upon them. It was however a fortunate circumstance for us as by 8 o'clock the ship had more fish on board than all hands could eat in 2 or 3 days, and before night so many that every mess who could raise salt corned as many as will last them this month or more. **(Banks)**

In the afternoon, canoes came alongside the ship. In one were forty-seven people, in all appearance come with a hostile intention, being completely armed with pikes, darts and stones. They began to sell some of their arms and one man offered for sale a *Ha'ahow* – a square piece of cloth they wear made of black and white dog-skin. Lieutenant Gore, who at the time was commanding officer, sent into the canoe a piece of cloth that the man had agreed to take in exchange for his, but as soon as he had Gore's cloth in his possession he would not part with his own, and immediately they began to shake their paddles and sing their war song in defiance. This enraged the second lieutenant so much that he levelled a musket at the man who still had the cloth in his hand and shot him dead:

> The name of this unfortunate young man, we afterwards learned, was Otirreeoònooe. **(Parkinson)**

> I must own it did not meet with my approbation because I thought the punishment a little too severe for the crime, and we had now been long enough acquainted with these People to know how to chastise trifling faults like this without taking away their lives. **(Cook)**

However, those Māori who were ashore acknowledged that the dead man deserved his punishment. Here they witnessed the villagers taking their meal of fish, shellfish, lobsters and birds; these were dressed either by broiling them on a skewer, which was stuck into the ground leaning over the fire, or in earth ovens like those in Tahiti, which were holes in the ground filled with provision and hot stones and covered over with leaves and earth.

Here, as in Tahiti, Banks was again deeply disturbed to witness a woman in mourning. She sat on the ground, tears constantly trickling down her cheeks, repeating in a low but very mournful voice words that he could not understand, and at every sentence cutting her arms, face or breast with a shell she held in her hand, so that she was almost covered with blood.

10th November

> This day was employed in an excursion to view the large river at the bottom of the bay. We went up about a league where we agreed to stop our disquisition and go ashore to dine. A tree in the neighbourhood on which were many shags nests and old shags setting by them confirmed our resolution; an attack was consequently made on the Shags and about 20 soon killed and as soon broiled and eat, everyone declaring that they were excellent food as indeed I think they were. Hunger is certainly most excellent sauce, but since our fowls and ducks have been gone we find ourselves able to eat any kind of birds (for indeed we throw away none) without even that kind of seasoning.
>
> An oyster bank had been found at the river by the wooding place. Here the longboat was sent and soon returned deep loaded with I sincerely believe as good oysters as ever came from Colchester and about the same size. They were laid down under the booms and employed the ships company very well who I verily think did nothing but eat from the time they came on board till

> night, by which time a large part were expended, but that gave us no kind of uneasiness as we well knew that not the boat only but the ship might be easily loaded in one tide almost, as they are dry at half ebb. **(Banks)**

Relations were friendly, so they went to visit an Indian Fort or Hippa built on the end of a hill, which jutted out into the sea and was enclosed by a palisade about 10 feet high (3 metres). Over the palisade was built a fighting stage from which to throw darts or stones at any assailants, and out of danger of their weapons. Bundles of darts and heaps of stones were laid here, ready in case of attack. The young men staged a mock attack for them to see its function.

Here in Mercury Bay, Dr Solander and Banks collected an enormous number of plants. And the great number of crayfish they were given, 'are certainly the largest and best I have ever eat'. **(Banks)**

<u>15th November</u>

> Several canoes were on board and in one of them Torava who sayd that as soon as ever we are gone he must go to his Hippa or fort, for the friends of the man who was killed on the 9th threatened to revenge themselves upon him as being a friend to us. **(Banks)**

> Before we left this Bay [Mercury Bay] we cut out upon one of the trees near the watering place, the Ships Name, date &c. and after displaying the English Colours I took formal possession of the place in the name of His Majesty. **(Cook)**

Five days later, on <u>20th November</u>, a river was discovered. It was named the 'Thames'.

> The banks of the river were completely clothed with the finest timber my eyes ever beheld [White pine/kahikatea – *Dacrycarpus dacrydioides*]. **(Banks)**

The one they measured:

> was as straight as an arrow and tapered but very little in proportion to its length, so that I judged that there was 356 solid feet [108 metres] of timber in this tree clear of the branches. We saw many others of the same sort several of which were taller than the one we measured. **(Cook)**

> We cut down a young one of these trees; the wood proved heavy and solid, too much so for mast but would make the finest plank in the world. **(Banks)**

22nd November

When one of the visitors stole the half hourglass out of the binnacle (a case that supports and protects the ship's compass, located near the helm) and was caught red handed, Lieutenant Hickes, who was the commanding officer, brought him to the gangway and gave him a dozen lashes with the cat o' nine tails. The rest of his people seemed to agree with the punishment when they knew the cause, and one old man beat the fellow after he had got into his canoe.

On <u>23rd November</u> they departed the <u>Firth of Thames</u> and kept standing along shore to the north-west.

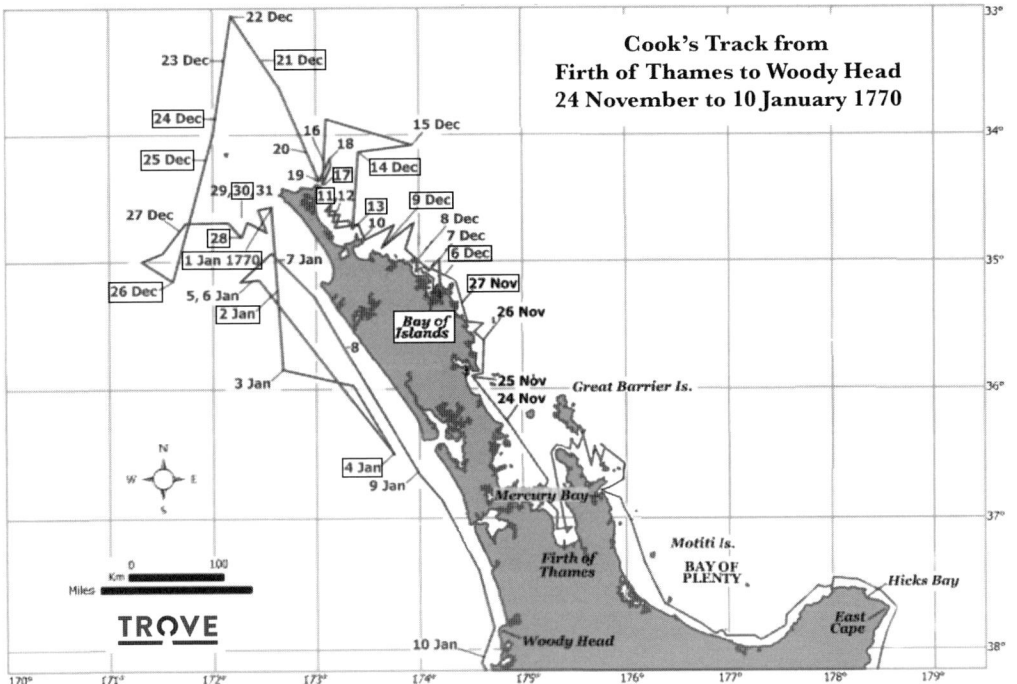

27th November

Two small canoes came off early in the morn and soon after two larger ones came from a distance. All came up together to the ship.

> The strangers were numerous and appeared rich: their Canoes were well carved and ornamented. The people themselves were browner than to the Southward and they had a much larger quantity of Amoco or black stains upon their bodies and faces; almost universally they had a broad spiral on each buttock and many had their thighs almost entirely black, small lines only being left untouched so that they looked like striped breeches.

> These people would not part with any of their arms &c. for any price we could offer; at last however one produced an axe of Talk and offered it for Cloth, it was given and the Canoe immediately put off with it. A musquet ball was fired over their heads on which they immediately came back and returned the cloth but soon after put off and went ashore.
>
> In the afternoon other Canoes came off and from some inattention of the officers were suffered to cheat unpunished and unfrightned. This put one of the Midshipmen who had suffered upon a droll though rather mischievous revenge. He got a fishing line and when the Canoe was close to the ship hove the lead at the man who had cheated, with so good success that he fastened the hook into his backside, on which he pulled with all his might and the Indian kept back, so the hook soon broke in the shank leaving its beard in his backside, no very agreeable legacy. **(Banks)**

Next morning, approaching the <u>Bay of Islands</u>, they were visited by people in canoes who began to pelt them with stones. One man, more active than the rest, took up a stick and threw it at one of their men on the taffrail. The captain went upon the poop where they immediately threw a stone at him. He levelled a gun loaded with small shot at the man who held a stone in his hand in the very action of throwing and struck him. The man clapped his hands to his face and fell flat in the canoe. They subsequently learned that the man had died. They were told three shots had struck his eye, 'and I suppose found there an easy passage to his brain'. **(Banks)**

On 30th November, they anchored in the Bay of Islands. Cook went with the pinnace and yawl, manned, and armed, and landed upon an island accompanied by Banks and Dr Solander. Soon they were surrounded by many hundreds of hostile warriors all bearing weapons. A serious situation arose in which the lives of the party hung by a thread. It was only the quick-thinking Lieutenant Hickes, who being in charge aboard the ship at the time, and seeing the direness of their predicament, brought her broadside to bear, and fired her great guns, saving their lives:

> The bay we were in [Bay of Islands] was indeed a most surprizing place: it was full of an innumerable quantity of Islands forming as many harbours, which must be as smooth as mill pools as they Landlock one another numberless times. Everywhere round us we could see large Indian towns, houses and cultivations: we had certainly seen no place near so populous as this one was. **(Banks)**

1st December

> Several Canoes were on board by Day break and sold some things chiefly for Indian Cloth and quart bottles. I have mentioned their custom of Eating

human flesh. I was loth a long time to believe that any human beings could have among them so brutal a custom. We have never failed wherever we went ashore and often when we conversed with canoes to ask the question; we have without one exception been answered in the affirmative, and several times as at Tolaga and today the people have put themselves into a heat by defending the Custom, which Tubia who had never before heard of such a thing takes every Occasion to speak ill of, exhorting them often to leave it off. They however as universally agree that they eat none but the bodies of those of their enemies who are killed in war, all others are buried. **(Banks)**

2nd December

Between 2 and 4 a.m. the gunner [**Stephen Forwood**] having the Charge of the watch, he together with Alexr Simpson Richd Littleboy and [Thomas Rossiter] found means to take out of the Spirit Cask, on the quarter deck between 10 and 12 Gallons of Rum being the whole that was in the Cask. They were caught in the very fact and part of the Rum was found in their possession. The three men I punished with 12 lashes each, but as to the Gunner who really deserved the whole upon his back is from his Drunkenness become the only useless person on board the Ship. I not only punished them with 12 Lashes each but stopped the allowance until it amounts to the quantity stolen. **(Cook)** [Forwood being a warrant officer, he could not be flogged.]

4th December

After breakfast we went ashore at a large Indian fort or heppa; a great number of people immediately crowded about us and sold almost a boat load of fish in a very short time. They then went and showed us their plantations which were very large of Yams, Cocos, and sweet potatoes; and after having a little laugh at our seine [long fishing net], which was a common kings seine, showed us one of theirs which was 5 fathom deep [9 metres] and its length we could only guess, as it was not stretched out, but it could not from its bulk be less than 4 or 500 fathom [914 metres]. **(Banks)**

5th December

By the evening all our empty Casks were filled with water and had at the same time got on board a large quantity of celery which is found here in great plenty. This I still continue to be boiled every morning with Oatmeal and Portable Soup for the ships companies breakfast – At 4 a.m. weighed with a light breeze at S.E. **(Cook)**

6th December

We were all happy in our breeze and fine clear moonlight; myself went down to bed and sat upon my cot undressing myself when I felt the ship strike upon a rock, before I could get upon my legs she struck again. I ran upon deck but before I could get there the danger was over; fortunately the rock was to wind ward of us so she went off without the least damage and we got into the proper channel, where the officers who had examined the bay declared there to be no hidden dangers – much to our satisfaction as the almost certainty of being eat as soon as you come ashore adds not a little to the terrors of shipwreck. **(Banks)**

9th December

Many canoes came off, Tupia persuaded them to come under the stern and after having bought of them some of their cloths, which they sold very fairly, began to enquire about the country:

Have you no hoggs among you? said Tupia. – No. – And did your ancestors bring none back with them? – No. – You must be a parcel of Liars then, said he, and your story a great lye for your ancestors would never have been such fools as to come back without them. Thus much as a specimen of Indian reasoning. **(Banks)**

11th December

Wind as hard hearted as ever: we turned all day without losing anything, [without losing ground] much to the credit of our old Collier, who we never fail to praise if she turns as well as this. **(Banks)**

13th December

At Noon had strong gales and hazy weather – tacked and stood to the westward. No land in sight, for the first time since we have been upon the Coast. **(Cook)**

14th December

Strong gales at west and WSW with Squalls at times attended with rain. In the evening brought the ship under her Courses, having first split the fore and Mizzen Topsails. **(Cook)**

This meant work for sailmaker John Ravenhill and his mates.

John Ravenhill, 'an old Man about 70 or 80 Years of age … generally more or less drunk every day.' **(Cook)**

> Saw land bearing S.W. which I take to be the northern extremity of this Country as we have now a large swell rolling in from the westward which could not well be was we covered by any land on that point of the compass. **(Cook)**

17th December

> The people at work repairing the Sails. The most of them having been split in the late blowing weather. **(Cook)**

21st December

> At noon clear weather, no land in sight. **(Cook)**

For three days deviations of S W winds prevented any westerly progress, and the land remained out of sight.

24th December

> Land in sight, an Island or rather several small ones most probably 3 Kings [Three Kings Islands, discovered by Abel Tasman], so that it was conjectured that we had Passed the Cape [North Cape, 'the northern most extremity of this Country' – **Cook**] which had so long troubled us. Calm most of the Day: myself in a boat shooting in which I had good success, killing chiefly several Gannets or Solan Geese so like European ones that they are hardly distinguishable from them. As it was the humour of the ship to keep Christmas in the old-fashioned way it was resolved of them to make a Goose pie for tomorrows dinner. **(Banks)**

25th December

> Christmas day: Our Goose pie was eat with great approbation and in the Evening all hands were as Drunk as our forefathers used to be upon the like occasion. **(Banks)**

26th December

> This morn all heads ached with yesterdays debauch. **(Banks)**

28th December

It began to blow very hard and increased in such a manner that by 8 [a.m.] o'clock it was a mere hurricane attended with rain and the Sea run prodigious high. **(Cook)**

30th December

At 6 [a.m.] Saw the land bearing N.E. distant about 6 Leagues [33 kilometres] which we judge to be the same as Tasman calls Cape Maria Van Diemen. **(Cook)**

1st January 1770

I cannot help thinking but what will appear a little strange that at this season of the year we should be three weeks in getting 10 Leagues [56 kilometres] to the westward and five weeks in getting 50 Leagues, [279 kilometres] but it will hardly be credited that in the midst of summer and in the Latitude of 35° such a gale of wind as we have had could have happened, which for its strength and continuance was such as I hardly was ever in before – Fortunately at this time we were a good distance from land otherwise it would have proved fatal to us. **(Cook)**

2nd January

Having no land in sight not daring to go near it as the wind blow'd fresh right on Shore and a high rolling Sea from the same quarter, and knowing that there was no harbour that we could put into in case we were caught upon a lee shore. **(Cook)**

4th January

The great sea which the prevailing westerly winds impel upon the Shore must render this a very dangerous Coast, this I am so fully sensible of that was we once clear of it I am determined not to come so near again if I can possible avoid it unless we have a very favourable wind indeed. **(Cook)**

The continuing hard gales, high seas and onshore winds during the two weeks they ran down the dangerous west coast of the north island of New Zealand forbade any landing.

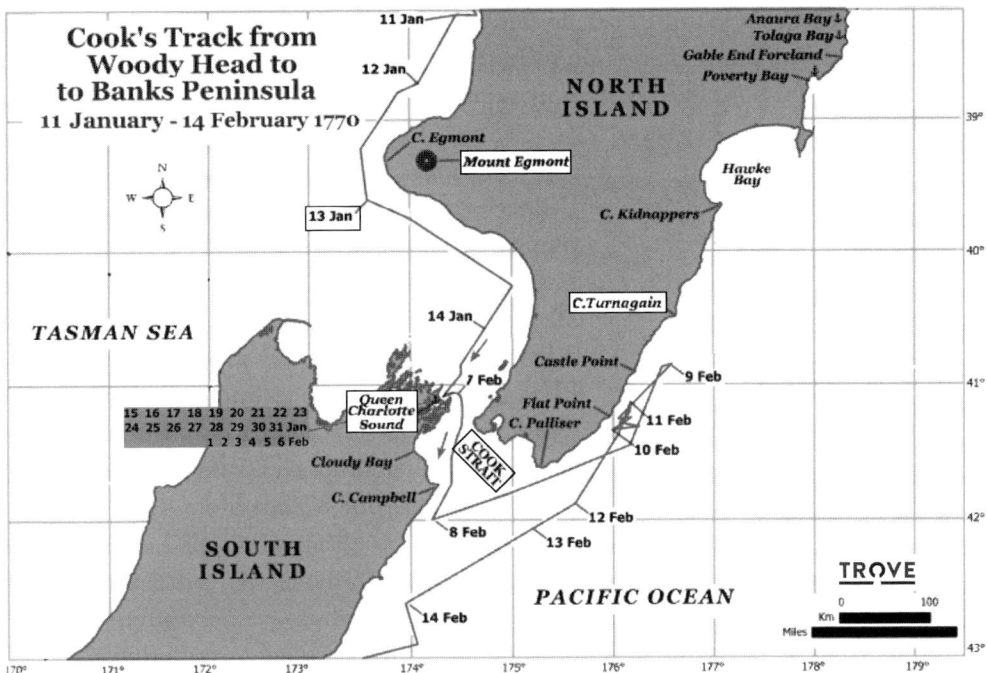

On 13th January, they turned south-east around the snow-covered peak of Mount Egmont:

> This morn soon after daybreak we had a momentary view of our great hill the top of which was thick covered with snow, though this month answers to July in England. How high it maybe I do not take upon me to judge, but it is certainly the noblest hill I have ever seen and it appears to the utmost advantage rising from the sea without another hill in its neighbourhood one 4th part of its height. **(Banks)**

> I have named it Mount Egmont in honour of the Earl of Egmont. **(Cook)**

Three days later, 'anchored in a very snug Cove' **(Cook)** [Ship Cove in Queen Charlotte Sound] they found a fine stream of excellent water, plenty of wood and in a few hauls of the seine caught 300 pounds of different sorts of fish, which were equally distributed to the ship's company. The ship was careened to clean her bottom, and here they confirmed the Māori were as good as their word about their cannibalistic habit. They met some men and Cook, 'got from one of them the bone of the forearm of a man or a woman which was quite fresh and the flesh had been but lately picked off which they told us they had eat'. **(Cook)**

The following day, four heads of the men who had been lately slain were brought to the ship and Banks purchased one.

During their three-week stay in Ship Cove, they replenished their supplies of wood and water, carried out the necessary repairs, and Cook climbed a hill from which he concluded there was a passage to the east. They set up a post, hoisted the Union flag, 'and I dignified this Inlet with the name of Queen Charlottes Sound and took formal possession of it and the adjacent lands in the name and for the use of his Majesty'. **(Cook)**

Having sailed through the passage (Cook Strait) they turned north-east and, 'when the weather clearing up we saw Cape Turn-again. I then called the officers upon deck and asked them if they were now satisfied that this land was an Island to which they answered in the affirmative.' **(Cook)**

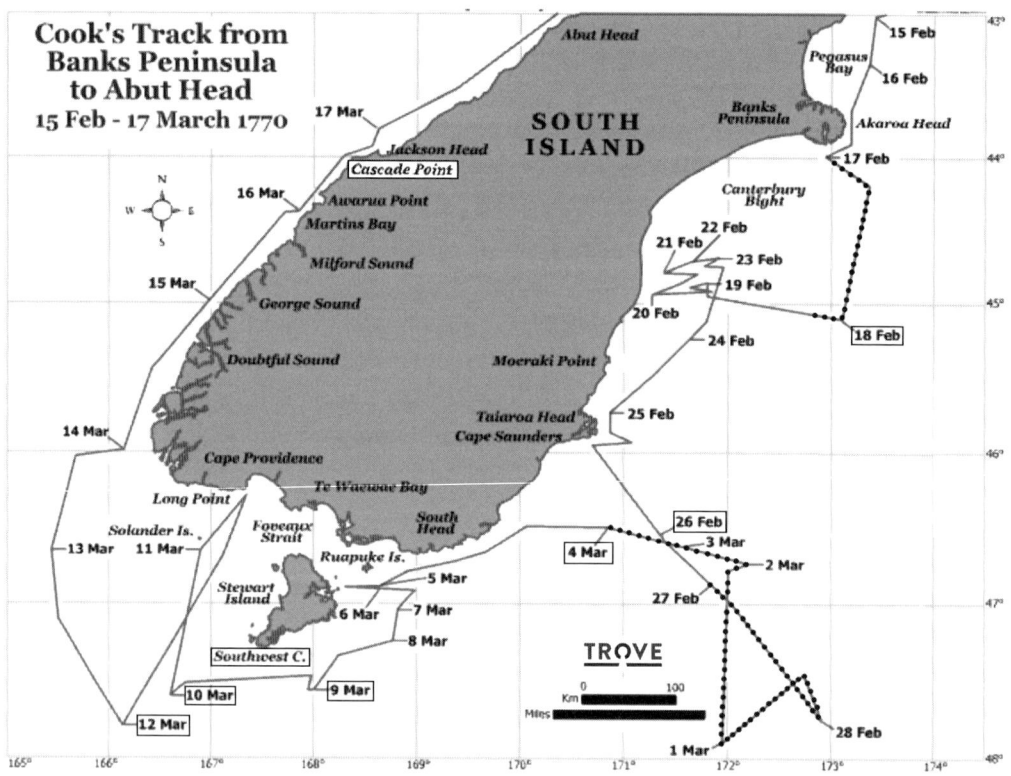

Now they turned their attention to the land to the south. The hunt for the Great South Continent was still alive in the minds of some, and when Gore imagined that he saw land to the SSE and SEBE the ship was put on a course to investigate:

18th February

> PM stood ESE in search of Mr Gores imaginary land until 7 o'clock at which time we had run 28 Miles since noon, but seeing no land but that we had left, or signs of any, we bore away SBW and continued upon that Course until noon, Seeing no signs of land, I thought it to no purpose standing any farther to the Southward, and therefore hauled to the Westward. **(Cook)**

Still Banks held firm to a Great South Continent, 'myself and one poor midshipman, the rest begin to sigh for roast beef'. **(Banks)**

26th February

> In the p.m. had the wind whifling all round the Compass, sometimes blowing a fresh gale and at other times almost calm. At 5 o'clock it fixed at WSW and soon blowed so hard as to put us past our topsails, and to split the fore sail all to pieces: after getting another to the yard we continued standing to the southward under two Courses. **(Cook)**

This long-winded gale lasted for six days, driving them far out of sight of land. On <u>4th March</u> they got sight of land again. It looked little populated.

On the <u>9th</u>:

> At day light we discovered under our lee bow a ledge of rocks. **(Cook)**

> They were luckily discovered by the midshipman's going to the mast head. The breeze being moderate, we put the helm a-lee, and were delivered from this imminent danger by the good providence of God. **(Parkinson)**

> I have named them the Traps because they lay as such to catch unweary strangers. **(Cook)**

On the <u>10th</u>:

> At sunset the Southermost point of land which I afterwards named South Cape [<u>Southwest Cape on map</u>] bore N 38° E^t distant 4 Leagues. I began now to think that this was the southermost land. **(Cook)**

> To the total demolition of our aerial fabric called continent. **(Banks)**
> They had reached the southernmost land of New Zealand.

> Fresh gales still and wind that will not let us get to the northward. We stood in with the shore which proved very high and had a most romantic appearance from the immense steepness of the hills, many of which were conical and most had their heads covered with snow. **(Banks)**

On <u>12th March</u> the wind finally shifted to the SWbW, and they stood to the NNW along the treacherous west coast.

> There was not the smallest signs of inhabitants, nor indeed have we seen any since we made this land except the fire on the 4th of March. **(Banks)**

> As we sailed along we passed a broken point, that had a flat top, from which the water poured down into the sea, and formed three grand natural cascades. This point we named <u>Cascades Point</u>. **(Parkinson)**

> Once we were very near the shore on which we saw that there was a most dreadful surf, occasioned by the S and SW swell which has reigned without intermission ever since we have been upon this side of the land. **(Banks)**

> No country upon earth can appear with a more rugged and barren aspect than this doth from the sea for as far inland as the eye can reach nothing is to be seen but the summits of these rocky mountains which seem to lay so near one another as not to admit any valleys between them. **(Cook)**

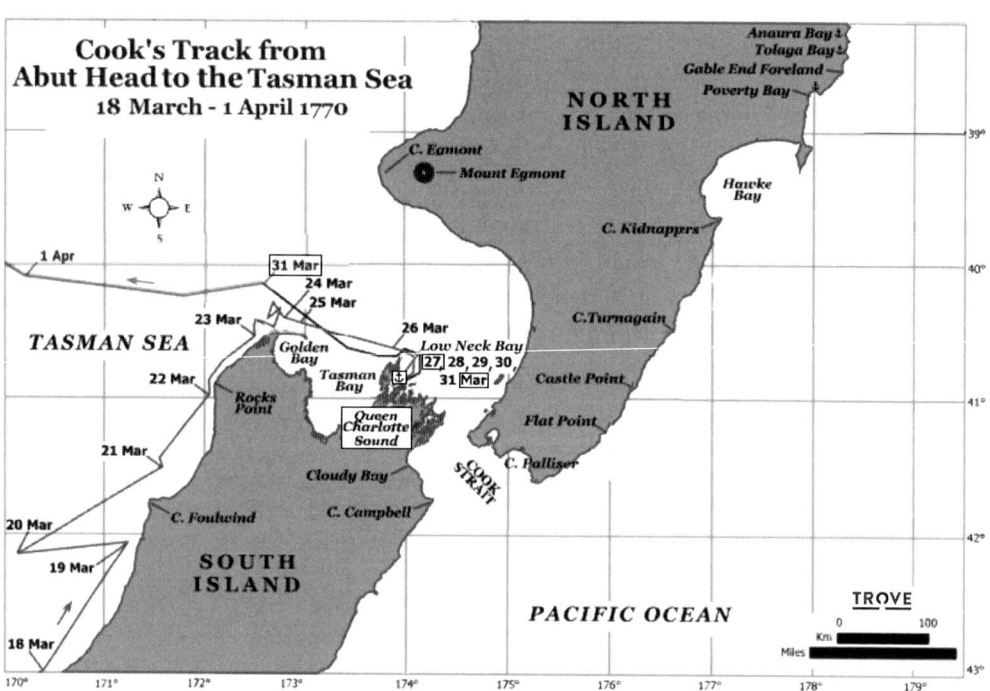

On 27th March, after having endured the dangers of foul winds, and the tedious suspense of many calms, they anchored in a bay. This they found to be about ten leagues NW of Queen Charlotte Sound.

'The sea is certainly an excellent school for patience.' **(Banks)**

They anchored in the shelter of D'urville Island, not far from their old snug anchorage of Ship Cove. Here they replenished their water and wood and caught an abundance of fish.

They had circumnavigated and charted the entirety of New Zealand.

By now, they had been more than a year and a half at sea and Cook called a meeting of his officers to decide the route home. The poor condition of the ship and provisions decided them to run for the East Indies and on the way to explore the uncharted coast of New Holland.

31st March

'We this morn weighed and sailed with a fair breeze of wind inclined to fall in with Van Diemen's Land as near as possible to the place where Tasman left it.' **(Banks)**

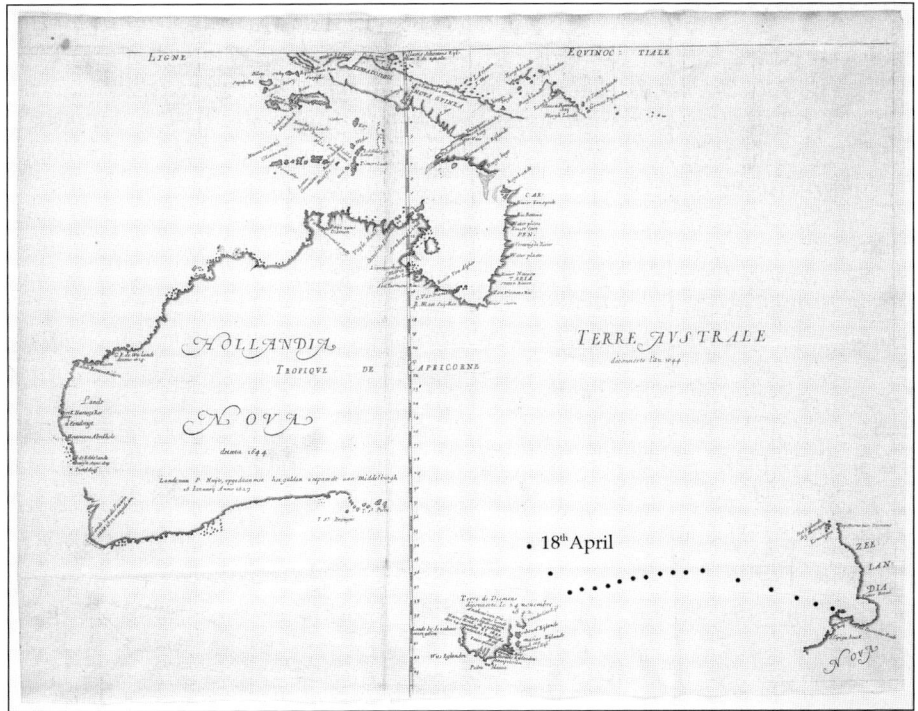

New Holland as it was known to Europeans before Cook. This was the first appearance of Abel Tasman's discoveries (*Endeavour's* route across the Tasman Sea is added).[6]

Continent of Smoke

Chapter 1

A Certain Sign the Country is Inhabited

Lieutenant James Cook (1728–1779).[1]

(Dates are as Cook gives them for whom the new day begins at noon. Underlined or bold text refers to marked positions on associated charts. Some expressions and spellings are left as original, and supported by a Glossary of Terms.)

18th April 1770

COOK: At 6 o'clock the gale increased to such a height as to oblige us to take in the fore topsail and main sail, and to run under the fore sail and mizzen all night, sounding every 2 hours, but found no ground with 120 fathom.[2]

A Certain Sign the Country is Inhabited 3

HM Bark Endeavour in a rough sea by Sydney Parkinson.³

Sydney Parkinson – Natural
History Artist (1745–1771).⁴

PARKINSON: We had a broken sea that caused the ship to pitch and roll very much; at the same time we shipped a sea fore and aft, which deluged the decks, and had like to have washed several of us overboard.[5]

Joseph Banks – Gentleman Naturalist (1743–1820).[6]

Port Egmont Hen.[7] (Representation only)

BANKS: In the morn a Port Egmont hen [Brown Skua *Catharacta skua antarctica*] and a Pintado bird were seen, at noon two more of the former.[8]

COOK: The first of these birds are certain signs of the nearness of land.

19th April

BANKS: A shoal of porpoises were about the ship which leaped out of the water like salmons, often throwing their whole bodies several feet high above the surface.

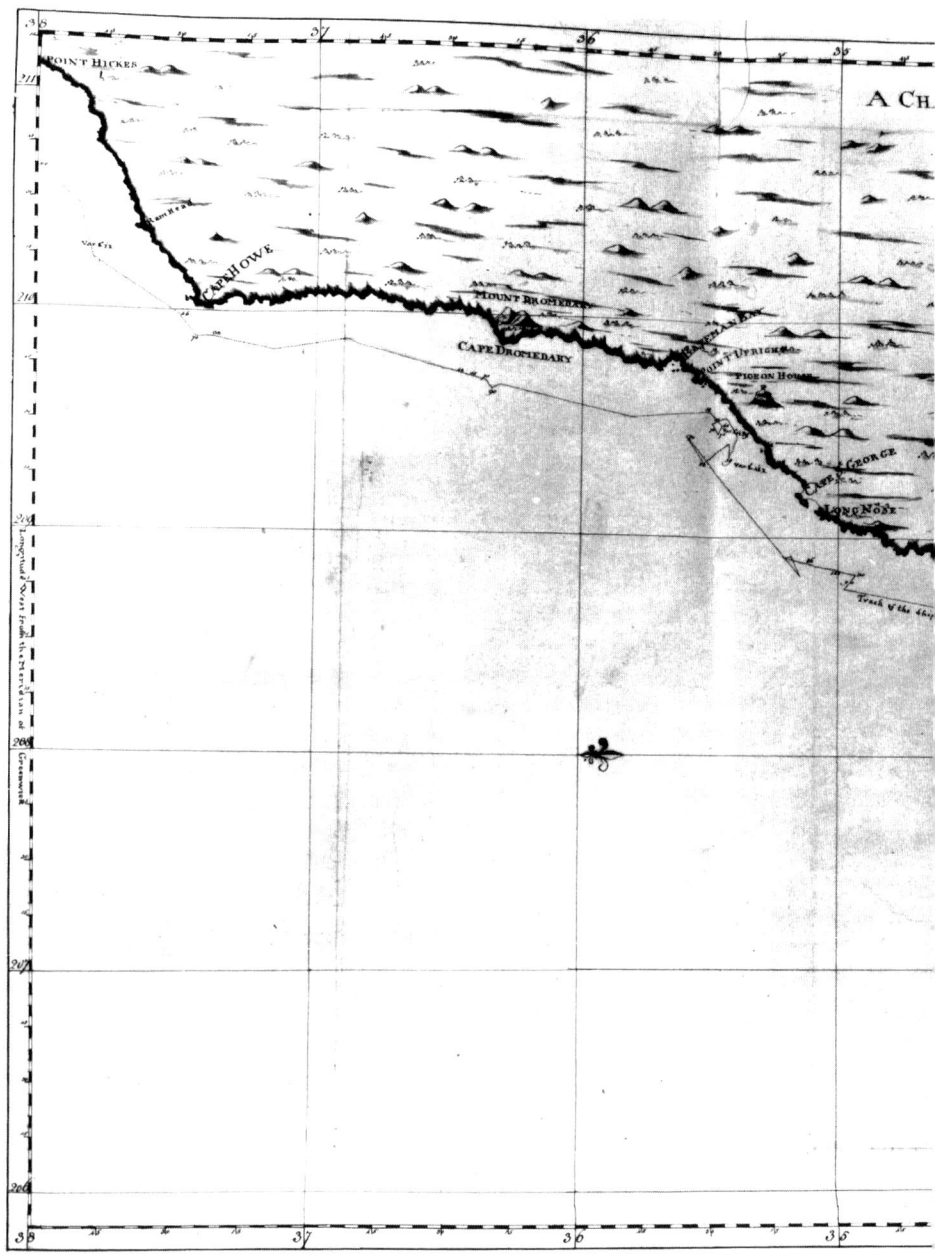

A Chart of the East Coast of New Holland from Point Hickes to Smoaky Cape by Lieutenant J. Cook Commander of the *Endeavour*.⁹ [Sixteen-year-old Isaac Smith, Able Seaman, became adept at drawing and copying charts surveyed by Cook. This is one of them.] (The following charts in Chapter 1 are enlargements of sections of this chart.)

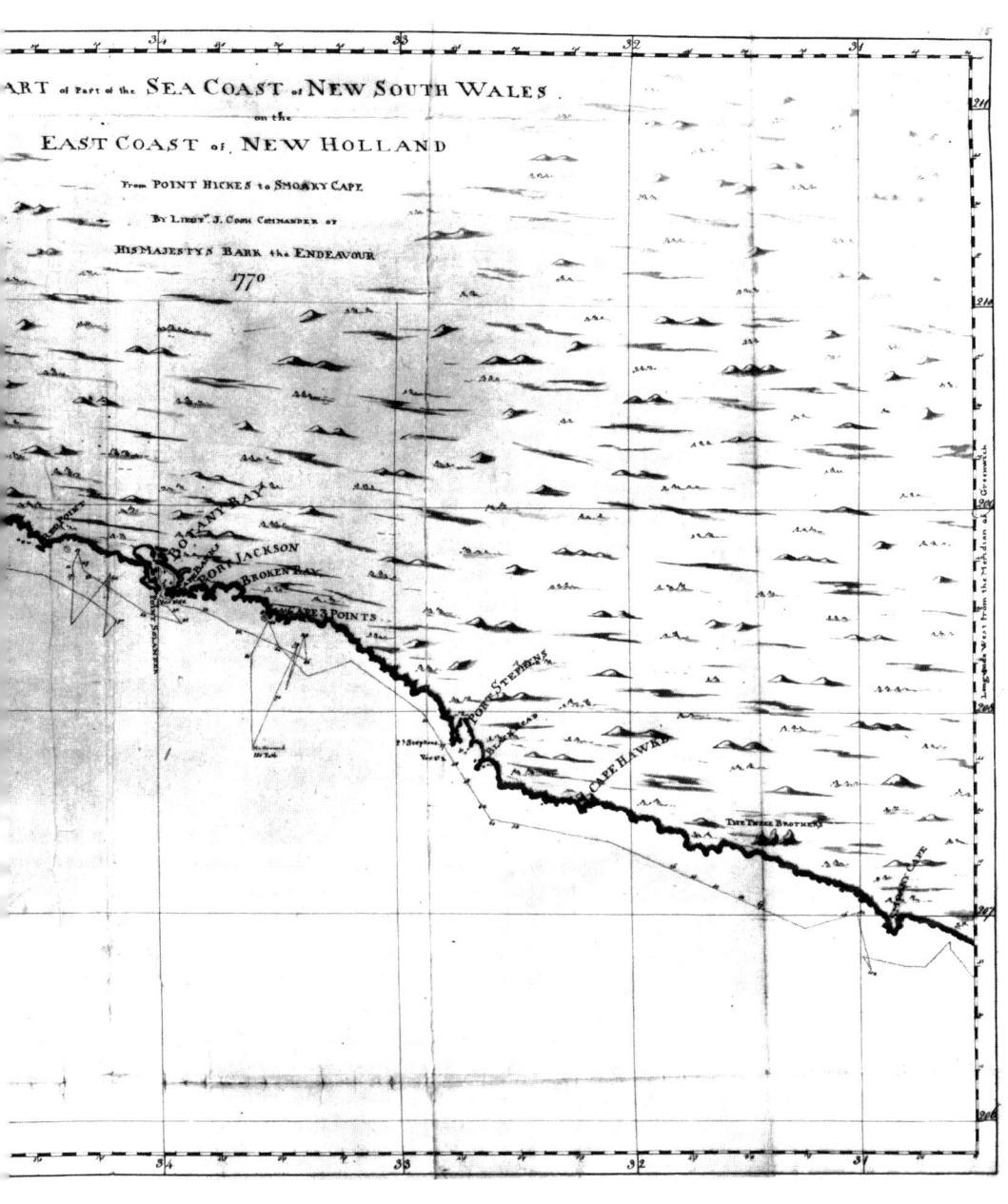

Portion of Cook's Chart Point Hickes to Smoaky Cape.[10]

19th April (cont'd)

COOK: At 5 [in the a.m.] set the topsails close reefed and 6 saw land. I have named it <u>Point Hickes</u>, because Lieutenant Hickes was the first who discovered this land. [Marked on Cook's chart above. Bidwell country.] *Some authorities believe that what Lieutenant Hickes saw and Cook named Point Hickes was in fact a cloud bank. The first true landmark named by Cook was Ram Head. See www.jamescookheritagetrail.com.au*

PARKINSON: It is moderately high: part of it appeared to be flat, and covered with sand; but, the weather being foggy, we had not a good view of it.

A Certain Sign the Country is Inhabited 9

BANKS: At ten it was pretty plainly to be observed; it made in sloping hills covered in part with trees and bushes but interspersed with large tracts of sand.

COOK: At noon a remarkable point bore N 20 degrees East. This point rises to a round hillock, very much like the <u>Ram Head</u> going into Plymouth Sound on which I called it by the same name. [Marked on Cook's chart above.]

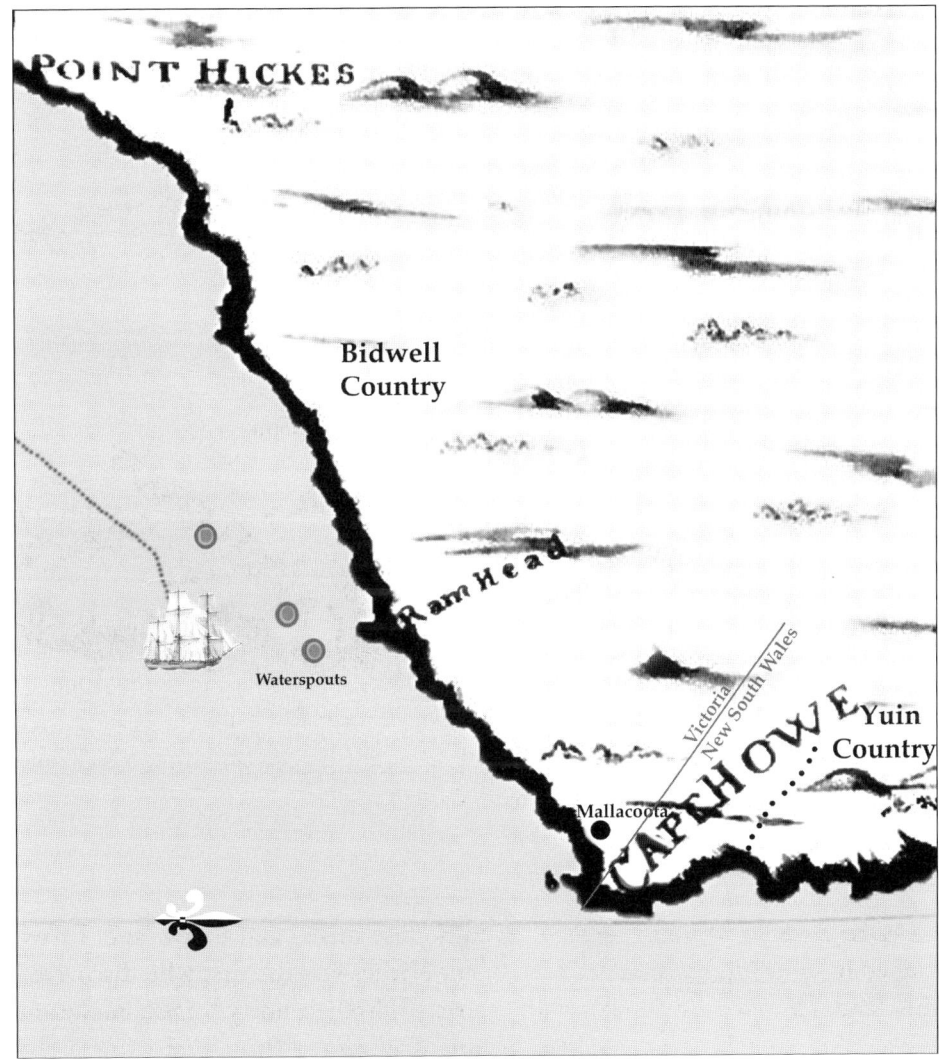

20th April

COOK: At 1 in the p.m. saw three water spouts at once.

BANKS: Two soon disappeared but the third which was about a League from us lasted full a quarter of an hour. It was a column which appeared to be of about the thickness of a mast or a middling tree, and reached down from a smoke coloured cloud about two thirds of the way to the surface of the sea; under it the sea appeared to be much troubled for a considerable space and from the whole of that space arose a dark coloured thick mist which reached to the bottom of the pipe.

Waterspouts.[11] (Representation only)

BANKS: When it was at its greatest distance from the water the pipe itself was perfectly transparent and much resembled a tube of glass or a column of water, if such a thing could be supposed to be suspended in the air. It very frequently contracted and dilated, lengthened and shortened itself and that by very quick motions; it very seldom remained in a perpendicular direction but generally inclined either one way or the other in a curve as a light body acted upon by wind is observed to do. During the whole time that it lasted smaller ones seemed to attempt to form in its neighbourhood; at last one did about as thick as a rope close by it and became longer than the old one which at that time was in its shortest state; upon this they Joined together in an instant and gradually contracting into the Cloud disappeared.

A Certain Sign the Country is Inhabited 11

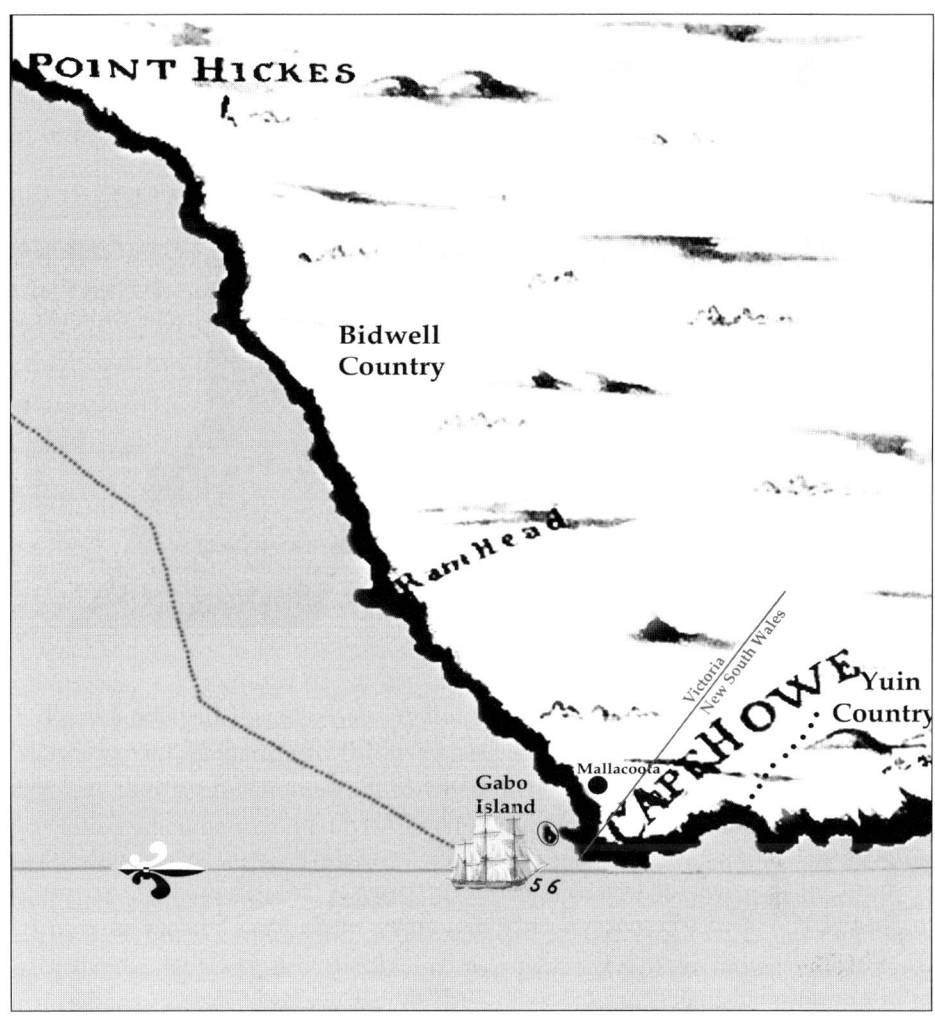

COOK: At 6 o'clock shortened sail and brought too for the night having 56 fathom a fine sandy bottom. A small Island lying close to a point on the Main bore west distant 2 leagues. [Gabo Island – marked but not named on chart above.] The point I have named <u>Cape Howe</u>. [Marked on chart above.]

12 The Endeavour Journals

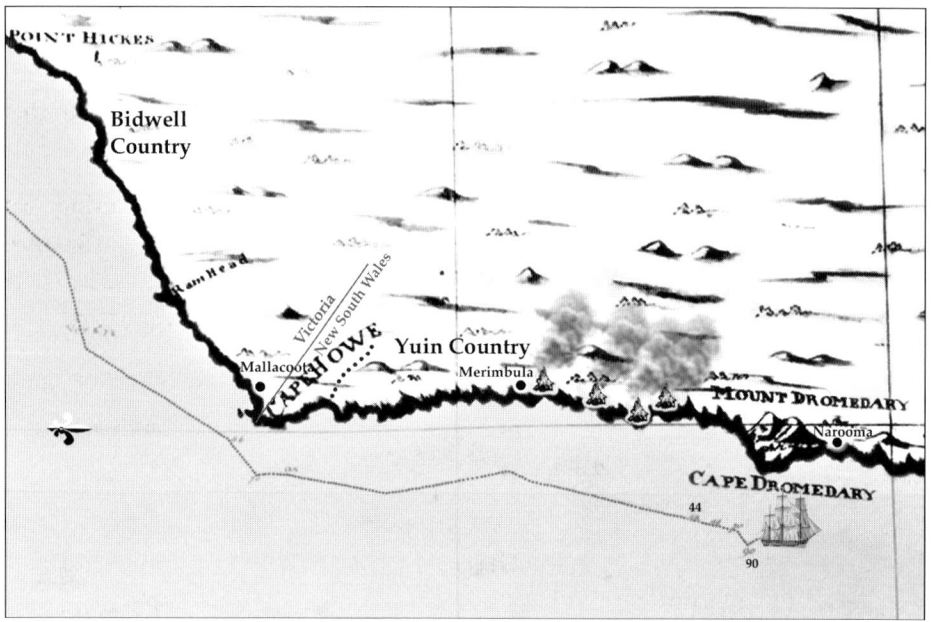

COOK: We at 10 o'clock wore and lay her head in until 4 a.m. at which time we made sail along shore to the Northward. The wind at SW with serene weather.

BANKS: The country this morn rose in gentle sloping hills which had the appearance of the highest fertility, every hill seemed to be clothed with trees of no mean size.

COOK: Very agreeable.

PARKINSON: About noon we saw some smoke ascending out of a wood near the seaside. [First smoke sighted at present-day Merimbula – Yuin country.]

BANKS: At noon a smoke was seen a little way inland...

21st April

BANKS: ... and in the evening several more.

COOK: A certain sign the country is inhabited.

PARKINSON: We saw some clouds of smoke rising a good way up the country, but we found no harbour.

COOK: At 6 o'clock [p.m.] being about 2 or 3 Leagues from the land we shortened sail and sounded and found **44** fathom water; stood on under an easy sail until midnight at which time we brought too until 4 a.m. when we made sail again having **90** fathom water.

BANKS: In the morn the land appeared much as it did yesterday but rather more hilly.

COOK: At 6 o'clock [a.m.] we were a breast of a pretty high mountain laying near the shore which on account of its figure I named <u>Mount Dromedary</u>. [Marked on chart above. Culturally significant spot for Yuin people – 'Gulaga'.] The shore underfoot of this mountain forms a point which I called <u>Cape Dromedary</u>.

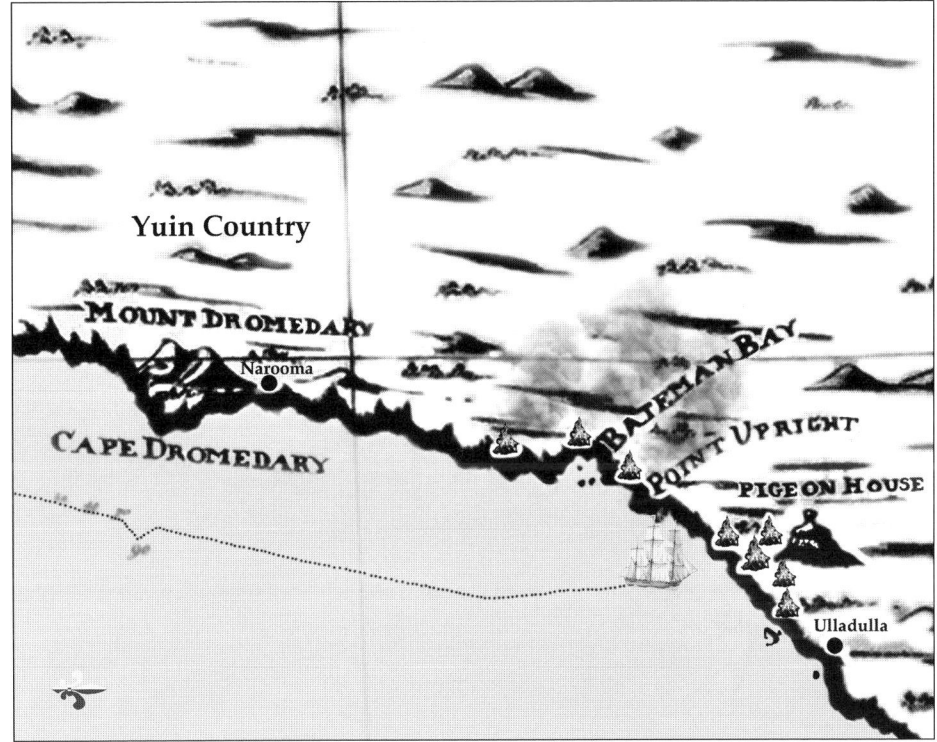

COOK: At Noon an open Bay wherein lay three or 4 small Islands bore NWBW. [Bateman Bay – marked on chart above.] This Bay seemed to be but very little sheltered from the sea winds and yet it is the only likely anchoring place I have yet seen upon the Coast.

22nd April

BANKS: In the even again it became flatter.

COOK: In the p.m. we saw the smoke of a fire in several places near the sea beach [vicinity of Bateman Bay].

14   The Endeavour Journals

BANKS: ... from whence we concluded it to be rather more populous.

COOK: At 5 o'clock we were abreast of a point of land which on account of its perpendicular cliffs I called <u>Point Upright</u>. [Marked on chart above.]

BANKS: At night five fires.

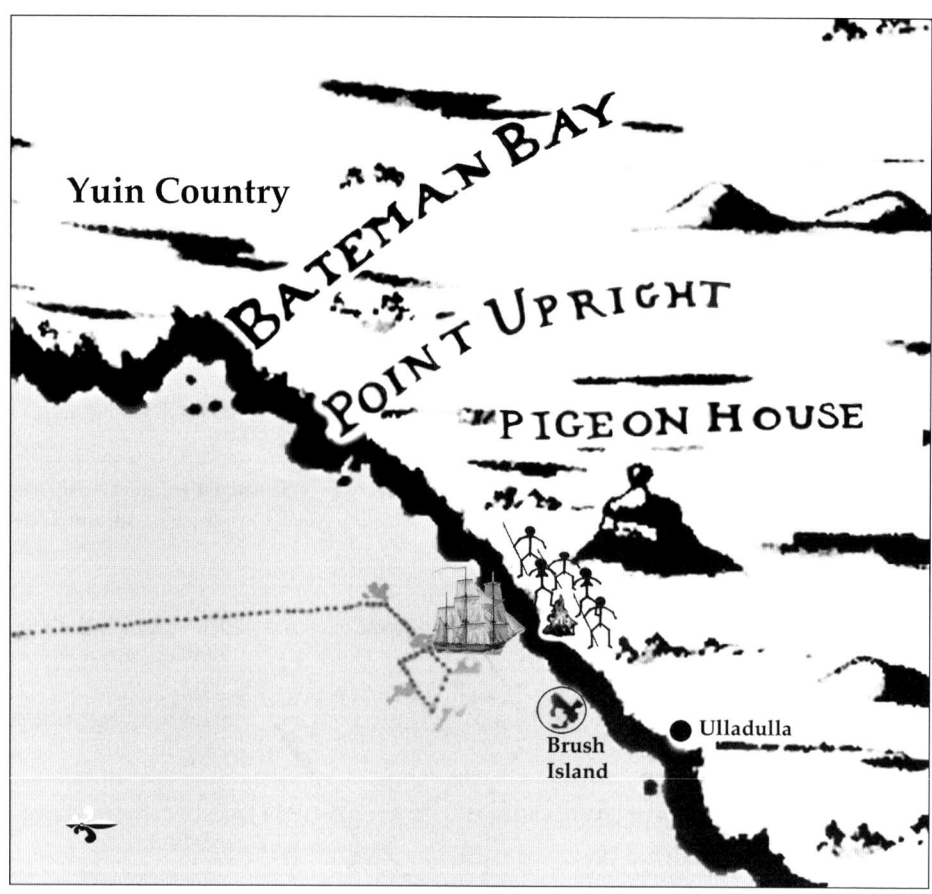

Detail of Cook's chart 'Point Hickes to Smoaky Cape'.

BANKS: In the morn we stood in with the land near enough to discern 5 people who appeared through our glasses to be enormously black. [Yuin people, on one of the beaches just south of Brush Island, the latter marked but not named on chart above.]

COOK: They appeared to be of a very dark or black colour but whether this was the real colour of their skins or the clothes they might have on I know not.

BANKS: So far did the prejudices which we had built on Dampier's account influence us that we fancied we could see their colour when we could scarce distinguish whether or not they were men.

PARKINSON: It is probable they live upon the produce of the earth, as we did not see any canoes, and the coast seems to be unfavorable for fishing.

BANKS: Since we have been on the coast we have not observed those large fires which we so frequently saw in the Islands and New Zealand made by the Natives in order to clear the ground for cultivation; we thence concluded not much in favour of our future friends.

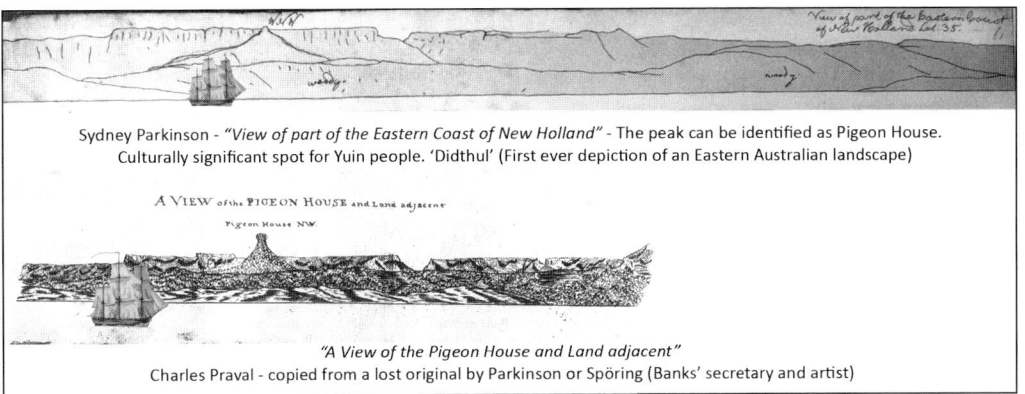

Sydney Parkinson - *"View of part of the Eastern Coast of New Holland"* - The peak can be identified as Pigeon House. Culturally significant spot for Yuin people. 'Didthul' (First ever depiction of an Eastern Australian landscape)

"A View of the Pigeon House and Land adjacent"
Charles Praval - copied from a lost original by Parkinson or Spöring (Banks' secretary and artist)

Pigeon House Hill. [Charles Praval joined the expedition later in Batavia (Jakarta) and was employed making copies of drawings of coastal views made earlier by Sydney Parkinson and Herman Spöring.]¹²

BANKS: A hill was in sight which much resembled those dove houses which are built four square with a small dome at the top.

COOK: Which looked like a <u>Pigeon House</u> and occasioned my giving it that name.

PARKINSON: The hills within land were remarkably flat.

COOK: Remarkably flat atop with steep rocky cliffs all round them as far as we could see.

BANKS: It has long been an observation among us that the air in this Southern hemisphere was much clearer than in our northern, these some days at least it has appeared remarkably so. The hill like a pigeon house was seen at a very great distance; the little dome on the top of it was first thought to be a rock standing up in the sea long before any other part was seen, and when we came up with it we found it to be several miles inland.

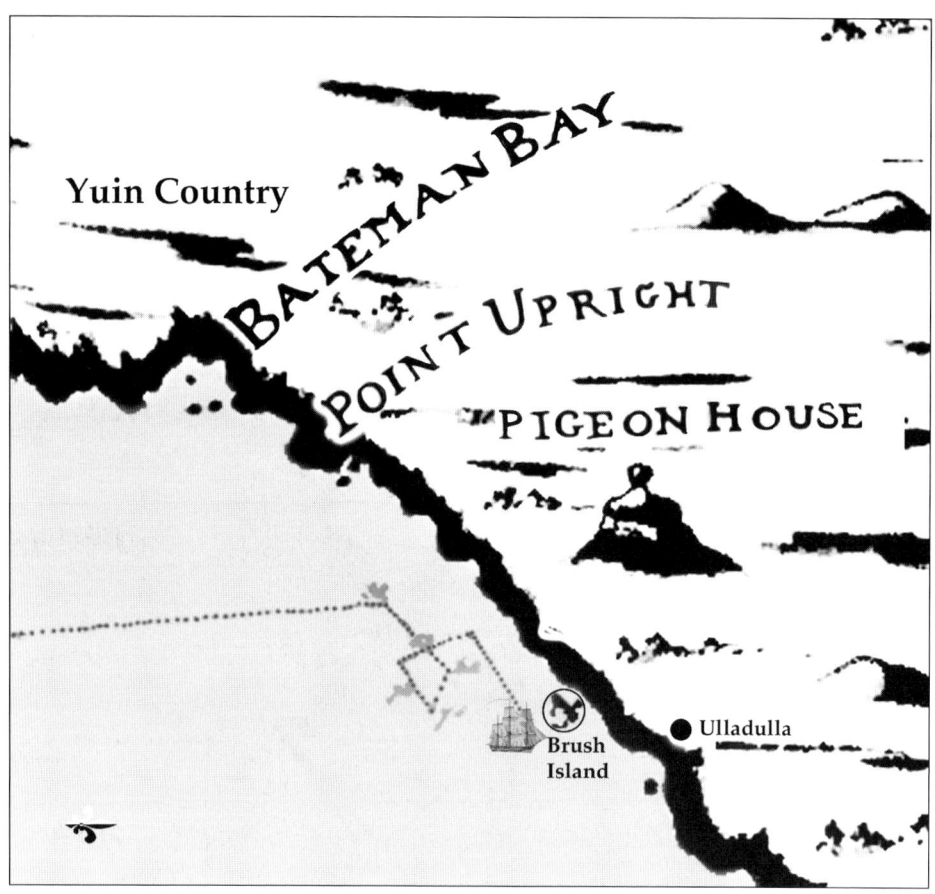

COOK: A small low Island laying close under the shore bore NW. [Brush Island – marked but not named on chart above.] When we first discovered this Island in the morning I was in hopes from its appearance that we should have found shelter for the Ship behind it but when we came to approach it near I did not think that there was even security for a boat to land, but this I believe I should have attempted had not the wind come on shore after which I did not think it safe to send a boat from the ship as we had a large hollow sea from ye SE rowling in upon the land which beat every were very high upon the Shore and this we have had ever since we came upon the Coast.

A Certain Sign the Country is Inhabited 17

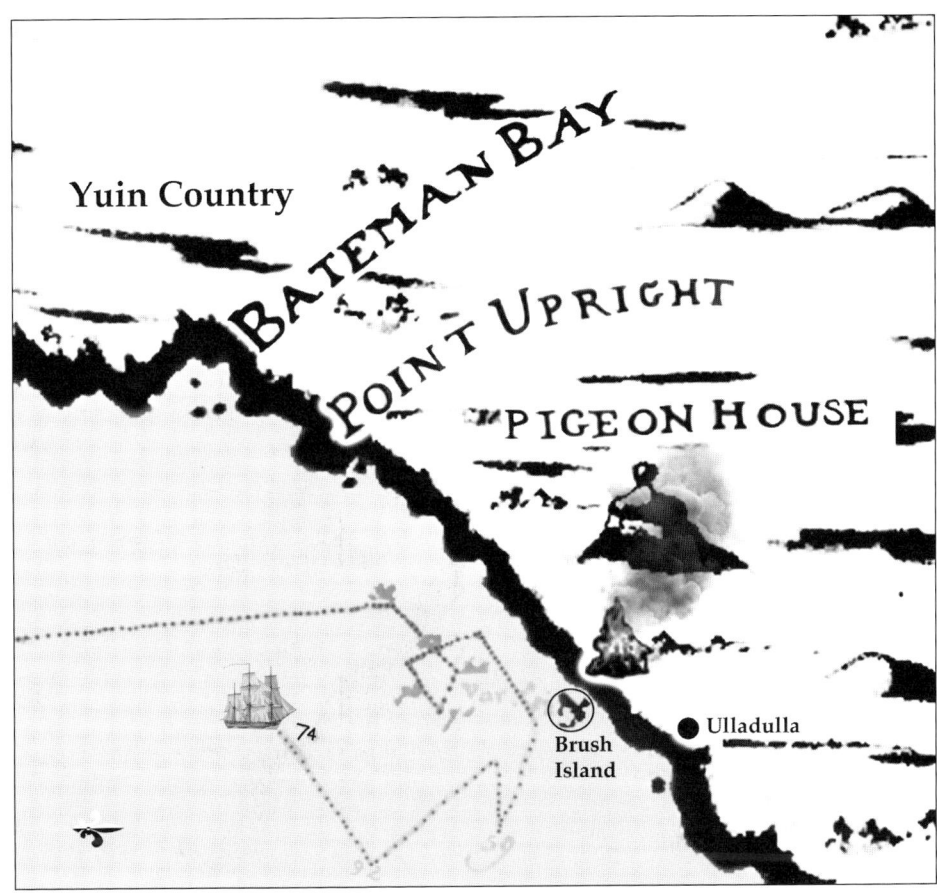

23rd April

BANKS: The ship was too far from the shore to see much of it. A larger fire was however seen than any we have seen before.

COOK: We were about Six Leagues from the land. In this situation had 74 fathom water.

BANKS: Calm today, myself in small boat but saw few or no birds. [Banks had his own skiff for his private use.] Took with the dipping net 1. *Holothuria obtusata* [now *Physalia physalis*], 2. *Mimus volutator* [now *Glacus atlantius*], 3. *Medusa pelagica* [now *Pelagia noctiluca*]. (Colour Plate No. 1)

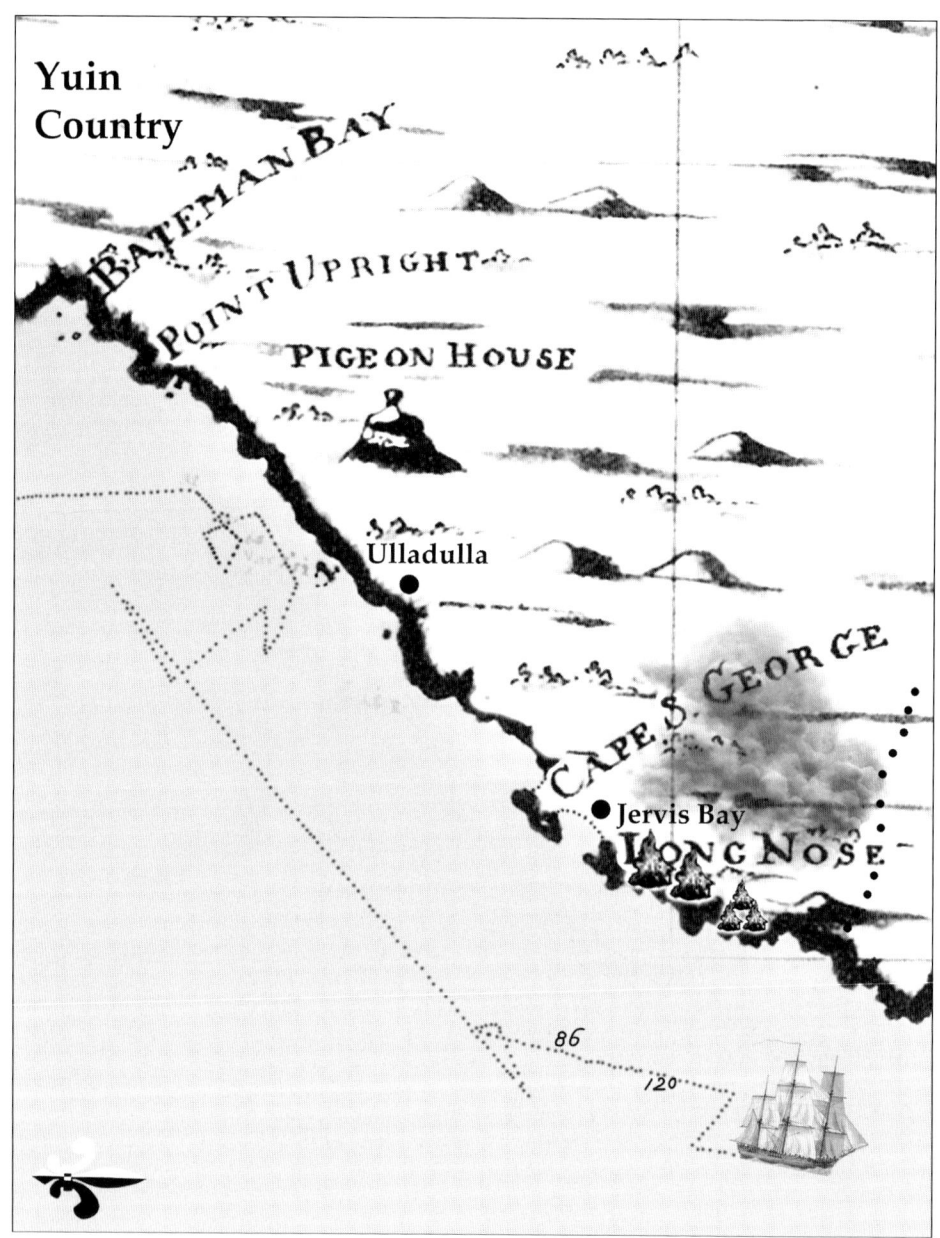

24th April

COOK: In the p.m. had variable Light airs and calms until 6 o'clock at which time a breeze sprung up. Stood to the NE. At noon a point of land I named <u>Cape St George</u> we having discovered it on that Saints day, bore west 19 miles. [Marked on chart above.]

BANKS: Two large fires were seen and several smaller.

25th April

COOK: At 5 o'clock [p.m.] sounded and had **86** fathom water. At 8 o'clock being very squally with lightning we close reefed the top sails and brought too in **120** fathom. At 3 a.m. we made sail again to the northward.

About 2 leagues to the northward of Cape St George the shore seems to form a bay [Jervis Bay] which appeared to be sheltered from the NE winds but as we had the wind it was not in my power to look into it. The north point of this bay on account of its figure I named <u>Long Nose</u>. [Marked on chart above.]

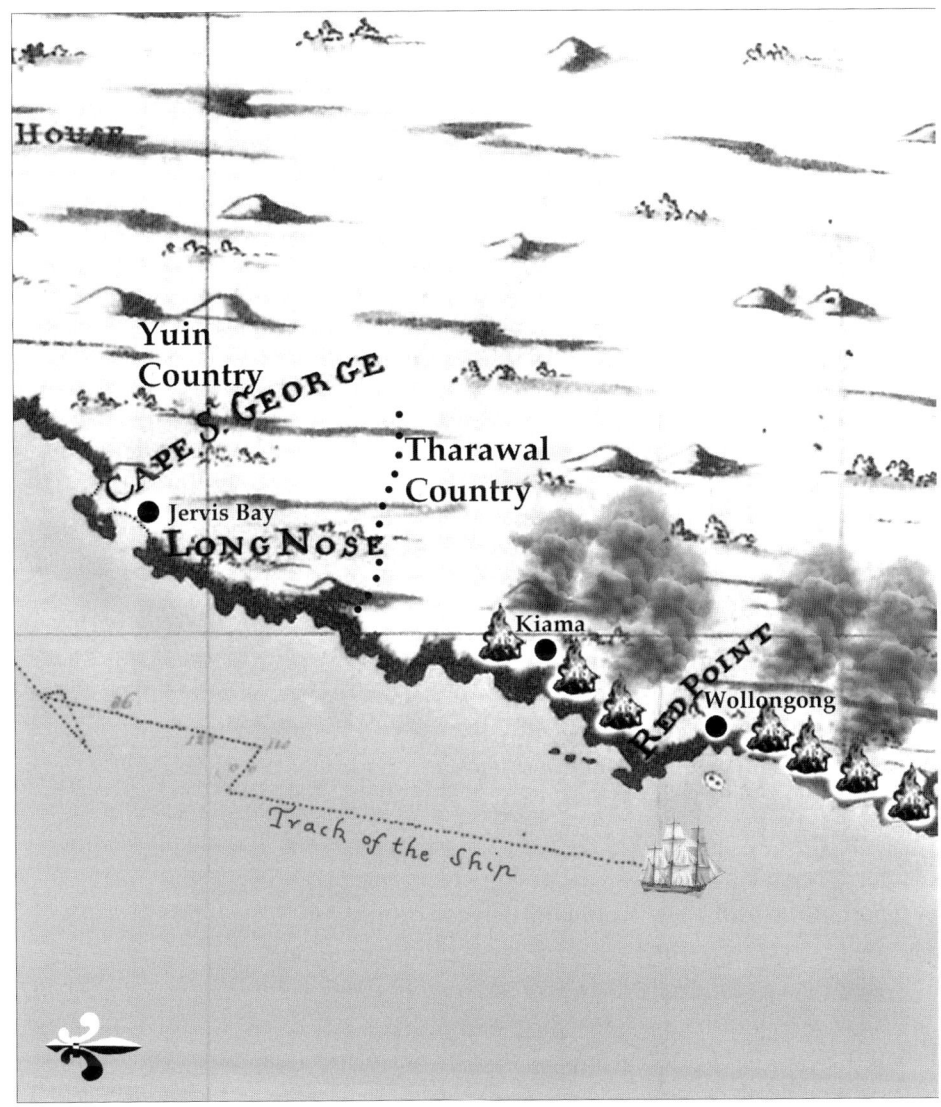

COOK: Eight Leagues to the northward of this is a point which I called Red Point, some part of the land about it appeared of that colour. [Marked on chart above.] A little way inland to the NW of this point is a round hill the top of which looks like a crown of a hatt [Mount Kembla or Hat Hill]. Course and distance Sailed since yesterday noon is NBE 49 Miles. In the Course of this days run we saw the smoke of fire in several places near the sea beach.

BANKS: Large fires were lighted this morn about 10 o'clock, we supposed that the gentlemen ashore had a plentiful breakfast to prepare.

PARKINSON: We saw several fires along the coast lit up one after another, which might have been designed as signals to us [Vicinity of Red Point marked on chart above – near Wollongong. Tharawal country].

BANKS: The country tho in general well enough clothed appeared in some places bare; it resembled in my imagination the back of a lean cow, covered in general with long hair, but nevertheless where her scraggy hip bones have stuck out farther than they ought accidental rubs and knocks have entirely bard them of their share of covering.

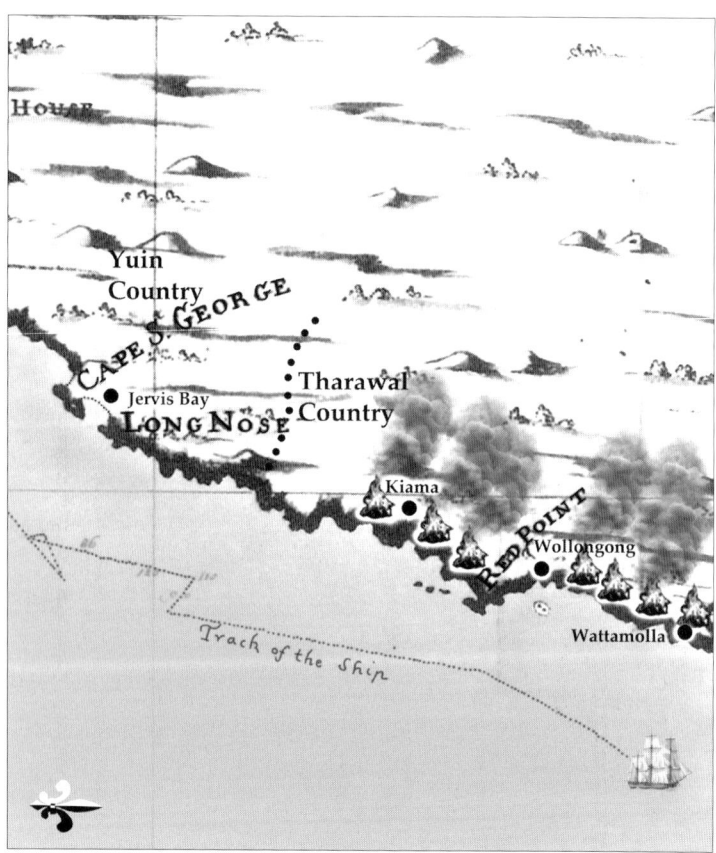

26th April

BANKS: In the evening it was calm.

COOK: Saw several smokes along shore before dark.

BANKS: All the fires were put out about 5 O'clock.

COOK: Two or 3 times a fire in the night.

BANKS: Fires were seen during the day the same as yesterday but none so large. Land today more barren in appearance that we had before seen it.

COOK: At Noon in the Latitude of 34°10' about 5 Leagues from the land are some white cliffs which rise perpendicularly from the sea to a moderate height [vicinity of Wattamolla].

BANKS: Chalky cliffs something resembling those of old England; within these it was flat and might be no doubt as fertile.

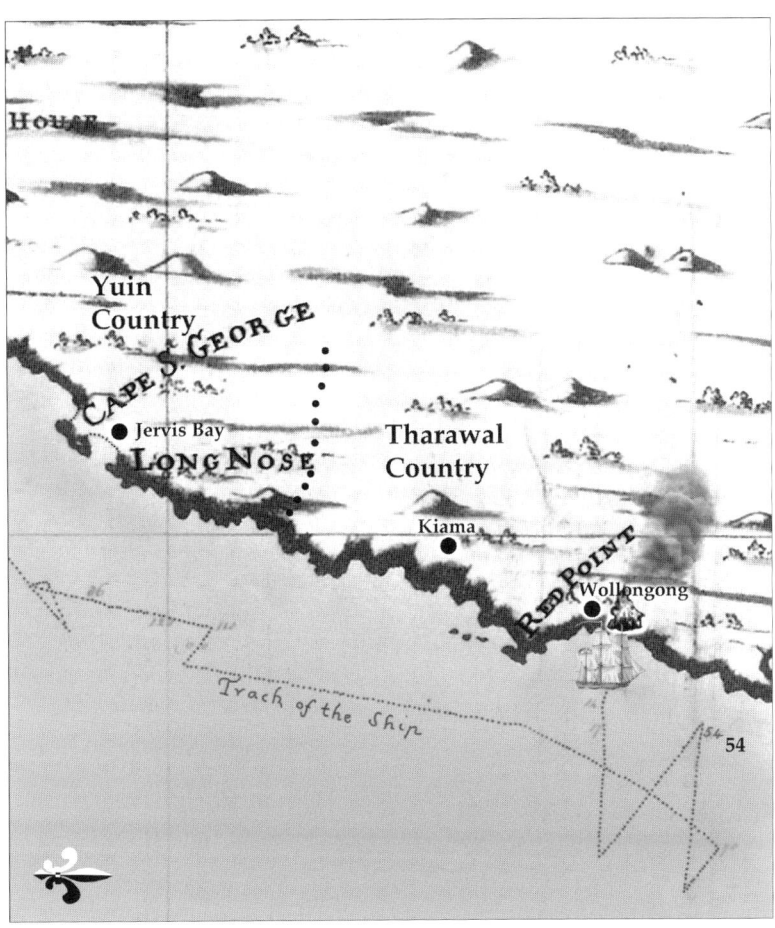

22   The Endeavour Journals

27th April

COOK: In the PM stood off shore until 2 o'clock then tacked and stood in until 6 [p.m.] at which time we tacked and stood off being then in **54** fathom and about 4 or 5 miles from land.

PARKINSON: In the morning, the wind being against us, we stood off and on shore.

BANKS: At noon we were very near it; one fire only was in sight [just north of Red Point – Tharawal country]. Some bodies of 3 feet long and half as broad floated very boyant past the ship; they were supposed to be cuttle bones which indeed they a good deal resembled but for their enormous size [Giant Cuttlefish – *Sepia apama*. Maximum size 80 cm]. (Colour Plate No. 2)

The Pinnace by Sydney Parkinson.[13]

28th April

BANKS: After dinner the Captain proposed to hoist out our boats and attempt to land, which gave me no small satisfaction.

COOK: The Pinnace took in water so fast that she was obliged to be hoisted in again to stop her leaks.

BANKS: Four men were at this time observed walking briskly along the shore, two of which carried on their shoulders a small canoe; they did not however attempt to put her in the water so we soon lost all hopes of their intending to come off to us, a thought with which we once had flattered ourselves. To see something of them however we resolved.

A Certain Sign the Country is Inhabited   23

The Yawl by Sydney Parkinson.[14]

COOK: Being now not above two miles from the shore Mister Banks, Doctor Solander, Tupia and myself put off in the yawl. [Tupia, a Tahitian High Priest, skilled navigator and Polynesian speaker, and his boy-servant Taiyota were taken on board in Tahiti.]

Taiyota – portrait by Sydney Parkinson.[15]

24   The Endeavour Journals

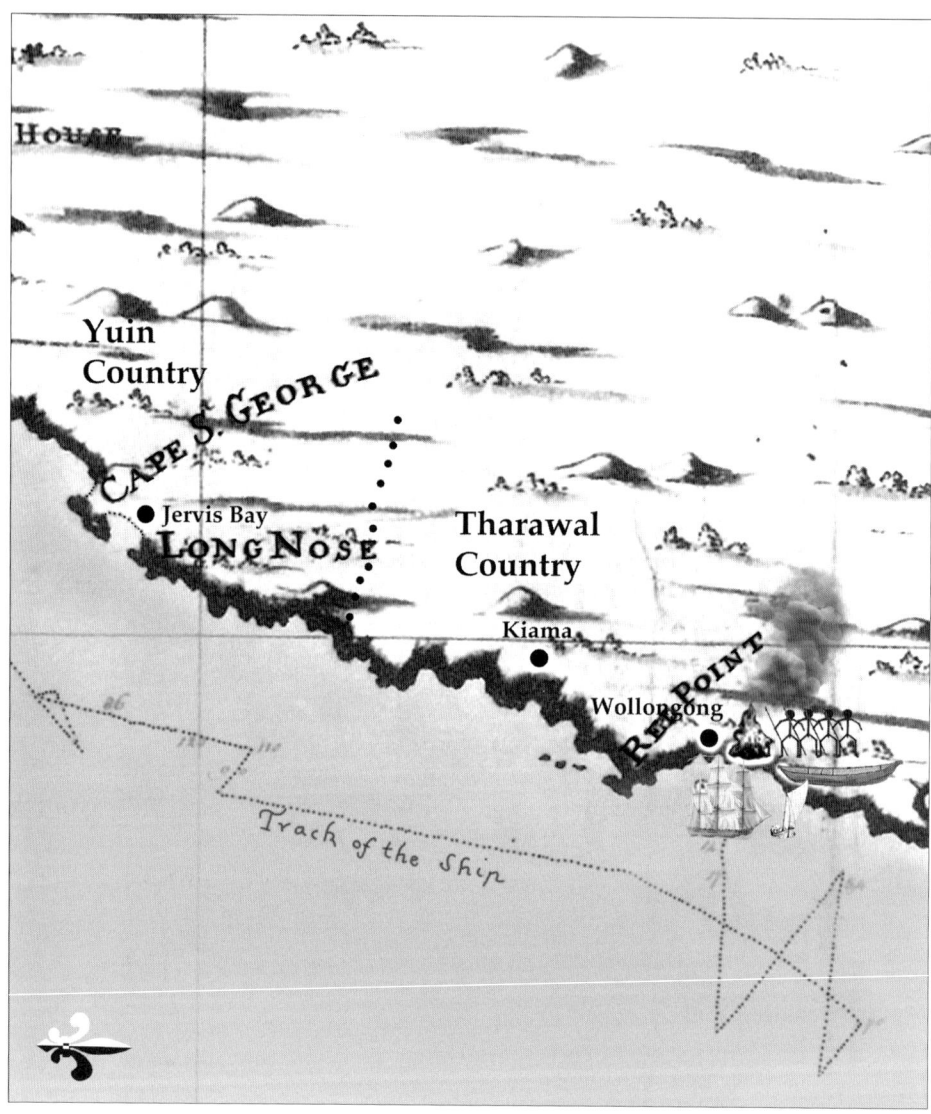

BANKS: The men sat on the rocks expecting us but when we came within about a quarter of a mile they ran away hastily into the country.

COOK: … which disappointed us in the expectation we had of getting a near view of them if not to speak to them.

BANKS: They appeared to us as well as we could judge at that distance exceedingly black. Near the place were four small canoes which they left behind.

COOK: Our disappointment was heightened when we found that we nowhere could affect a landing by reason of the great surf which beat everywhere upon the shore.

BANKS: We were obliged to content ourselves with gazing from the boat at the productions of nature which we so much wished to enjoy a nearer acquaintance with. Among them we could discern many cabbage trees [*Livistona australis*] but nothing else which we could call by any name. (Colour Plate No. 3)

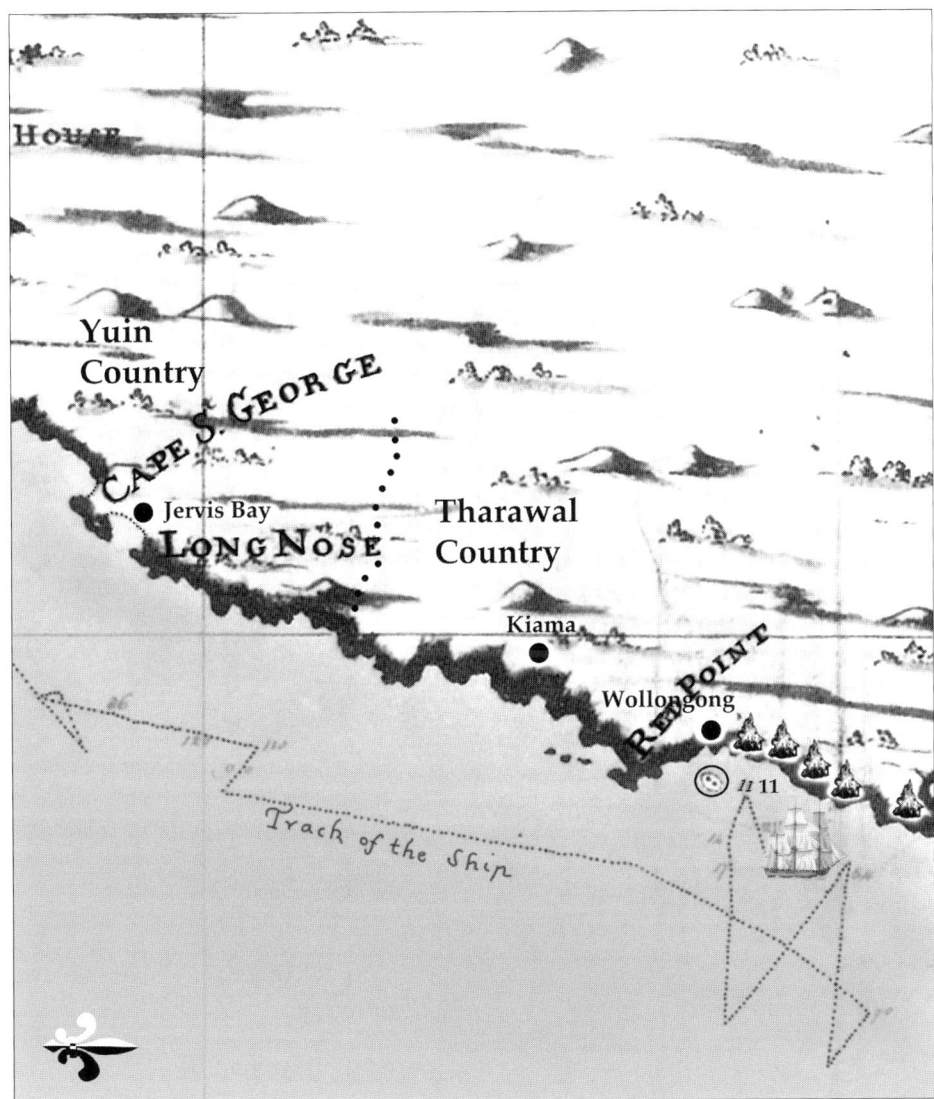

COOK: We returned to the Ship about 5 in the evening. At this time it fell calm and we were not above a mile and a half from shore in **11** fathom water and within some **breakers** that lay to the southward of us, but luckily a light breeze came off from the land which carried us out of danger and with which we stood to the northward.

PARKINSON: From the ship, the country looked very pleasant and fertile; and the trees, quite free from underwood, appeared like plantations in a gentleman's park.

BANKS: In the course of the night many fires were seen.

Chapter 2

Resolved to Dispute our Landing to the Utmost

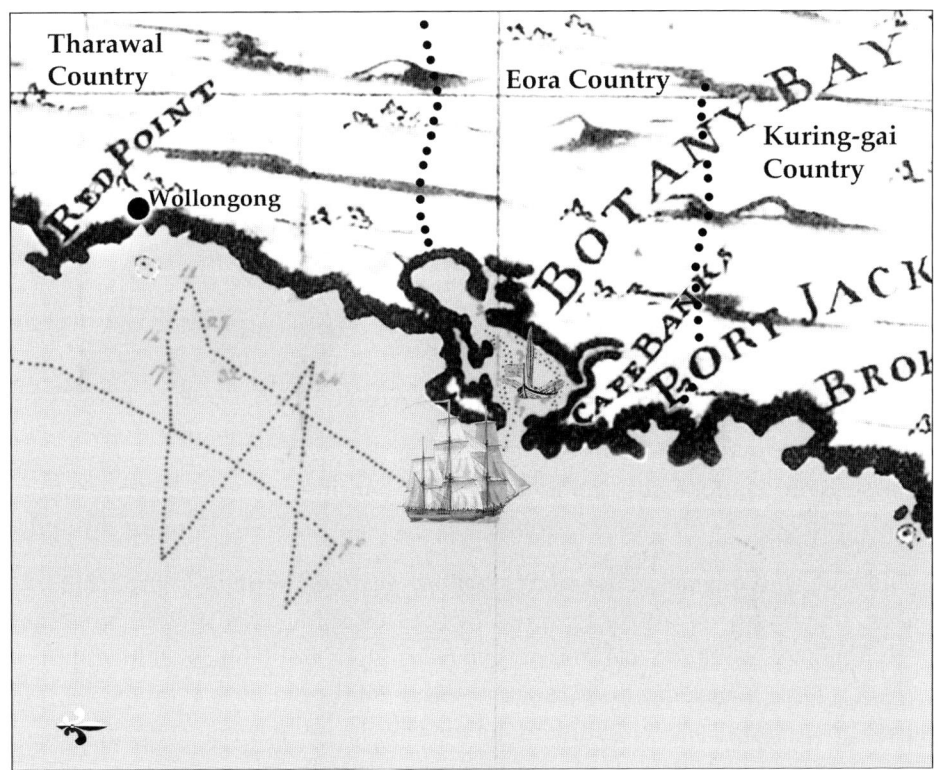

Detail of Cook's chart 'Point Hickes to Smoaky Cape.'¹

28th April (cont'd)

COOK: At day light in the morning we discovered a Bay [Botany Bay] which appeared to be tolerably well sheltered from all winds into which I resolved to go with the Ship and with this view sent the Master in the Pinnace to sound the entrance while we kept turning up with the Ship having the wind right out.²

28   The Endeavour Journals

Ship's Master – Robert Molineux (1746–1771)[3]

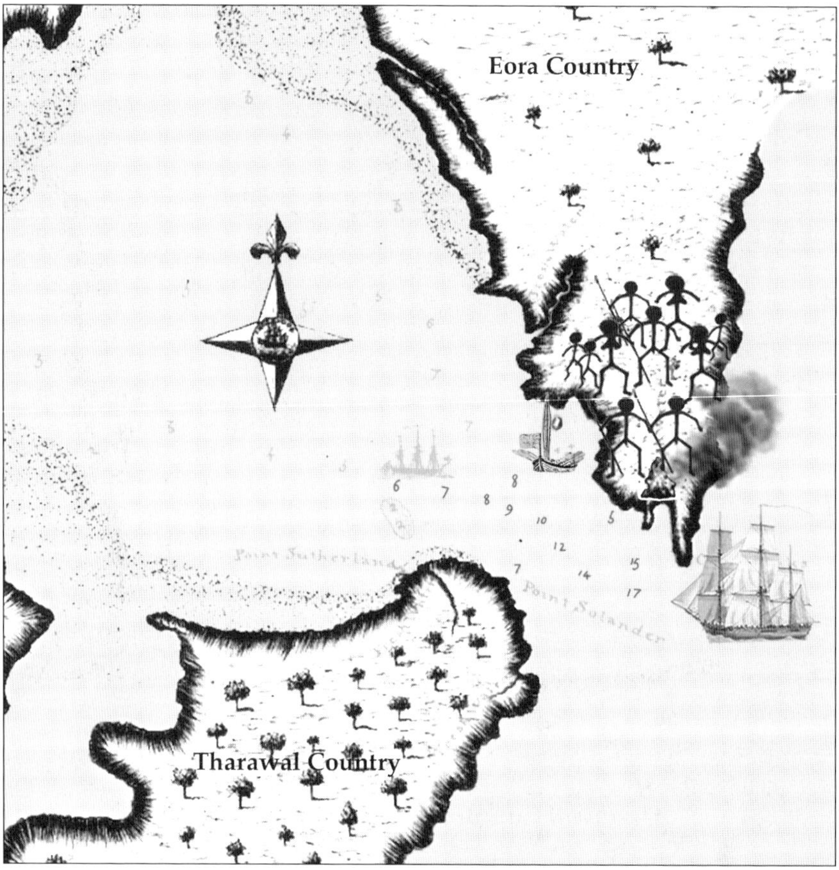

Portion of a Sketch of Botany Bay – JAMES COOK – Isaac Smith.[4] (Sixteen-year-old Isaac Smith, Able Seaman, became adept at drawing and copying charts surveyed by Cook.)

COOK: At Noon the entrance bore NNW distance 1 mile.

BANKS: A small smoke arising from a very barren place directed our glasses that way and we soon saw about 10 people, who on our approach left the fire and retired to a little eminence where they could conveniently see the ship [Eora people – Kameygal clan – on the north head of Botany Bay]. Soon after this two Canoes carrying 2 men each landed on the beach under them, the men hauled up their boats and went to their fellows upon the hill. Our boat which had been sent ahead to sound now approached the place and they all retired higher up on the hill. Our boat proceeded along shore [north shore] and the Indians followed her at a distance. When she came back the officer [Robert Molineux] who was in her told me that in a cove a little within the harbour they came down to the beach and invited our people to land by many signs and words which he did not at all understand; all however were armed with long pikes and a wooden weapon made something like a short scimitar.[5]

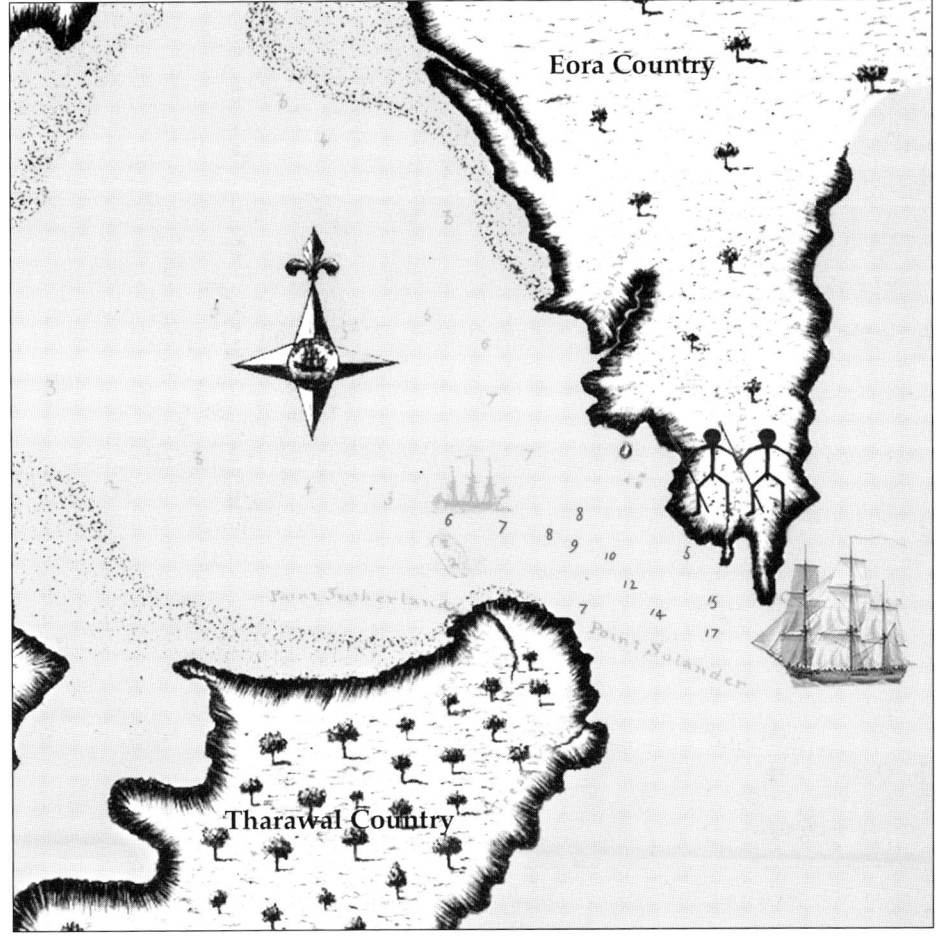

30 The Endeavour Journals

Sydney Parkinson's sketch of the two Eora warriors seen on the north penisular at the entrance to Botany Bay.[6] (This pencil drawing from Sydney Parkinson's hand is one of only two original images of New Holland Aborigines known to exist. It's a field drawing. Done on the spot and during the event as it took place.)

BANKS: During this time a few of the Indians who had not followed the boat remained on the rocks opposite the ship, threatening and menacing with their pikes and swords – two in particular who were painted with white, their faces seemingly only dusted over with it, their bodies painted with broad strokes drawn over their breasts and backs resembling much a soldiers cross belts, and their legs and thighs also with such like broad strokes drawn round them which imitated broad garters or bracelets. Each of these held in his hand a wooden weapon about 2 and a half feet long, in shape much resembling a scimitar; the blades of these looked whitish and some thought shining insomuch that they were almost of opinion that they were made of some kind of metal, but myself thought they were no more than wood smeared over with the same white pigment with which they paint their bodies. These two seemed to talk earnestly together, at times brandishing their crooked weapons at us as in token of defiance.

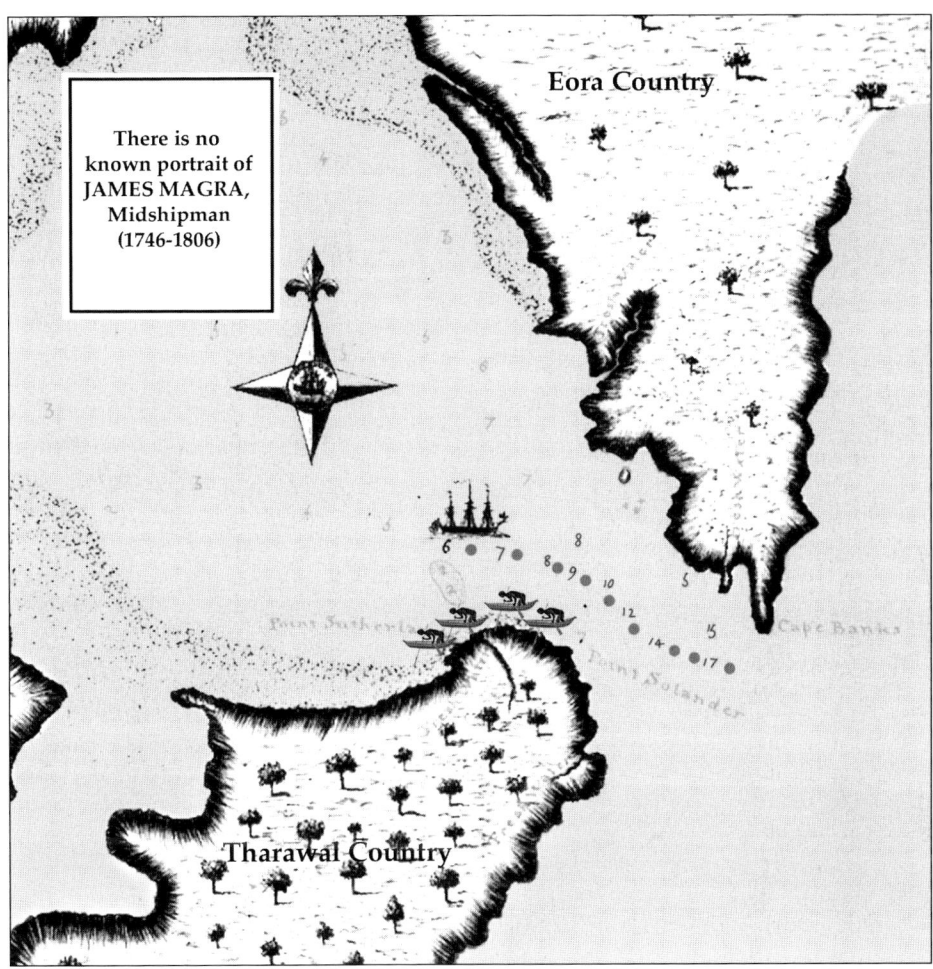

29th April

COOK: In the p.m. wind southerly clear weather with which we stood into the bay.

BANKS: Under the South head of it were four small canoes; in each of these was one man who held in his hand a long pole with which he struck fish, venturing with his little embarkation almost into the surf. These people seemed to be totally engaged in what they were about: the ship passed within a quarter of a mile of them and yet they scarce lifted their eyes from their employment; I was almost inclined to think that attentive to their business and deafened by the noise of the surf they neither saw nor heard her go past them [Tharawal men – Gweagal clan].

MAGRA: At half past one in the afternoon we **anchored** in 6 fathom and a half, sandy ground.[7]

BANKS: We came to an anchor abreast of a small village consisting of about 6 or 8 houses. Soon after this an old woman followed by three children came out of the wood; she carried several pieces of stick and the children also had their little burthens; when she came to the houses 3 more younger children came out of one of them to meet her. She often looked at the ship but expressed neither surprise nor concern [Tharawal people – Gweagal clan – present-day Kurnell].

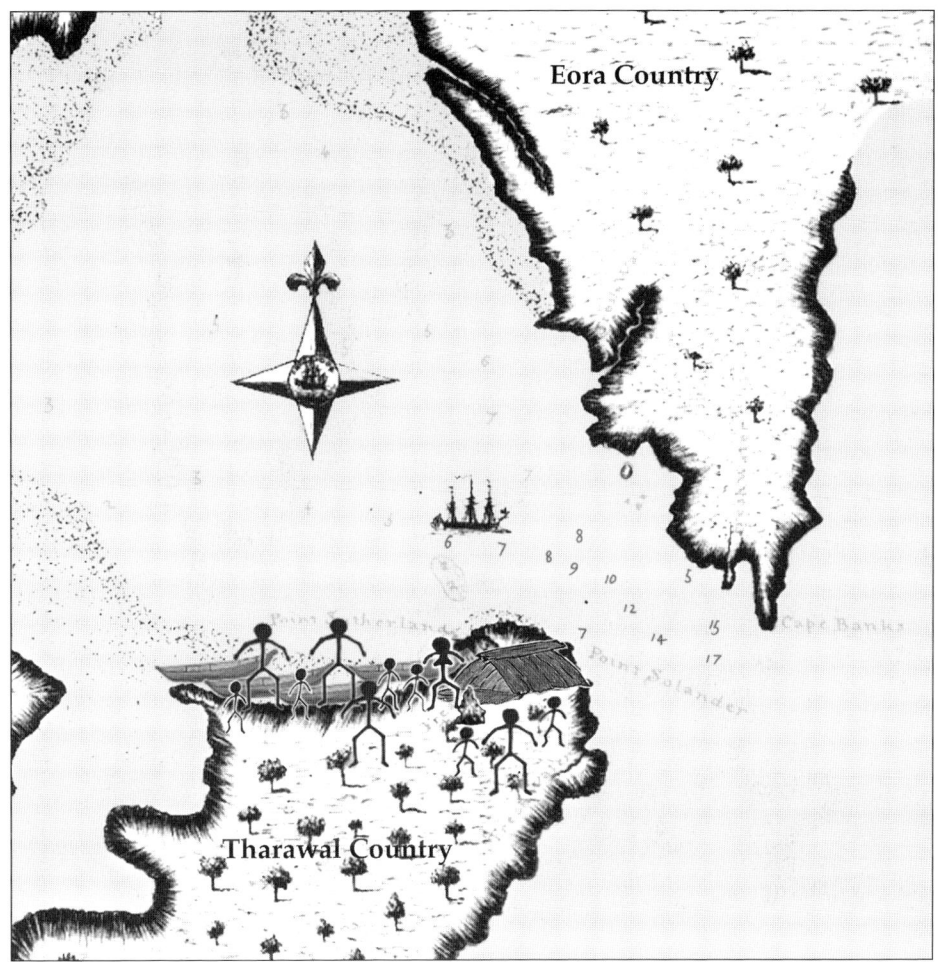

BANKS: Soon after this she lighted a fire and the four Canoes came in from fishing; the people landed, hauled up their boats and began to dress their dinner to all appearance totally unmoved at us, tho we were within a little more than a quarter of a mile of them.

Of all these people we had seen so distinctly through our glasses we had not been able to observe the least signs of clothing: myself to the best of my judgement plainly discerned that the woman did not copy our mother Eve even in the fig leaf.

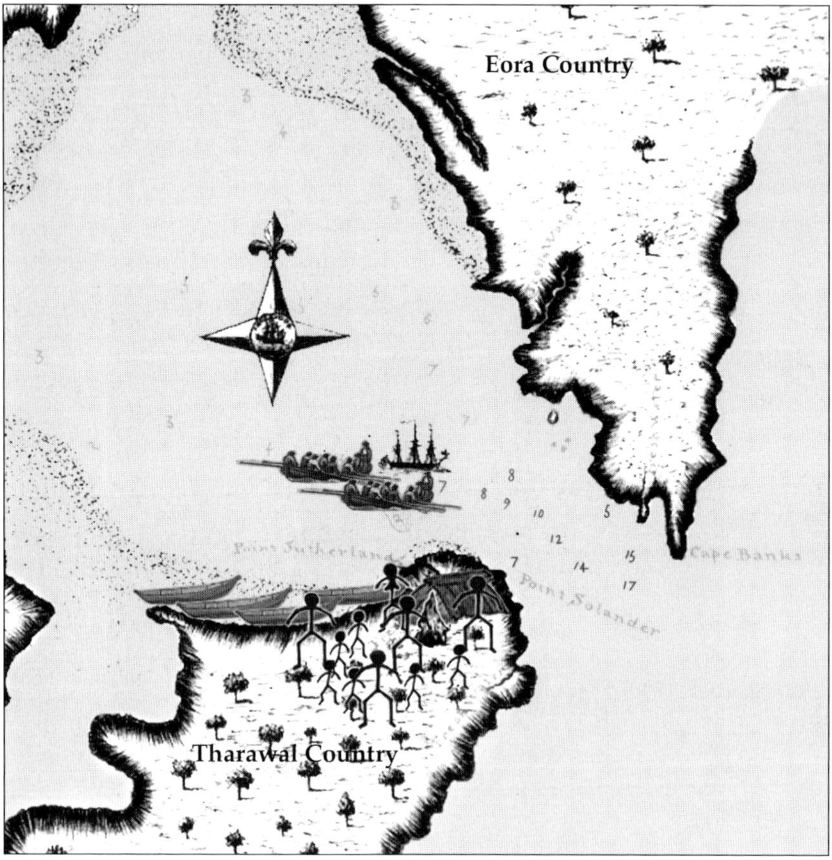

![Map showing Eora Country and Tharawal Country with Point Sutherland, Point Solander, and Cape Banks]

BANKS: After dinner, the boats were manned and we set out from the ship intending to land at the place where we saw these people, hoping that as they regarded the ships coming in to the bay so little they would as little regard our landing.

EARL OF MORTON: The following hints, hastily put together; and probably very incorrect, are however humbly submitted to the consideration of Captain Cook and the other Gentlemen, by their hearty well-wisher and Most obedient Servant James Douglas, 14th Earl of Morton and President of the Royal Society.

Earl of Morton.[8]

> To exercise the utmost patience and forbearance with respect to the natives of the several Lands where the ship may touch.

> To check the petulance of the sailors, and restrain the wanton use of Fire Arms.

To have it still in view that shedding the blood of those people is a crime of the highest nature. They are human creatures, the work of the same omnipotent Author, equally under his care with the most polished Europeans; perhaps being less offensive, more entitled to his favour.

They are the natural, and in the strictest sense of the word, the legal possessors of the several Regions they inhabit.

No European Nation has a right to occupy any part of their country, or settle among them without their voluntary consent. Conquest over such people can give no just title; because they could never be the Aggressors.

They may naturally and justly attempt to repel intruders, whom they may apprehend are come to disturb them in the quiet possession of country, whether that apprehension be well or ill founded.

Therefore should they in a hostile manner oppose a landing, and kill some men in the attempt, even this would hardly justify firing among them, 'till every other gentle method had been tried.[9]

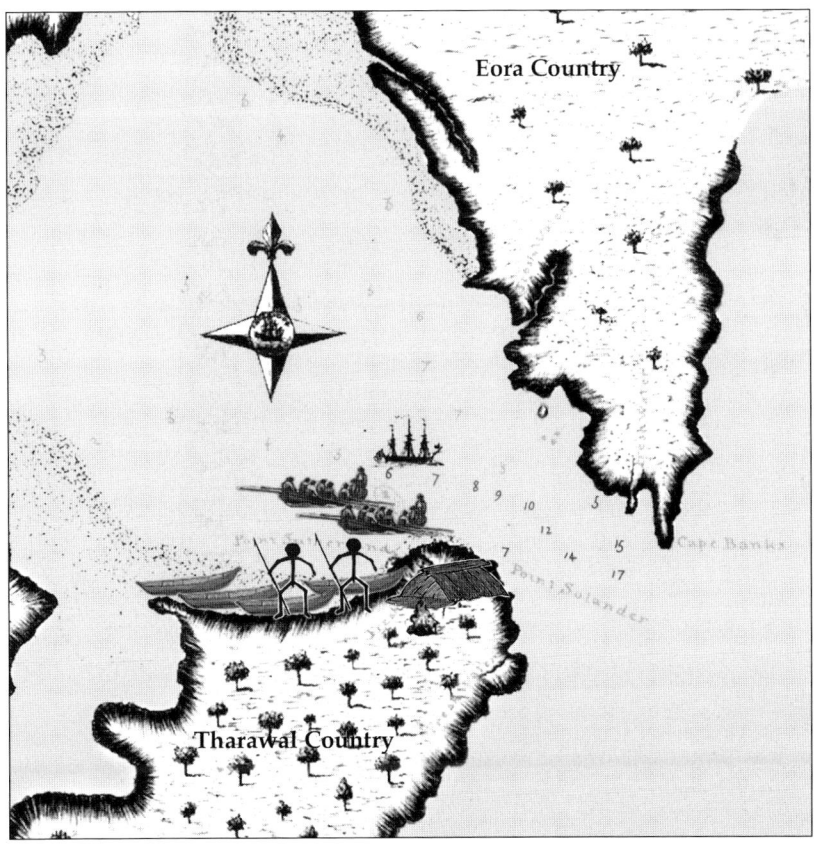

BANKS: As soon as we approached the rocks two of the men came down upon them, each armed with a lance of about 10 feet long and a short stick which he seemed to handle as if it was a machine to throw the lance.

PARKINSON: Their countenance bespoke displeasure; they threatened us, and discovered hostile intentions, often crying to us, 'Warra warra wai'.

MAGRA: On their breasts we observed rude figures of men, darts &c. done with a kind of white paint.

Two of the Natives of New Holland, Advancing in Combat[10] Engraving by Thomas Chambers 1773. [With the death of Sydney Parkinson later in the voyage, Stanfield Parkinson was left to compile his brother's journal for publication. Stanfield realized he had no developed drawing of New Holland Aborigines to illustrate the journal. To remedy this omission, he commissioned this engraving to be made. This was based on written descriptions of people recorded in Sydney's journal from both Botany Bay and Endeavour River. Reference could not be made to his brother's sketch of the two Eora warriors above. These were in Banks's hands. For this reason, much of what was authentic in Sydney's drawing was left out or wrongly imagined in the engraving. Using these slim resources Stanfield Parkinson instructed Thomas Chambers to assemble this invented image. (Banks to Hawkesworth Mitchel Library Sydney, MS A80.4, 12th January 1773.)]

BANKS: They called to us very loud in a harsh sounding Language, shaking their lances and menacing, in all appearance resolved to dispute our landing to the utmost tho they were but two and we 30 or 40 at least.

COOK: As soon as I saw this I ordered the boats to lay upon their oars in order to speak to them but this was to little purpose for neither us nor Tupia could understand one word they said.

BANKS: In this manner we parleyd [to negotiate by word of mouth] with them for about a quarter of an hour, they waving to us to be gone, we again signing that we wanted water and that we meant them no harm.

COOK: We then threw them some nails beads &c. a shore which they took up and seemed not ill pleased in so much that I thought that they beckoned to us to come ashore but in this we were mistaken for as soon as we put the boat in they again came to oppose us.

PARKINSON: They kept aloof and dared us to come on shore.

BANKS: They remained resolute so a musquet [musket] was fired over them.

COOK: I fired a musquet between the two which had no other effect than to make them retire back where bundles of their darts lay and one of them took up a stone and threw at us which caused my firing a second Musquet load with small shot.

BANKS: Small shot was fired at the Eldest of the two who was about 40 yards from the boat; it struck him on the legs but he minded it very little so another was immediately fired at him. On this he ran up to the house about 100 yards distant and soon returned with a shield.

PARKINSON: And also a wooden sword, and then they advanced boldly, gathering up stones as they came along, which they threw at us.

BANKS: In the mean time we had landed on the rock. He immediately threw a lance at us and the young man another which fell among the thickest of us.

PARKINSON: One of which fell between my feet!

BANKS: Two more musquets with small shot were then fired at them on which the Eldest threw one more lance and then ran away as did the other.

PARKINSON: They took the alarm and were very frantic and furious, shouting for assistance, calling 'Hala, hala, mae'; that is, (as we afterwards learned,) Come hither; while their wives and children set up a most horrid howl. We endeavoured to pacify them, but to no purpose, for they seemed implacable.

MAGRA: Unsupported by any of their countrymen, they retreated slowly to their houses within the bushes, but constantly faced us the whole way.

COOK: But not in such haste but what we might have taken one, but Mr Banks being of opinion that the darts were poisoned made me cautious how I advanced into the woods.

Sydney Parkinson's sketch of one of the huts and the shield.[11]

COOK: We found here a few Small huts made of the bark of trees in one of which were four or five small children.

MAGRA: Their huts were wretchedly built, as they consisted of nothing more than pieces of bark of trees loosely spread over a few cross spars, about four feet above the ground.

BANKS: We found the children hid behind the shield and a piece of bark in one of the houses.

We were conscious from the distance the people had been from us when we fired that the shot could have done them no material harm; we therefore resolved to leave the children on the spot without even opening their shelter. We therefore threw into the house to them some beads, ribbands, cloths &c. as presents and went away.

The original shield now in the British Museum, London. Drawing by John Frederick Miller (1771) – sometimes ascribed to Sydney Parkinson.[12]

Three of the 'forty or fifty' spears confiscated at Botany Bay, held in the Cambridge University Museum of Archaeology and Anthropology, currently on loan to Chau Chak Wing Museum, University of Sydney. Soon to be returned to the Gweagal people at Botany Bay.[13]

BANKS: We however thought it no improper measure to take away with us all the lances which we could find about the houses, amounting in number to forty or fifty. They were of various lengths, from 15 to 6 feet in length; both those which were thrown at us and all we found except one had 4 prongs, headed with very sharp fish bones, which were besmeared with a greenish coloured gum that at first gave me some suspicions of Poison. Upon examining the lances we had taken from them we found that the very most of them had been used in striking fish, at least we concluded so from sea weed which was found stuck in among the four prongs.

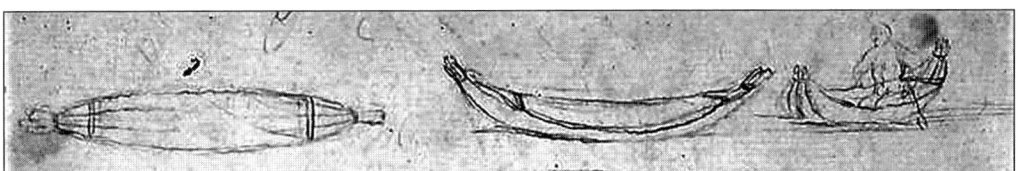

Sydney Parkinson's sketch of the canoes.[14]

COOK: Three Canoes lay upon the beach the worst I think I ever saw. They were about 12 or 14 feet long made of one piece of the bark of a tree drawn or tied up at each end and the middle kept open by means of pieces of sticks by way of Thwarts.

BANKS: The people were blacker than any we have seen in the voyage tho by no means negroes; their beards were thick and bushy and they seemed to have a redundancy of hair upon those parts of the body where it commonly grows; the hair of their heads was bushy and thick but by no means woolly like that of a Negro; they were of a common size, lean and seemed active and nimble; their voices were coarse and strong.

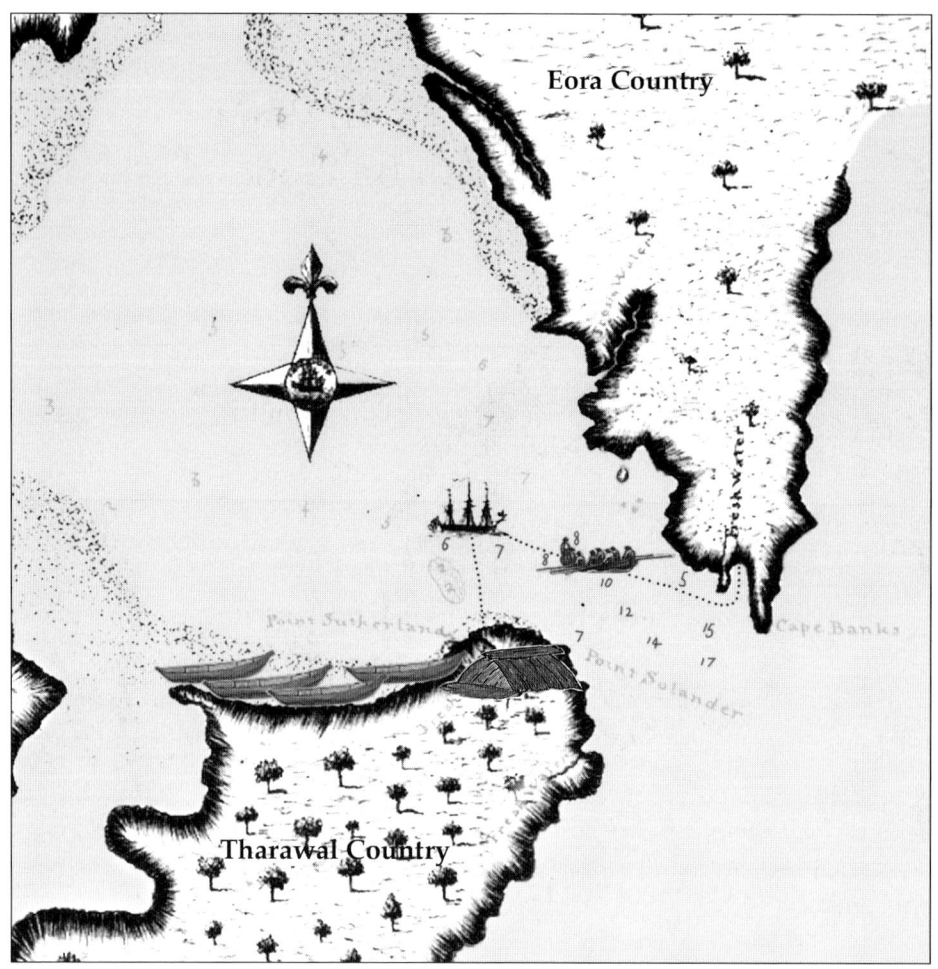

COOK: After searching for fresh water without success except a little in a small hole dug in the sand, we embarked.

BANKS: We returned to the ship in order to get rid of our load of lances, and having done that went to that place at the mouth of the harbour where we had seen the people in the morn.

COOK: When we landed now there were no body to be seen – we found here some fresh water which came trinkling down and stood in pools among the rocks: but this was troublesome to come at.

BANKS: At night many moving lights were seen in different parts of the bay such as we had been used to see at the Islands; from hence we supposed that the people here strike fish in the same manner. The fires were seen during the greatest part of the night.

(END OF FIRST DAY BOTANY BAY – CIVIL TIME)

Chapter 3

All They Seemed To Want Was For Us To Be Gone

29th April (cont'd)

COOK: I sent a party of men a shore in the morning to the place where we first landed to dig holes in the sand by which means and a small stream they found fresh water sufficient to water the ship.

BANKS: We went ashore at the houses, but found not the least good effect from our present yesterday: No signs of people were to be seen; in the house in which the children were yesterday was left every individual thing which we had thrown to them.

COOK: Probably the natives were afraid to take them away.

Naturalist Dr Daniel Solander (1733–1782).[1]

BANKS: Dr Solander and myself went a little way into the woods [present-day Kurnell], and found many plants [one of which was *Epacris longiflora*] but saw nothing like people. (Colour Plate No. 4)

[With the death of fellow artist Alexander Buchan earlier in Tahiti, Sydney Parkinson was faced with double the artistic workload. To accommodate the great number of new plants discovered in New Zealand and New Holland, he adopted a colour-coding system, making only an **outline pencil drawing** of the plant, with **short colour notes.** These he intended to complete at a later date. Parkinson died on the voyage home to England and this completion work passed to other artists employed by Banks in London. Some of these **finished drawings** were turned into copper plates for an ambitious work Banks envisioned known as *Banks' Florilegium*. This remained unfinished in Banks's lifetime. The first complete full-colour edition of the Florilegium was published between 1980 and 1990 in thirty-four parts by Alecto Historical Editions and the British Museum – Natural History.]

All They Seemed To Want Was For Us To Be Gone

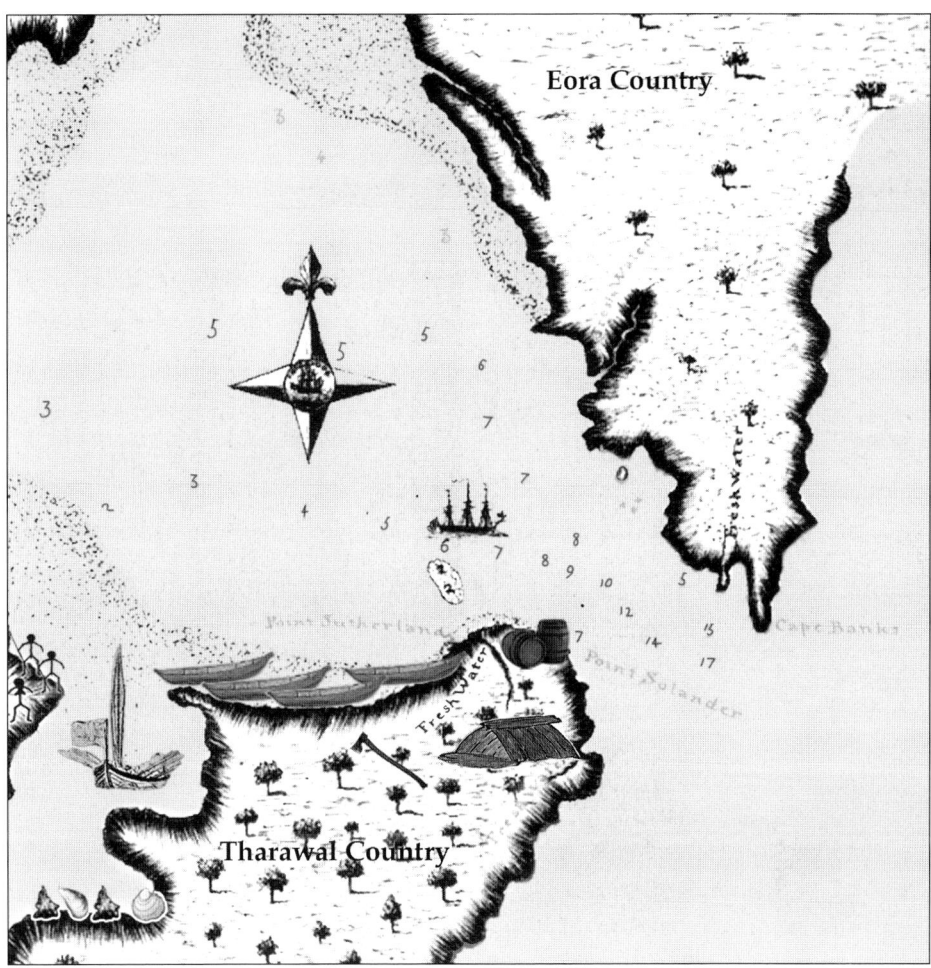

COOK: After breakfast we sent some empty casks a shore and a party of men to cut wood and I went myself in the Pinnace to sound and explore the Bay. In the doing of which I saw several of the natives but they all fled at my approach. I landed in two places one of which the people had but just left as there were small fires and fresh muscles [mussels] broiling upon them here likewise lay vast heaps of the largest oyster shells I ever saw.

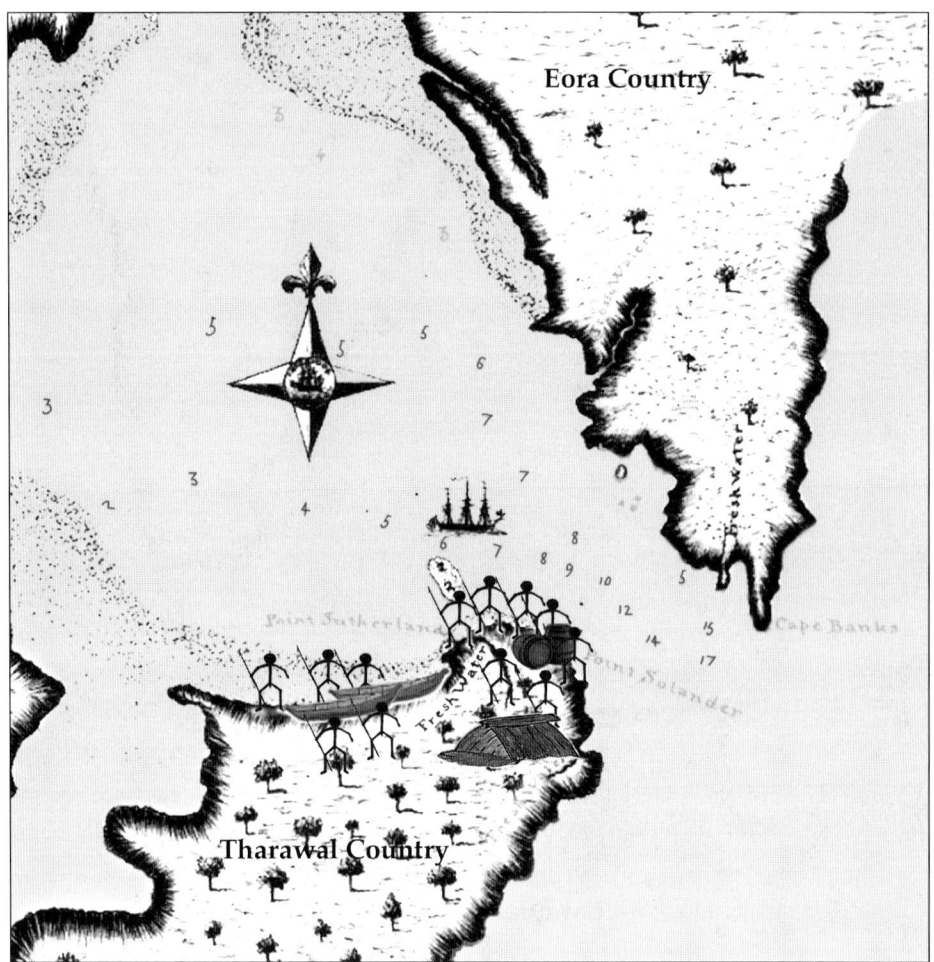

BANKS: At noon all hands came on board to dinner. The Indians, about 12 in number, as soon as they saw our boat put off came down to the houses. Close by these was our watering place at which stood our cask: they looked at them but did not touch them, their business was merely to take away two of four boats which they had left at the houses; this they did, and hauled the other two above high-water mark, and then went away as they came.

All They Seemed To Want Was For Us To Be Gone   45

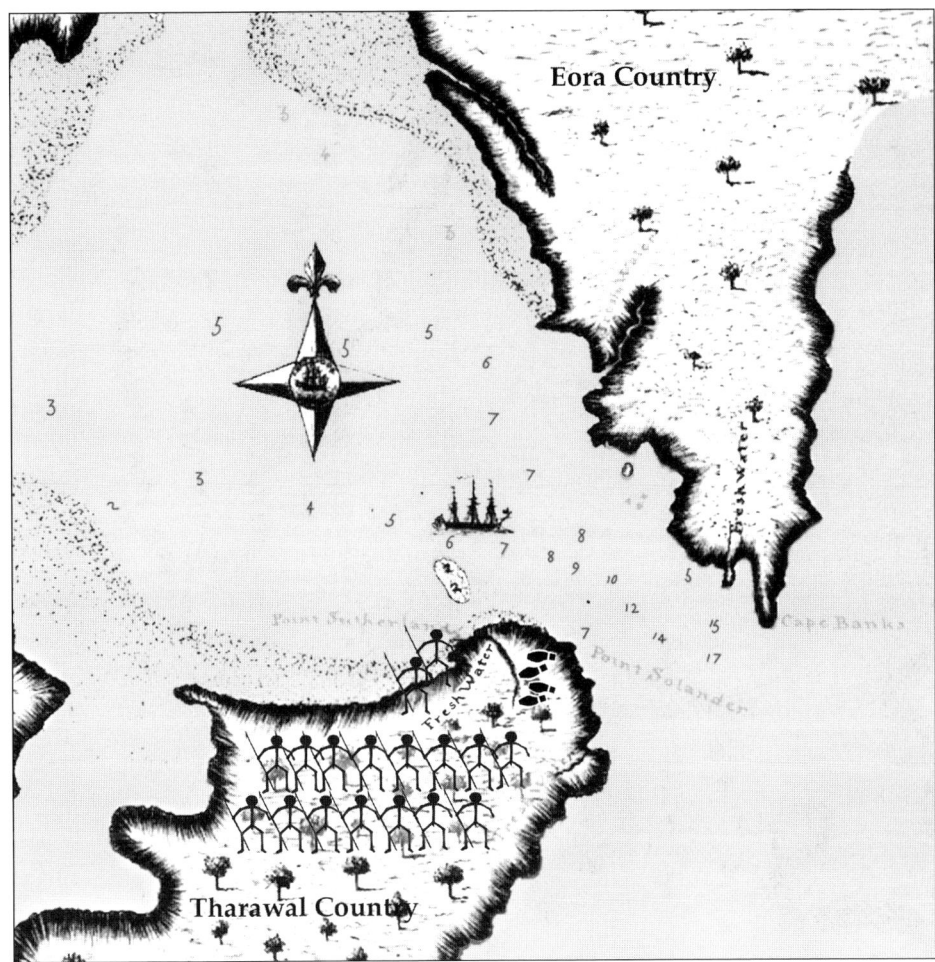

30th April

COOK: In the afternoon 16 or 18 of them came boldly up to within 100 yards of our people at the watering place and there made a stand.

BANKS: They sent two before the rest, our people did the same; they however did not wait for a meeting but gently retired. Our boat was about this time loaded so everybody went off in her, and at the same time the Indians went away.

COOK: Mr Hicks who was the officer ashore did all in his power to entice them to him by offering them presents &c. but it was to no purpose all they seemed to want was for us to be gone – they were all armed with darts and wooden swords.

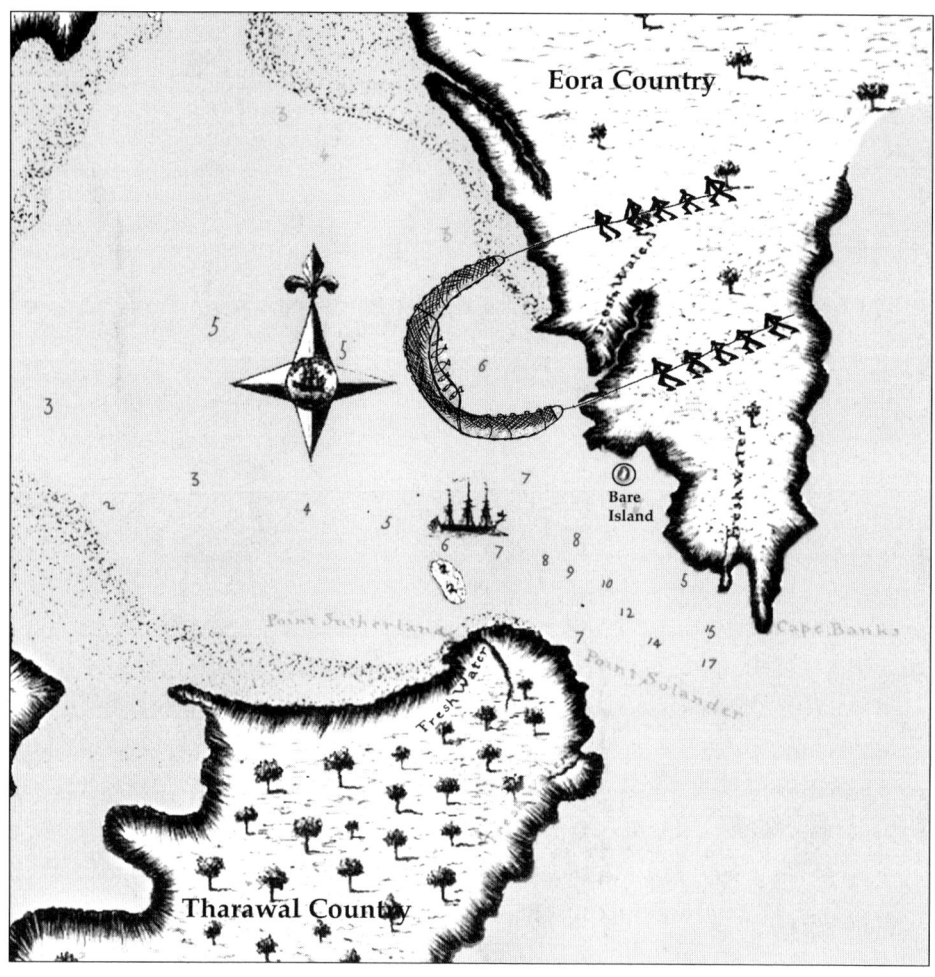

COOK: After I returned from sounding the bay I went to a cove on the north side [present-day Yarra Bay or Frenchman's Cove].

BANKS: Myself with the Captain were in a sandy cove on the Northern side of the harbour, where we hauled the seine and caught many very fine fish, more than all hands could eat.

COOK: In 3 or 4 hauls we caught about 300 pounds weight, which I caused to be equally divided among the Ships Company. I afterwards found a very fine stream of fresh water in the first sandy cove within the Island [Bare Island] before which a ship might lay almost land locked and wood for fuel may be got everywhere.

(END OF SECOND DAY – CIVIL TIME)

All They Seemed To Want Was For Us To Be Gone 47

BANKS: Before day break this morn the Indians were at the houses abreast of the Ship: they were heard to shout much. At sunrise they were seen walking away along the beach; we saw them go into the woods where they lighted fires about a mile from us. Dr Solander and myself into the woods.

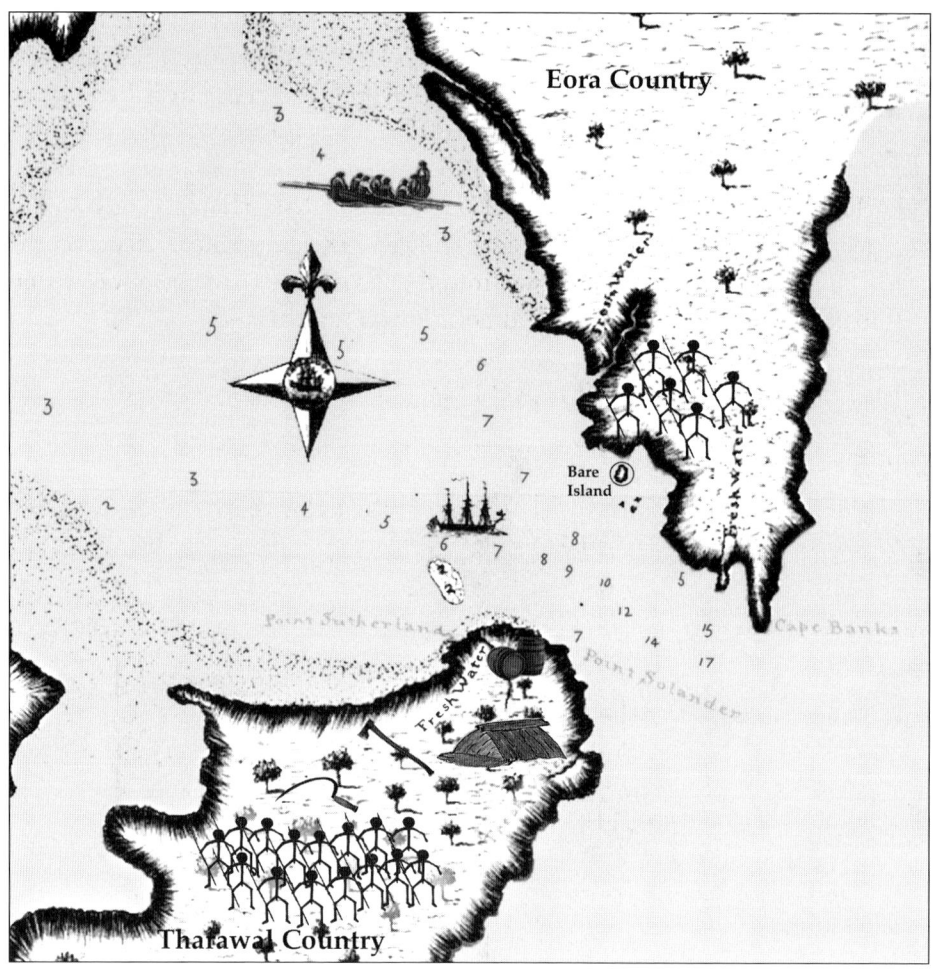

BANKS: Our people went ashore as usual. The grass cutters were farthest from the body of the people: towards them came 14 or 15 Indians having in their hands sticks that shone (sayd the Sergeant of marines) like a musquet. The officer on seeing them gathered his people together: the hay cutters coming to the main body appeared like a flight so the Indians pursued them, however but a very short way, for they never came nearer than just to shout to each other, maybe a furlong.

COOK: In the a.m. I went in the Pinnace to sound and explore the North side of the bay where I neither met with inhabitants or anything remarkable.

1st May

BANKS: At night they came again in the same manner and acted over again the same half pursuit.

COOK: I being on board at this time went immediately ashore, but before I got there they were going away, I followed them alone and unarmed some distance along shore but they would not stop until they got farther off than I choose to trust myself: these were armed in the same manner as those that came yesterday.

BANKS: Myself in the Even landed on a small Island [Bare Island] on the Northern side of the bay to search for shells; in going I saw six Indians on the main who shouted to us but ran away into the woods before the boat was within half a mile of them, although she did not even go towards them.

(END OF THIRD DAY – CIVIL TIME)

COOK: Last night Forby Sutherland seaman [ship's poulterer] departed this life. He died of consumption with which he had been afflicted ever since our departure from Strait of le Maire [Tierra Del Fuego]. In the a.m. his body was buried a shore at the watering place which occasioned my calling the south point of this Bay after his name [marked on plan].

BANKS: The Captain, Dr Solander, myself and some of the people, making in all 10 musquets, resolved to make an excursion into the countrey.

MAGRA: Hoping for an opportunity of taking some of the natives, intending to cloth and make them presents, and afterwards send them back to their friends; expecting that such proof of our pacific intentions would be sufficient to engage them to pay us a visit, and enter into some commerce and traffic.[2]

BANKS: We walked till we completely tired ourselves, which was in the evening, seeing by the way only one Indian who ran from us as soon as he saw us. We saw many Indian houses and places where they had slept upon the grass without the least shelter; in these we left beads ribbands &c.

MAGRA: But these presents were never carried away, though we had reason to believe the place had been visited several times after by the natives.

COOK: Dr Solander had a bad sight of a small Animal something like a rabbit.

BANKS: My Greyhound just got sight of him and instantly lamed himself against a stump which lay concealed in the long grass; we saw also the dung of a large animal that had fed on grass which much resembled that of a Stag; also the footsteps of an animal clawed like a dog or wolf and as large as the latter; and of a small animal whose feet were like those of a polecat or weasel.

COOK: I saw some trees that had been cut down by the natives with some sort of a blunt instrument and several trees that were barked the bark of which had been cut by the same Instrument, in many of the trees, especially the palms, were cut steps about 3 or 4 feet asunder for the convenience of climbing them.

BANKS: The trees over our heads abounded very much with Loryquets and Cockatoos; both these sorts flew in flocks of several scores together.

PARKINSON: We shot a few of them, which we made into a pie, and they ate very well. We also met with a black bird, very much like our crow, and shot some of them too, which also tasted agreeably.

2nd May

COOK: Between 3 and 4 o'clock we returned out of the country and after dinner went ashore to the watering place.

All They Seemed To Want Was For Us To Be Gone 51

COOK: In the morning I had sent Mister Gore with a boat up to the head of the bay to dredge for oysters. He saw some natives who by signs invited him ashore.

Third Lieutenant John Gore (1729–1790).³

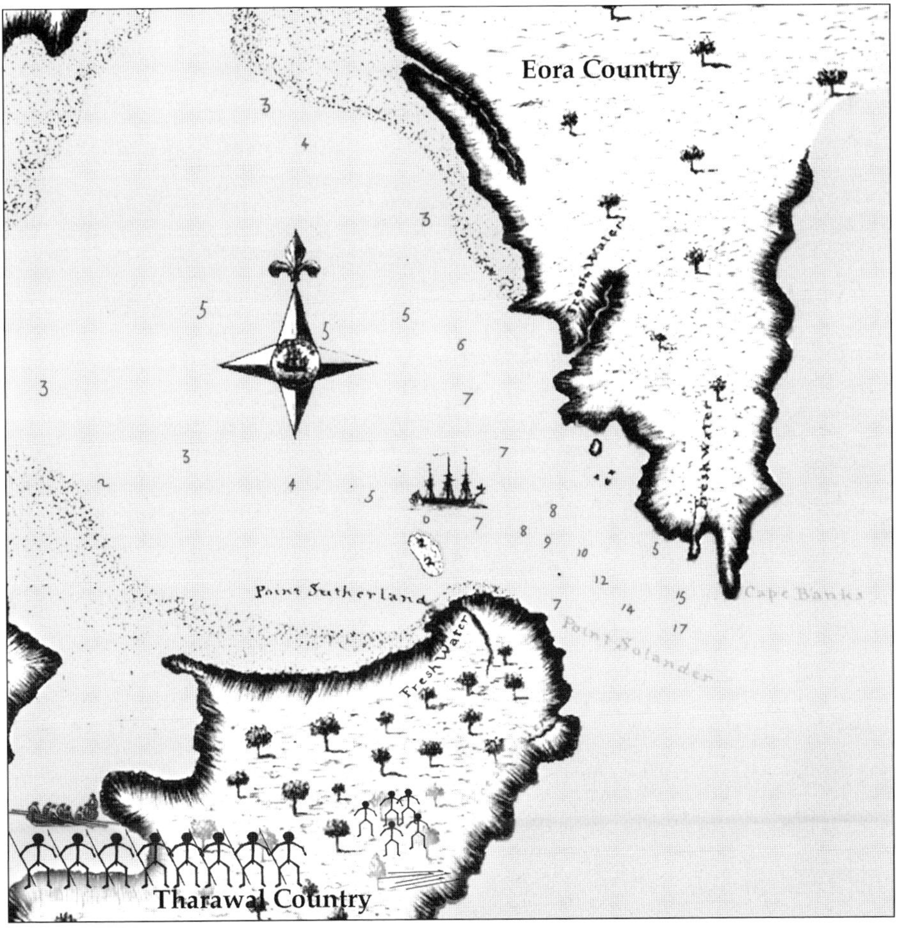

MAGRA: They singled out as many men from among themselves as they had counted in the boat, and then came down to the water's edge – their countrymen throwing down their arms and retiring a good distance – and there they challenged us to battle.

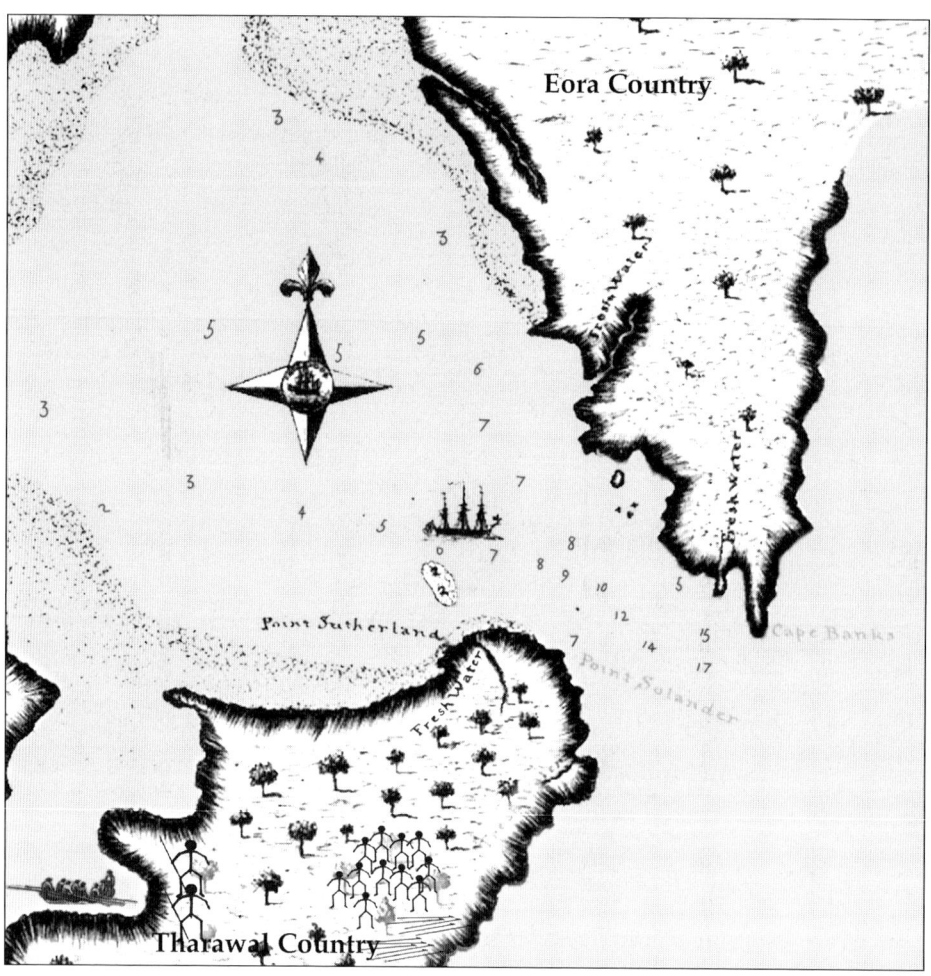

MAGRA: But this being refused, they selected two only, out of their number, and challenged as many of us to fight them, the others retiring to avoid any suspicion of treachery: but this offer being likewise rejected, they all retired.

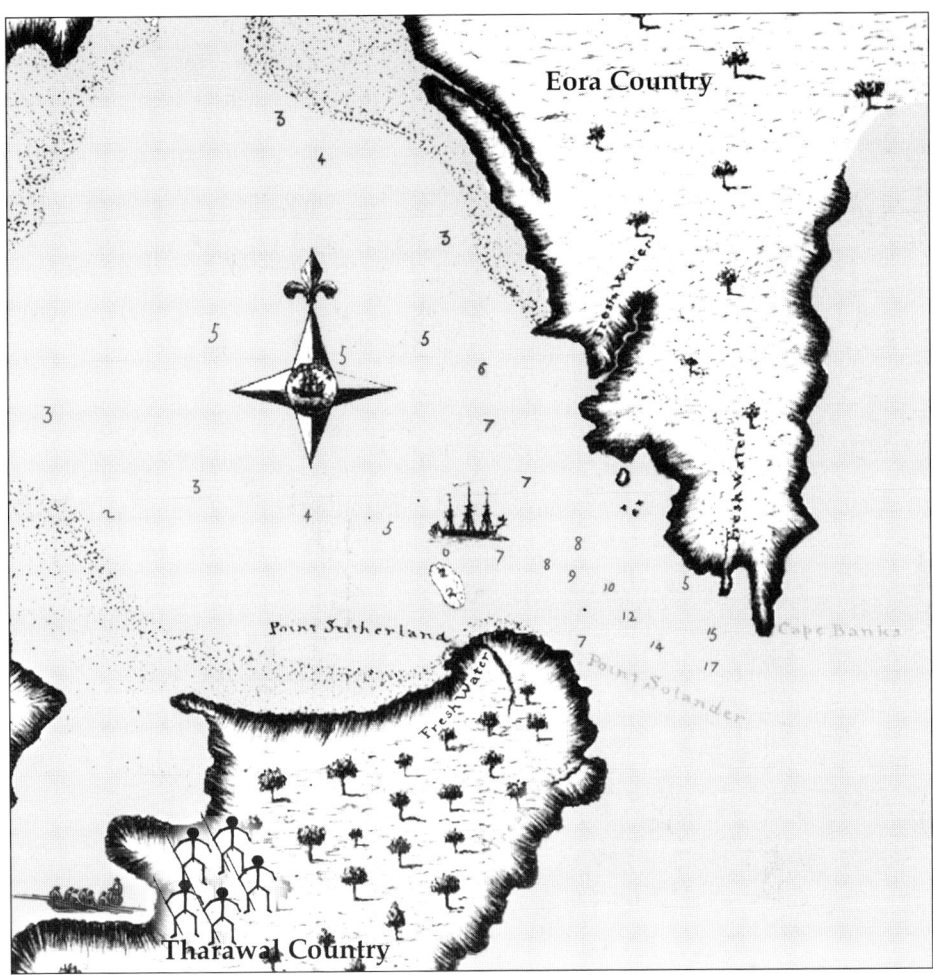

MAGRA: But soon after several others came to the shore, and an officer fired a musket loaded with a ball into a tree at some distance, that he might let them see how far it would carry; and being much pleased at the sight, they desired him, by signs, to let them see another discharge, which he did, and they soon after retired apparently well pleased.

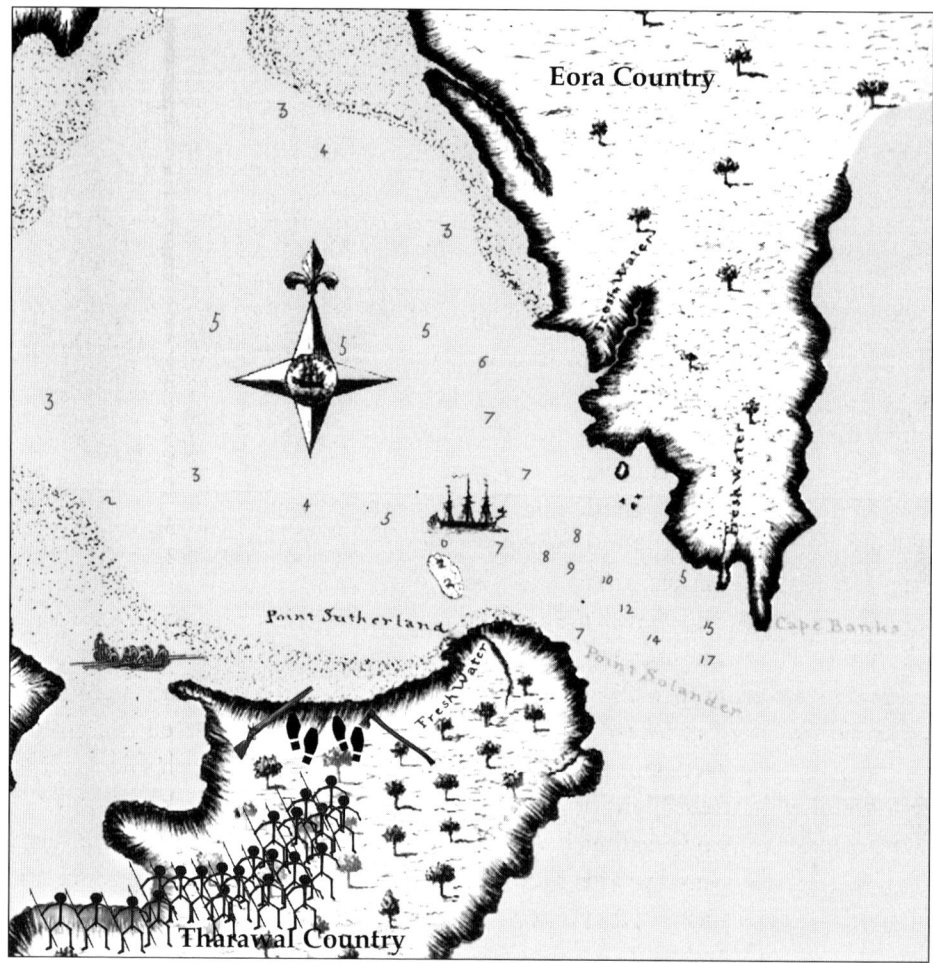

MAGRA: The officers then determining to return by land through the woods, dispatched the boat forwards, but they had not proceeded two miles on their way, before they were overtaken by two and twenty of the natives, all armed, who followed close at their heels, but stopped whenever the officers faced them, and retired if they began to advance towards them, but again followed them when they proceeded on their way to the place where the boat had been directed to wait: in this manner they continued their return, until they came near the place where a part of our crew was employed in cutting wood, when they were joined by several other gentlemen who had been shooting.

COOK: Doctor Munkhouse and one or two more.

MAGRA: One of them proposed a scheme to entrap some of the Indians, which had near proved fatal.

BANKS: More curious perhaps than prudent.

MAGRA: The design was to advance as near to the natives as they would permit, without retiring; and then feigning a fright, to turn suddenly and run from them, expecting in this manner to decoy them in a pursuit which might afford the working parties an opportunity of surrounding and taking some of them: but whether the Indians suspected the artifice or not, the gentlemen had not run above twenty yards after their pretended fright, before the natives, giving loud shrieks, advanced hastily, and threw their spears at them with great force. One of the gentlemen who was nearest, hearing their cry, suddenly turned his head, and seeing the spears in their flight, had scarce sufficient time to save himself behind a tree, though but at a few feet distant: one of the spears entered the ground which he had quitted, and another pierced deep into the tree behind which he had sheltered himself. Many others fell in different places, one sticking fast in the branch of a tree above the head of a gentleman who ran the farthest from them, and who was then at fifty yards distance; another passed between his legs into the ground. After this attack, they all precipitately retired to the woods.

Surgeon William Brougham Munkhouse (1732–1770).[4]

COOK: Doctor Solander, I, and Tupia made all the haste we could after them.

BANKS: Everyone else stayed behind; this however did not stop the Indians who walked leisurely away till our people were tired of following them.

COOK: We could by neither words nor actions prevail upon them to come near us.

MAGRA: We, collecting their spears, returned with them to our ship.

(END OF FOURTH DAY – CIVIL TIME)

BANKS: The morn was rainy and we who had got already so many plants were well contented to find an excuse for staying on board to examine them a little at least.

56   The Endeavour Journals

COOK: The wind at SE with rain which prevented me from making an excursion up the head of the bay as I intended.

3rd May

BANKS: In the afternoon however it cleared up and we returned to our old occupation of collecting, in which we had our usual good success. [*Lambertia formosa* was among those collected at Botany Bay.] (Colour Plate No. 5)

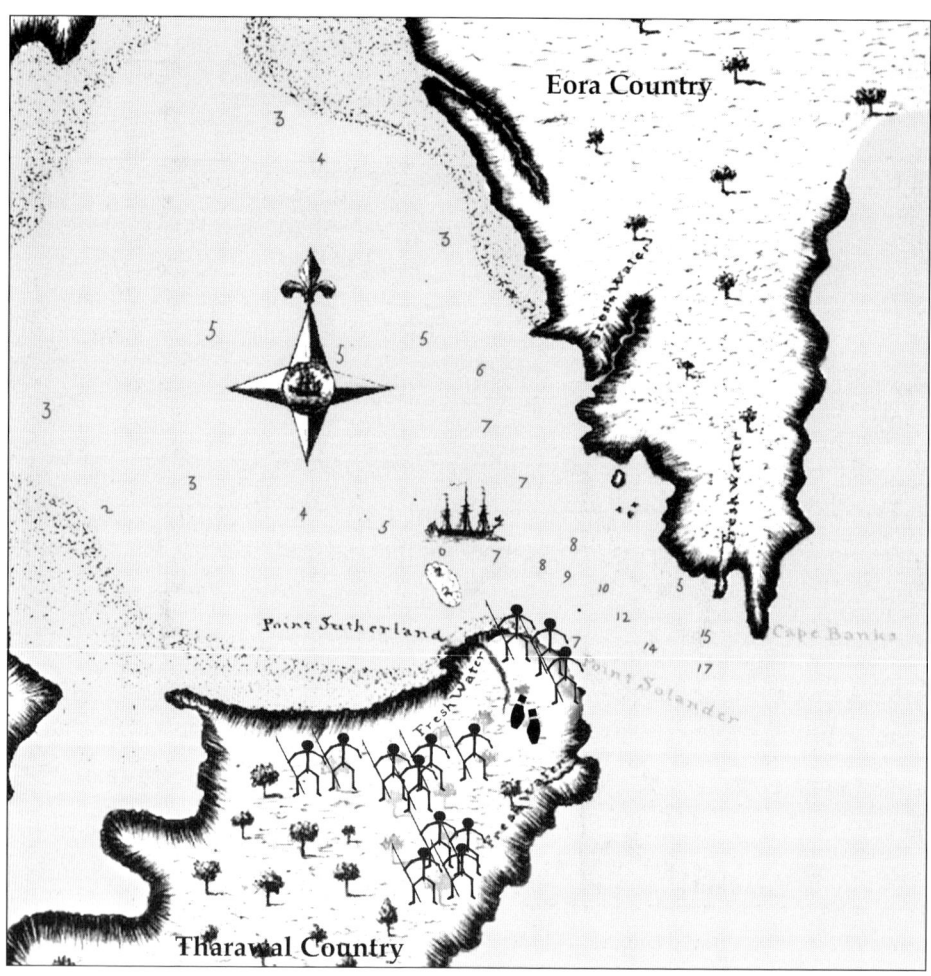

COOK: I made a little excursion along the Sea Coast to the southward accompanied by Mr Banks and Dr Solander. At our first entering the woods we saw 3 of the natives who made off as soon as they saw us – more of them were seen by others of our people who likewise made off as soon as they found they were discovered.

BANKS: Tupia who strayed from us in pursuit of parrots, of which he shot several, told us on his return that he had seen nine Indians who ran from him as soon as they perceived him.

[A Rainbow Lorikeet was captured at Botany Bay and kept by Tupia as a pet. The living specimen was brought back to England and painted by Moses Griffith in 1772. The earliest known European painting of an Australian bird.] (Colour Plate No. 6)

(END OF FIFTH DAY – CIVIL TIME)

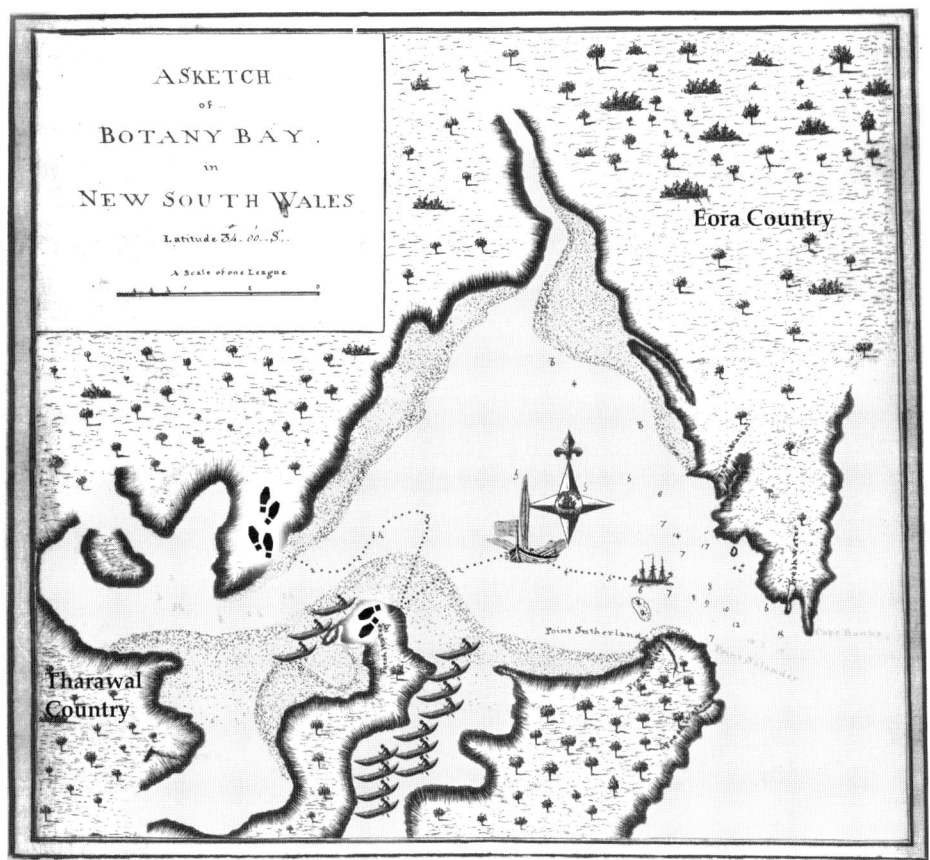

'A Sketch of Botany Bay' – JAMES COOK/Isaac Smith.[5]

COOK: In the a.m. I went in the Pinnace to the head of the Bay accompanied by Doctor Solander and Munkhouse in order to examine the Country and to try to form some connections with the natives.

In our way thither we met with 10 or 12 of them fishing each in a small canoe who retired in to shoal water upon our approach, others again we saw at the first place we landed at who took to their canoes and fled before we came near them. [They first landed at Towra Point, where they found Magenta Lilly Pilly – *Syzygium paniculatum* – later mentioned by Banks, p.60.] (Colour Plate No. 7)

After this we took water and went almost to the head of the inlet where we landed and travelled some distance inland. We found the face of the country much the same as I have before described but the land much richer for instead of sand I found in many places a deep black Soil [Sans Souci region].

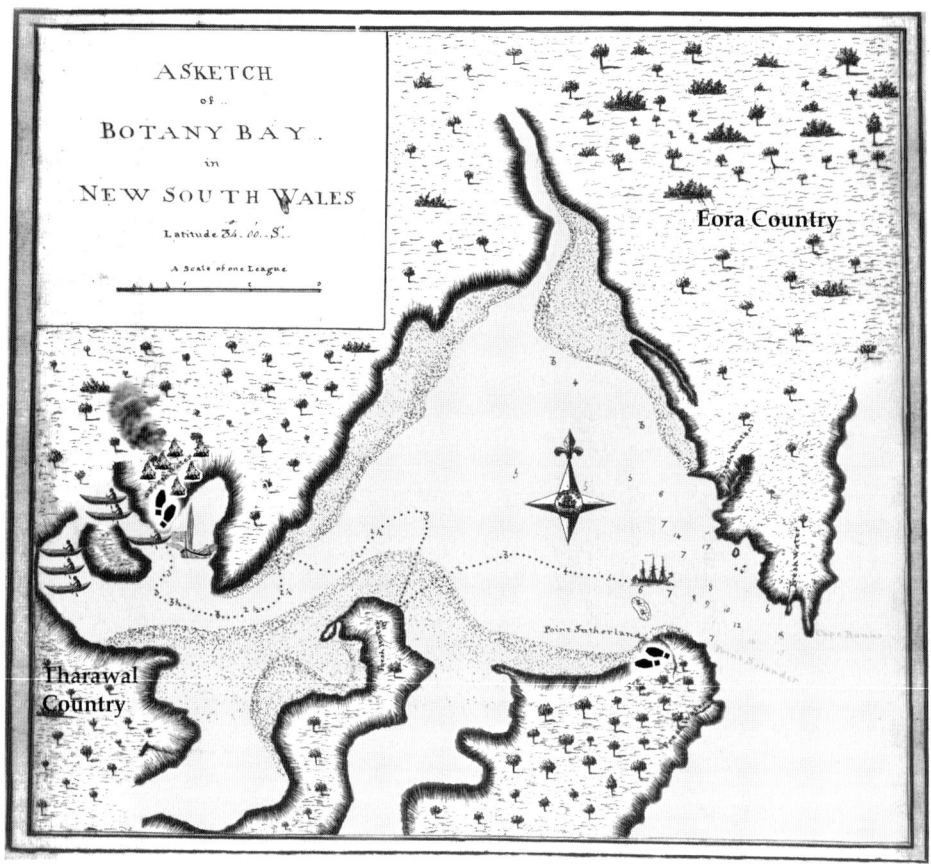

COOK: After we had sufficiently examined this part we returned to the boat and seeing some smoke and canoes at another part [Kogarah Bay] we went thither in hopes of meeting with the people but they made off as we approached – there were six canoes and six small fires near the shore and muscles roasting upon them, and a few oysters laying near, from this we conjectured that there had been just six people who had been out each in his canoe picking up the shell fish and come ashore to eat them where each had made his fire to dress them by – we tasted of their cheer and left them in return strings of beads. Near to this place at the foot of a tree was a small well or Spring of water.

BANKS: Our collection of Plants was now grown so immensely large that it was necessary that some extraordinary care should be taken of them least they should spoil in the books. I therefore devoted this day to that business and carried all the drying paper, near 200 Quires of which the larger part was full, ashore and spreading them upon a sail in the sun kept them in this manner exposed the whole day, often turning them and sometimes turning the Quires in which were plants inside out. By this means they came on board at night in very good condition.

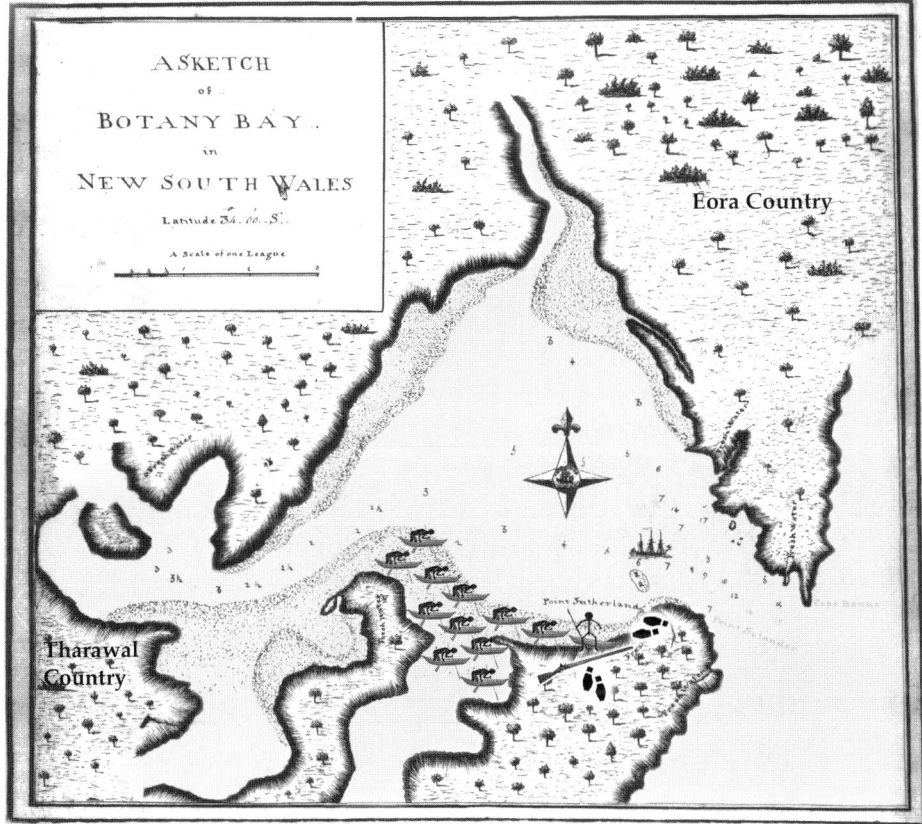

BANKS: During the time this was doing 11 Canoes, in each of which was one Indian, came towards us. We soon saw that the people in them were employed in striking fish; they came within about half a mile of us intent on their own employments and not at all regarding us. Opposite the place where they were several of our people were shooting; one Indian may be prompted by curiosity landed, hauled up his canoe and went towards them; he stayed about a quarter of an hour and then launched his boat and went off, probably that time had been spent in watching behind trees to see what our people did. I could not find however that he was seen by anybody.

4th May

BANKS: When the damp of the Even made it necessary to send my Plants and books on board I made a small excursion in order to shoot anything I could meet with and found a large quantity of Quails [Brown quail – *Synoicus ypsilophorus*] much resembling our English ones, of which I might have killed as many almost as I pleased had I given my time up to it, but my business was to kill variety and not too many individuals of any one species. (Colour Plate No.8)

COOK: The day being now fast spent we set out on our return to the ship. Upon my return to the ship in the evening I found that none of the natives had appeared near the watering place but about 20 of them had been fishing in their canoes at no great distance from us.

BANKS: The Captain and Dr Solander found several trees which bore fruit of the Jambosa kind [*Syzygium paniculatum* – Magenta Lilly Pilly found on Towra Point where they still grow], much in colour and shape resembling cherries; of these they eat plentifully and brought home also abundance, which we eat with much pleasure tho they had little to recommend them but a light acid. (Colour Plate No. 7)

(END OF SIXTH DAY – CIVIL TIME)

Chapter 4

To Try To Form Some Connections With The Natives

4th May (cont'd)

COOK: In the a.m. as the wind would not permit us to sail I sent out some parties into the Country to try to form some connections with the natives.

BANKS: Myself in the woods botanizing as usual, now quite void of fear as our neighbours have turned out such rank cowards.

One of our midshipmen straggling by himself a long way from anyone else met by accident with a very old man and woman and some children: they were setting under a tree and neither party saw the other till they were close together. They shewd signs of fear but did not attempt to run away. He had nothing about him to give to them but some parrots which he had shot: these they refused, withdrawing themselves from his hand when he offered them in token either of extreme fear or disgust. The people were very old and grey headed, the children young. The hair of the man was bushy about his head, his beard long and rough, the woman's was cropped short round her head; they were very dark coloured but not black nor was their hair woolly.

COOK: They were quite naked, even the woman had nothing to cover her nudity.

BANKS: He stayed however with them but a very short time.

COOK: They were close to the water side where several more were in their canoes gathering shell fish and he being alone was afraid to make any stay with the two old people least he should be discovered by those in the Canoes.

BANKS: He feared that the people in them might observe him and come ashore to the assistance of the old people, who in all probability belonged to them.

Tupia's watercolour of Tharawal canoeists in Botany Bay.[1] (The earliest known painting of east coast aborigines.)

[It was left to Tupia (Tupaia), the Polynesian high priest taken on board in Tahiti, to make the only other representation of the people of Botany Bay. Tupia used water colours given to him by Banks to paint this evocative depiction of canoeists at Botany Bay.]

BANKS: Seventeen canoes came fishing near our people in the same manner as yesterday only stayed rather longer, emboldened a little I suppose by having yesterday met with no kind of molestation.

Our surgeon [William Brougham Munkhouse] who had strayed a long way from the people with one man in his company, in coming out of a thicket observed 6 Indians standing about 50 yards from him; one of these gave a signal by a word pronounced loud, on which a lance was thrown out of the wood at him which however came not very near him. The 6 Indians on seeing that it had not taken effect ran away in an instant, but on turning about towards the place from whence the lance came he saw a young lad, who undoubtedly had thrown it, come down from a tree where he had been stationed probably for that purpose; he descended however and ran away so quick that it was impossible even to attempt to pursue him.

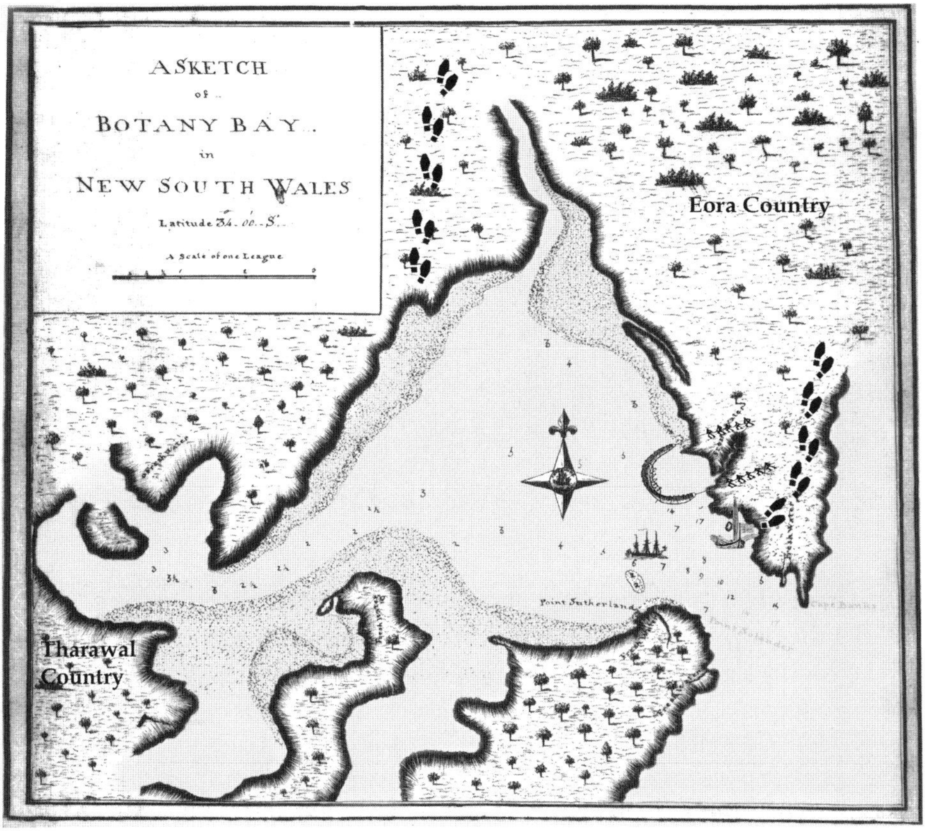

To Try To Form Some Connections With The Natives 63

'A Sketch of Botany Bay' – JAMES COOK/Isaac Smith.[2]

5th May

COOK: In the p.m. I went with a party of Men over to the North shore and while some hands were hauling the seine a party of us made an excursion of 3 or 4 Miles into the Country or rather along the Sea Coast. We met with nothing remarkable, great part of the Country for some distance in land from the sea Coast is mostly a barren heath diversified with marshes and Morasses.

BANKS: Myself in the afternoon ashore on the NW side of the bay, where we went a good way into the country which in this place is very sandy and resembles something our Moors in England, as no trees grow upon it

John Gore.[3]

but everything is covered with a thin brush of plants about as high as the knees. The hills are low and rise one above another a long way into the country by a very gradual ascent, appearing in every respect like those we were upon. While we were employed in this walk the people hauled the seine upon a sandy beach and caught great plenty of small fish.

COOK: Which the sailors call Leather Jackets on account of their having a very thick skin. They are known in the West Indies.

BANKS: On our return to the ship we found also that our 2nd lieutenant [John Gore*] who had gone out striking had met with great success: he had observed that the large sting rays of which there are abundance in the bay followed the flowing tide into very shallow water.

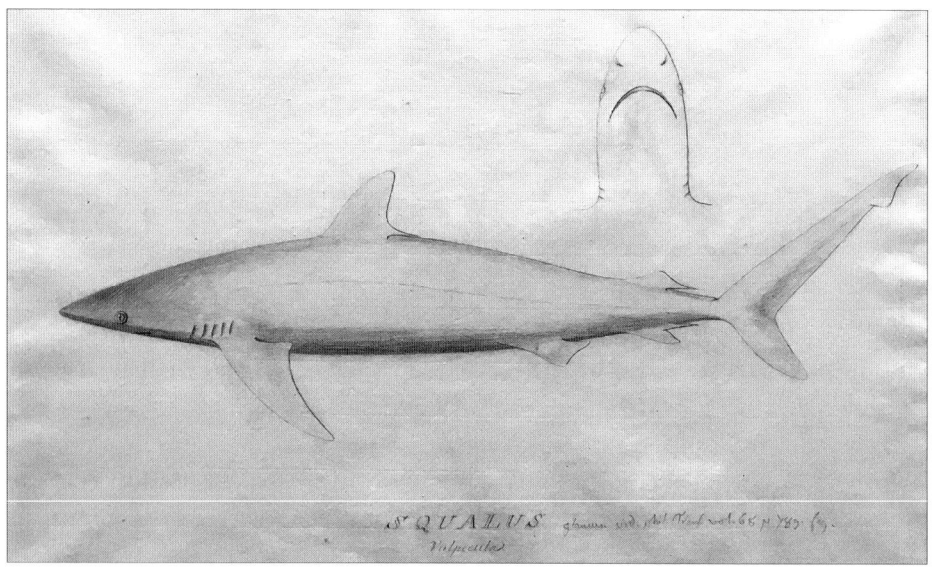

Herman Spöring.[4]

PARKINSON: On these shallows we found a great number of rays, some shell fish, and a few sharks. The rays are of an enormous size.

---

* More exactly John Gore was Third Lieutenant, Zachary Hickes was Second Lieutenant and James Cook First Lieutenant. But Cook was addressed as Captain, not Lieutenant, and in some cases, as in this, Gore was referred to as Second Lieutenant.

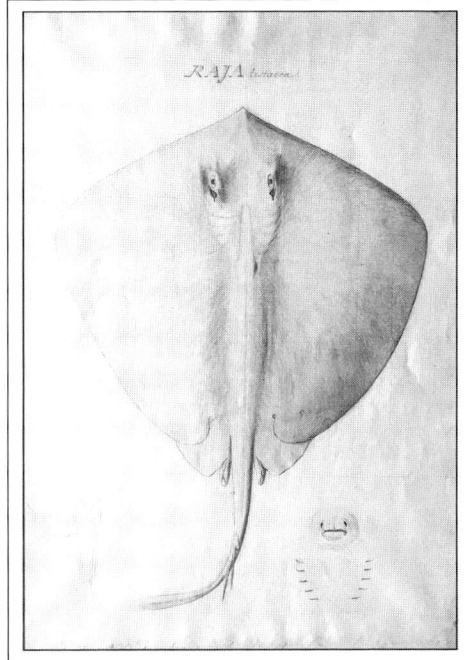 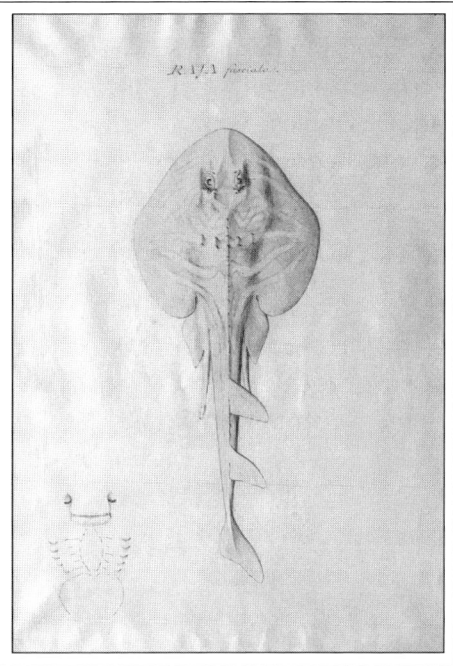

Common Stingray – *Urollophus testaceus*[5] Pencil sketch by Herman Spöring.

Fiddler Ray or Banjo Shark – *Trygonorrhina fasciata*[6] Pencil sketch by Herman Spöring.

BANKS: He [John Gore] therefore took the opportunity of flood and struck several in not more than 2 or 3 feet water; one that was larger than the rest weighed when his guts were taken out 239 pounds [108 kg].

As tomorrow was fixed for our sailing Dr Solander and myself were employed the whole day in collecting specimens of as many things as we possibly could to be examined at sea.

(END OF SEVENTH DAY – CIVIL TIME)

BANKS: The day was calm and the Mosquetos of which we have always had some more than usually troublesome. No Indians were seen by anybody during the whole day.

COOK: In the a.m. as the wind still continued northerly I sent the yawl again afishing and I went with a party of Men into the Country but met with nothing extraordinary.

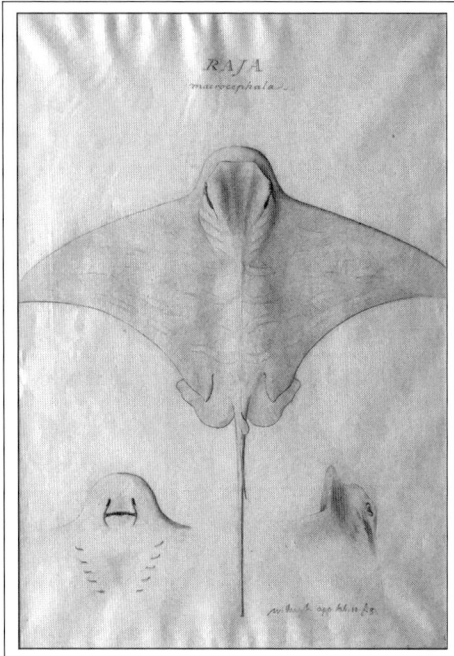 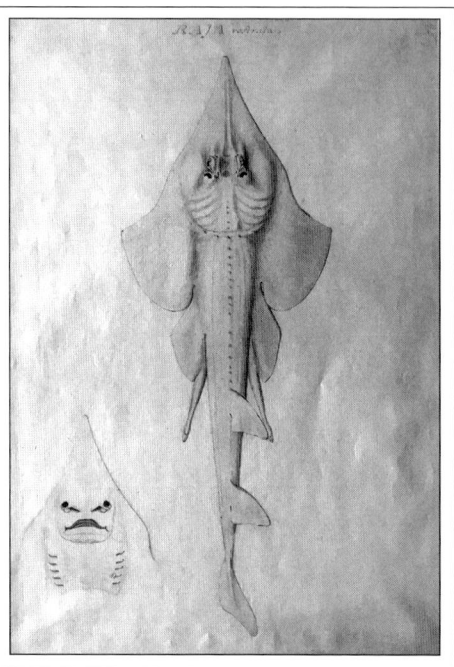

Eagle Ray – *Myliobatas australis*[7] Pencil sketch by Herman Spöring.

Shovelnose Ray – *Aptychaotrema banksii*[8] Pencil sketch by Herman Spöring.

6th May

COOK: In the evening the yawl returned from fishing having caught two Sting rays weighing near 600 pounds [272 kg].

BANKS: The biggest weighed without his guts 336 pounds [152 kg].

COOK: The great quantity of these sort of fish found in this place occasioned my giving it the name of Stingrays Harbour.
[Cook first named the bay Stingrays Harbour, but at a later date changed the name to Botany Bay, and wrote:]

> The great quantity of new plants Mister Banks and Doctor Solander collected in this place occasioned my giving it the name of Botany Bay.
> Sting rays I believe the Natives do not eat because I never saw the least remains of one near any of their huts or fire places. However, we could know but very little of their customs as we never were able to form any connections with them, they had not so much as touched the things we had left in their huts on purpose for them to take away. Oysters, Muscles, Cockles etc I believe are the chief support of the inhabitants who go into shoal water with their little canoes and pick them out of the sand and mud

with their hands and sometimes roast and eat them in the Canoe having often a fire for that purpose as I suppose for I know no other it can be for. The Natives do not appear to be numerous neither do they seem to live in large bodies but dispersed in small parties along by the water side. Although I have said that shell fish is their chief support yet they catch other sorts of fish, some of which we found roasting on the fire the first time we landed, some of these they strike with gigs and others they catch with hook and line we have seen them strike fish with their gigs & hooks and lines we found in their huts.

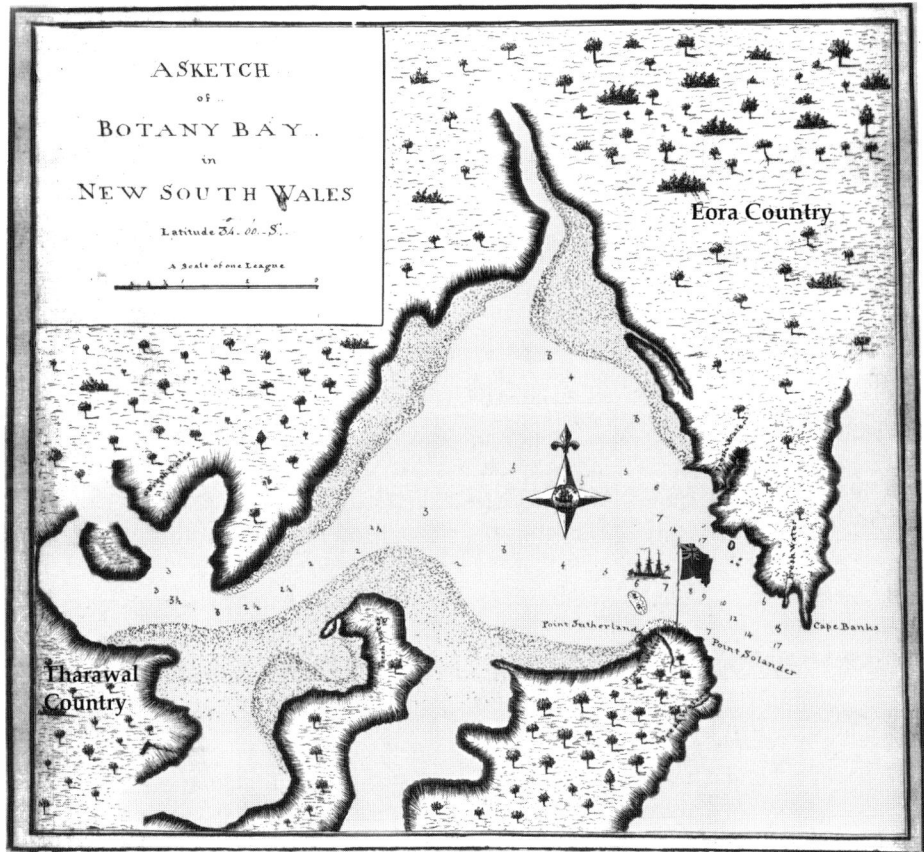

COOK: During our stay in this Harbour I caused the English Colours to be displayed a shore every day and an inscription to be cut out upon one of the trees near the watering place setting forth the Ships name, date &c.

Having seen everything this place afforded we at daylight in the Morning weighed with a light breeze at NW and put to sea and the wind soon after coming to the Southward we steered along shore NNE.

(END OF EIGHTH DAY – CIVIL TIME – Departure from Botany Bay)

[There is evidence that some friendly relations were established between the Europeans and the Aborigines at Botany Bay. Three word lists containing in all sixty indigenous words were collected by Zachary Hickes (First Lieutenant), Isaac Smith (Midshipman) and William Brougham Munkhouse (Ship's surgeon), but this evidence is not generally accepted. See *1770: The Endeavour Lists – Forgotten Words from Botany Bay* by Keith Vincent Smith.]

Chapter 5

A Great Quantity of Smoke

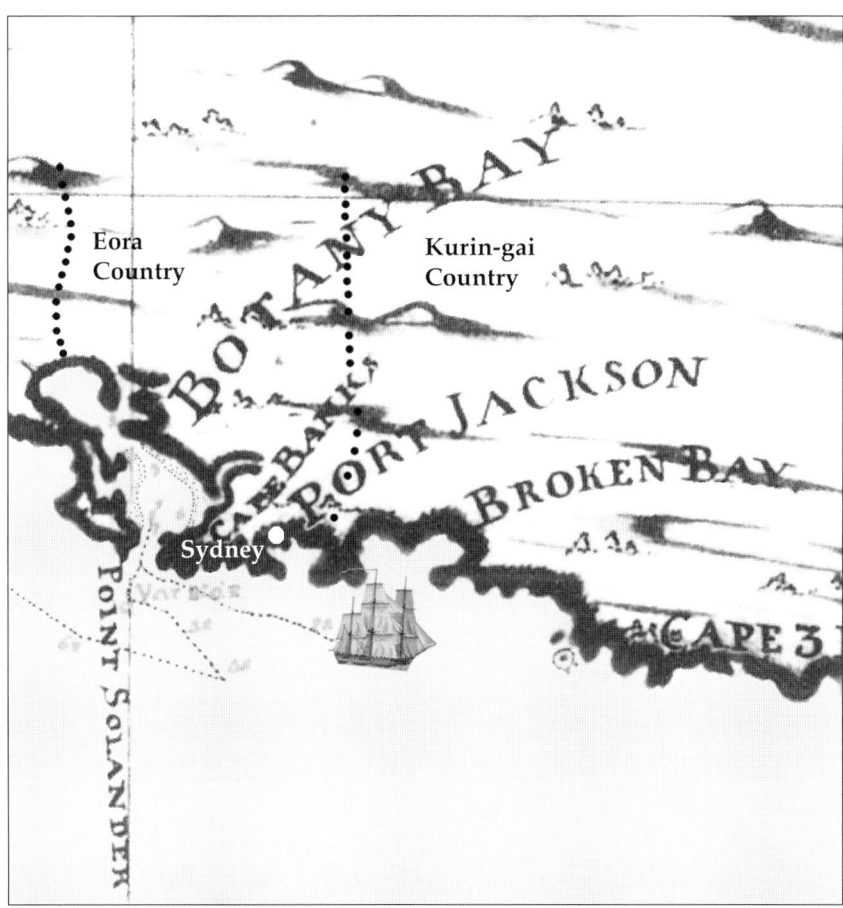

Portion of Cook's chart 'Point Hickes to Smoaky Cape'.[1]

6th May (cont'd)

BANKS: Went to sea this morn with a fair breeze of wind. The land we sailed past during the whole forenoon appeared broken and likely for harbours.

COOK: At Noon we were about 2 or 3 Miles from the land and abreast of a Bay or Harbour wherein there appeared to be safe anchorage which I called Port Jackson [present-day Sydney Harbour]. It lies 3 leagues to the northward of Botany Bay.

BANKS: We dined to day upon the sting-ray and his tripe: the fish itself was not quite so good as a skate nor was it much inferior.

PARKINSON: They tasted very much like the European rays, and the viscera had an agreeable flavour, not unlike stewed turtle.

BANKS: The tripe everybody thought excellent. We had with it a dish of the leaves of *Tetragonia cornuta* boiled which eat as well as spinage or very near it. (Colour Plate No. 9)

[The mariners were well acquainted with 'spinach' (Warrigal Greens) before discovering it around the shores of Botany Bay. They gathered it for their table earlier in the voyage at a number of locations in New Zealand. It would have helped contain the onset of scurvy as it contains a considerable amount of vitamin C.]

7th May

COOK: At sun set some broken land that appeared to form a Bay. This Bay I named <u>Broken Bay</u>. [Marked on chart below.]

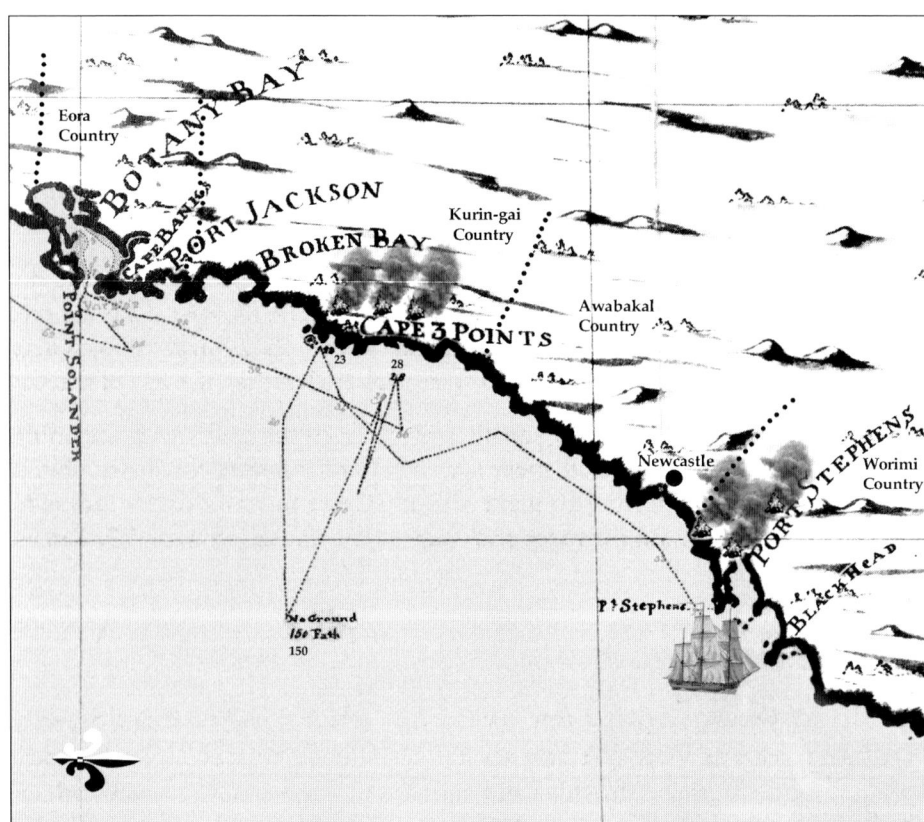

MAGRA: We sailed along the coast a few leagues from the shore, that we might be able to survey the land, and occasionally procure supplies of wood and water, or endeavour to establish a traffic with the natives.

BANKS: During last night a very large dew fell which wetted all our sails as completely as if they had been dipped overboard.

COOK: At Noon some pretty high land which projected out in three bluff points and occasioned my calling it Cape Three Points. [Marked on chart above.]

8th May

COOK: In the p.m. saw some smooks upon the shore [Cape Three Points – Kuring-gai country]. In the evening we were about 2 or 3 Miles from the land and had **28** fathom water.

BANKS: We had lost ground yesterday so that the land was what we had seen before; upon it however we observed several fires. At night a foul wind rose up much the same time and in much the same manner as yesterday.

9th May

COOK: We stood off shore until 12 at night. We had no ground at **150** fathom. Stood in shore until 8 o'clock a.m., and hardly fetched Cape Three Points.

10th May

COOK: In the p.m. we stood inshore until 4 o'clock when we tacked in **23** fathom water being about a mile from land.

PARKINSON: In the evening we saw two of the most beautiful rainbows my eyes ever beheld: the colours were strong, clear, and lively; those of the inner one were so bright as to reflect its shadow on the water. They formed a complete semicircle; and the space between them was much darker than the rest of the sky.

BANKS: Last night a very heavy squall came off from the land which according to the seamen's phrase made all sneer again; it pay'd however for the trouble it gave by bringing a fair wind. [To make all sneer again means 'to carry canvas to such an extent as to strain the ropes and spars to the utmost'. Smyth, *The Sailor's Word Book* 1867.]

11th May

COOK: At 4 p.m. we past a low rocky point which I named Point Stevens. [Marked on chart above.] We saw several smokes a little way in the Country rise up from the flat land, by this I did suppose that there were Lagoons which afforded subsistence for the natives [Worimi country].

72 The Endeavour Journals

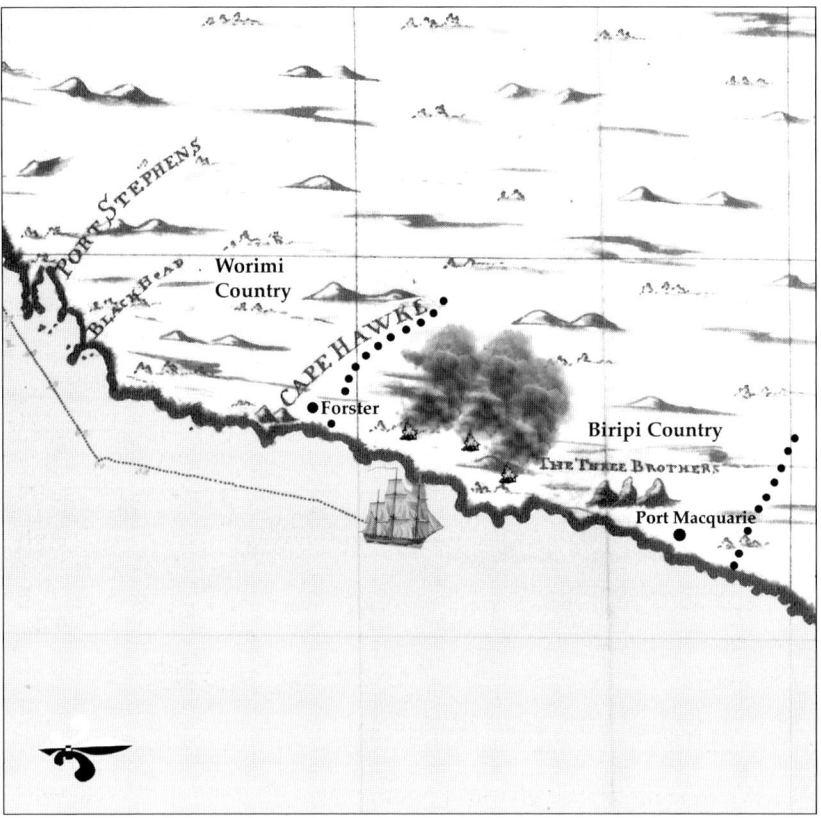

COOK: We run under an easy sail all night until 4 a.m. when we made all sail. At 8 o'clock [a.m.] we were abreast of a high point of land which made in two hillocks – this point I called <u>Cape Hawke</u>.

12th May

COOK: In the p.m. we saw several smokes a little way in land from the Sea and one upon the top of a hill which was the first we have seen upon elevated ground since we have been upon the coast [Biripi country]. At sun set three remarkable large high hills lying contiguous to each other and not far from the shore bore NNW [The Three Brothers].

BANKS: Behind them the countrey rose in gradual slopes carrying a great shew of fertility.

COOK: As these hills bore some resemblance to each other we called them <u>The Three Brothers</u>.

BANKS: This evening we finished drawing the plants got in the last harbour [Botany Bay] which had been kept fresh till this time by means of tin chests and wet cloths. In 14 days just, one draughtsman [Sydney Parkinson] has made 94 sketch drawings, so quick a hand has he acquired by use.

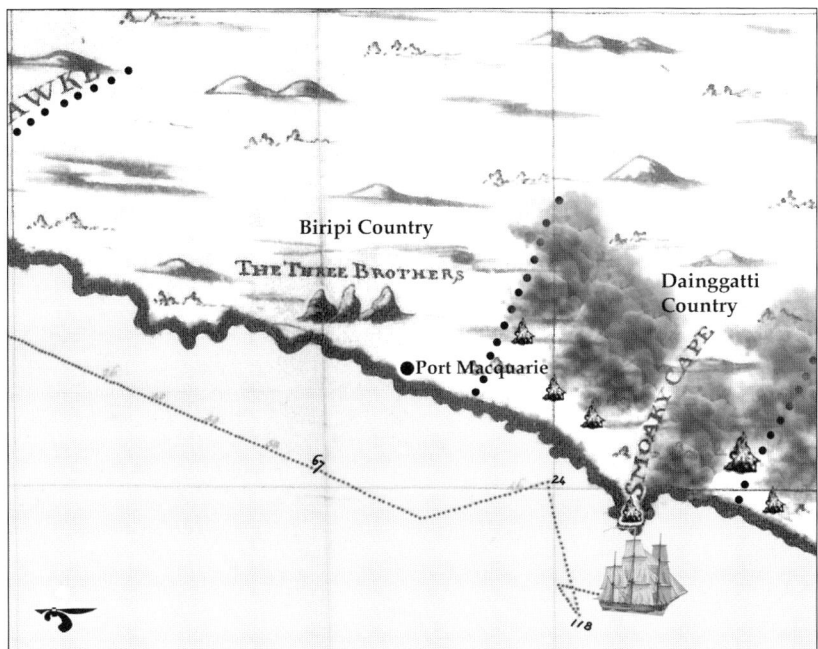

COOK: We steered NEBN all night having from 27 to **67** fathom water. In the a.m. several smokes we seen a little way in land.

13th May

COOK: In the p.m. stood in until 6 o'clock at which time we tacked being about 3 or 4 Miles from the land and in **24** fathom water. Stood off shore until midnight then tacked being in **118** fathom water.

BANKS: Many porpoises about the ship [8 a.m.].

COOK: At noon a point or head land on which were fires that caused a great quantity of smoke which occasioned my giving it the name of <u>Smoaky Cape</u> bore SW distant 4 leagues [Dainggatti country]. Besides the smoke seen upon this Cape we saw more in several places along the Coast.

BANKS: One very large which I judged to be at least a league inland.

14th May

BANKS: Innumerable shoals of fish about the ship in the afternoon and some birds of the Nectris kind [Wedge-tailed Shearwater – *Puffinus pacificus*]. (Colour Plate No. 10)

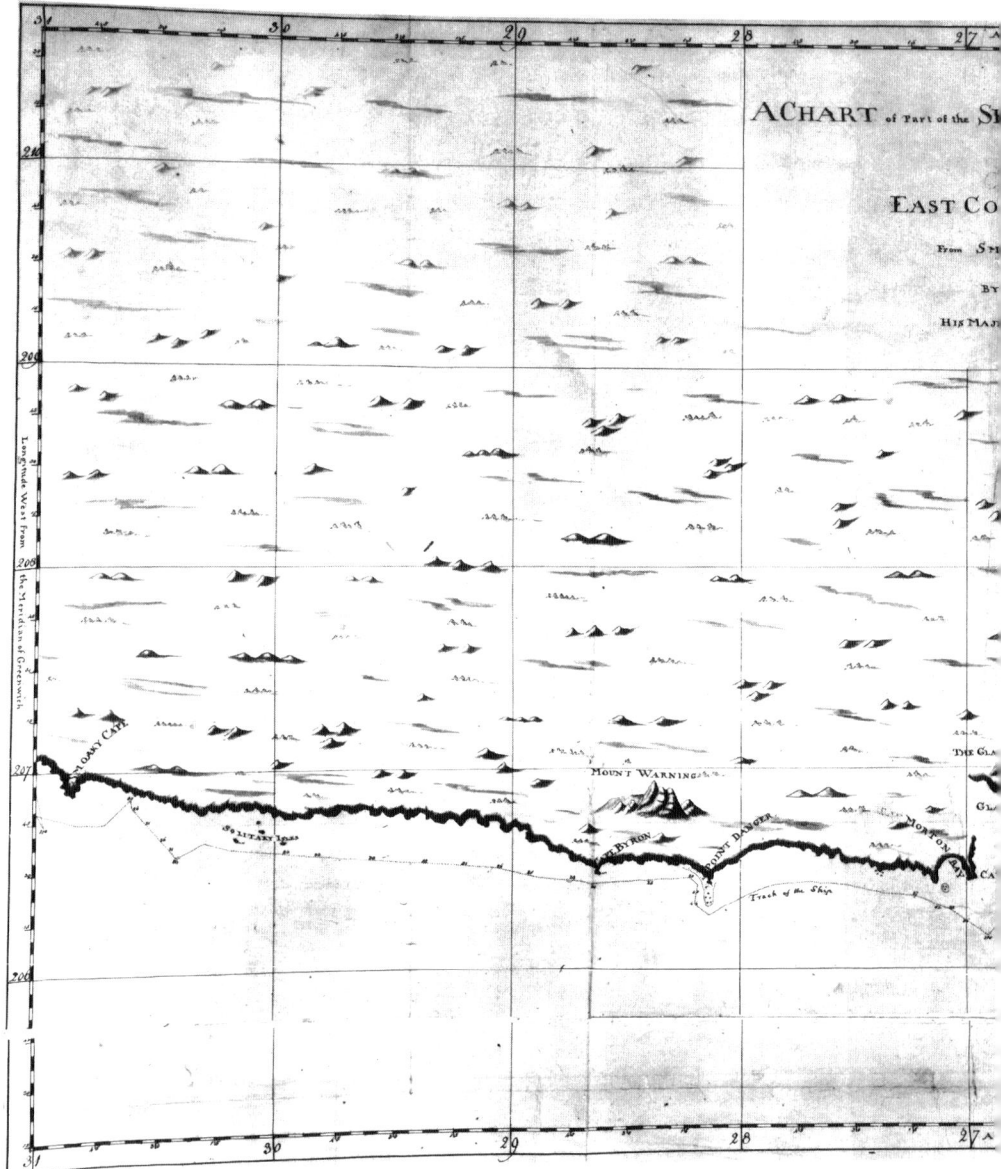

'A CHART of the SEA COAST of NEW SOUTH WALES on the EAST COAST of NEW HOLLAND from Smoaky Cape to Cape Townsend by Lieut.ᵗ J. Cook Commander of His Majestys Bark the *Endeavour* 1770.'[2] (The following seven charts in Chapter 5 are enlargements of sections of this chart.)

A Great Quantity of Smoke   75

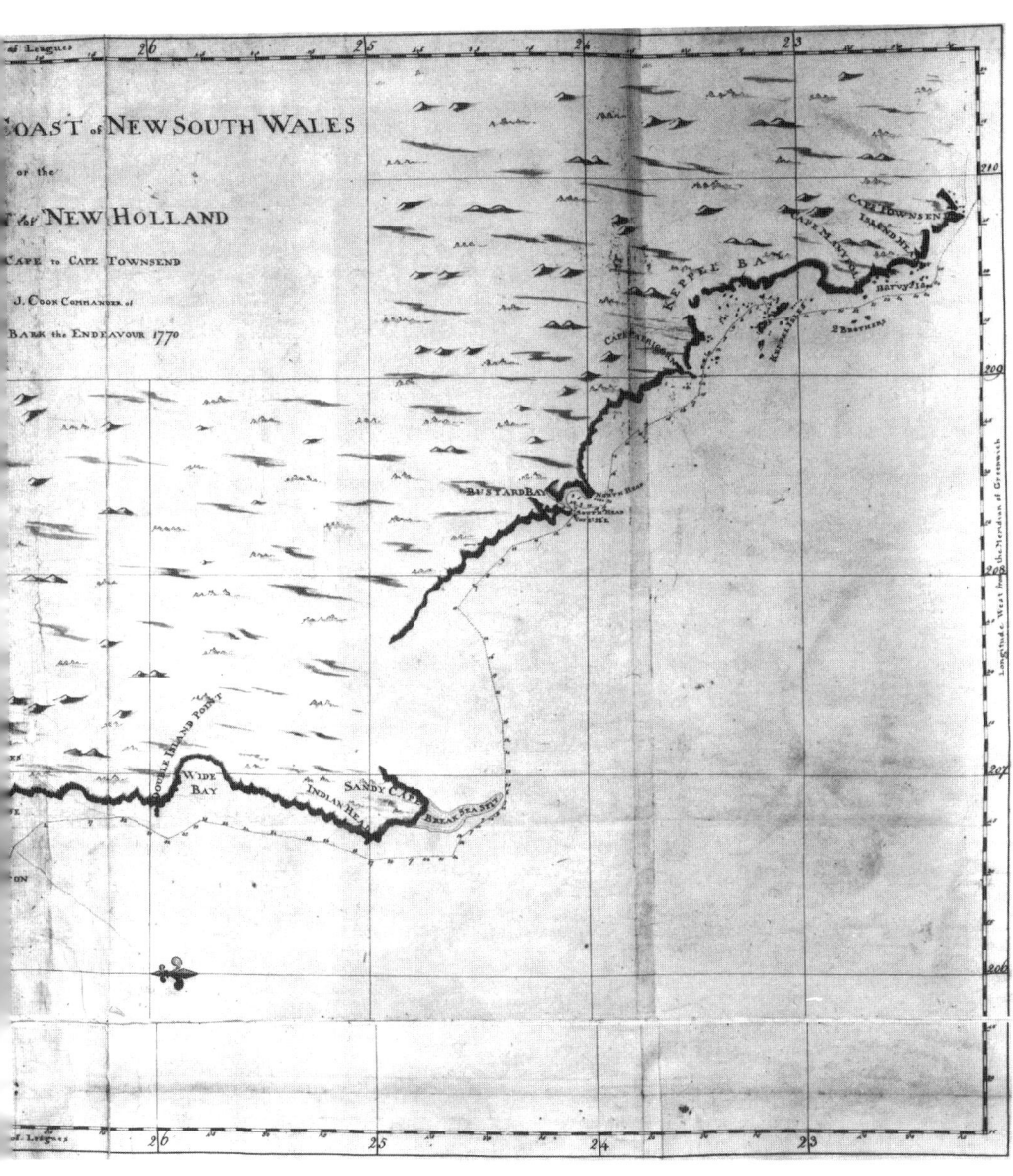

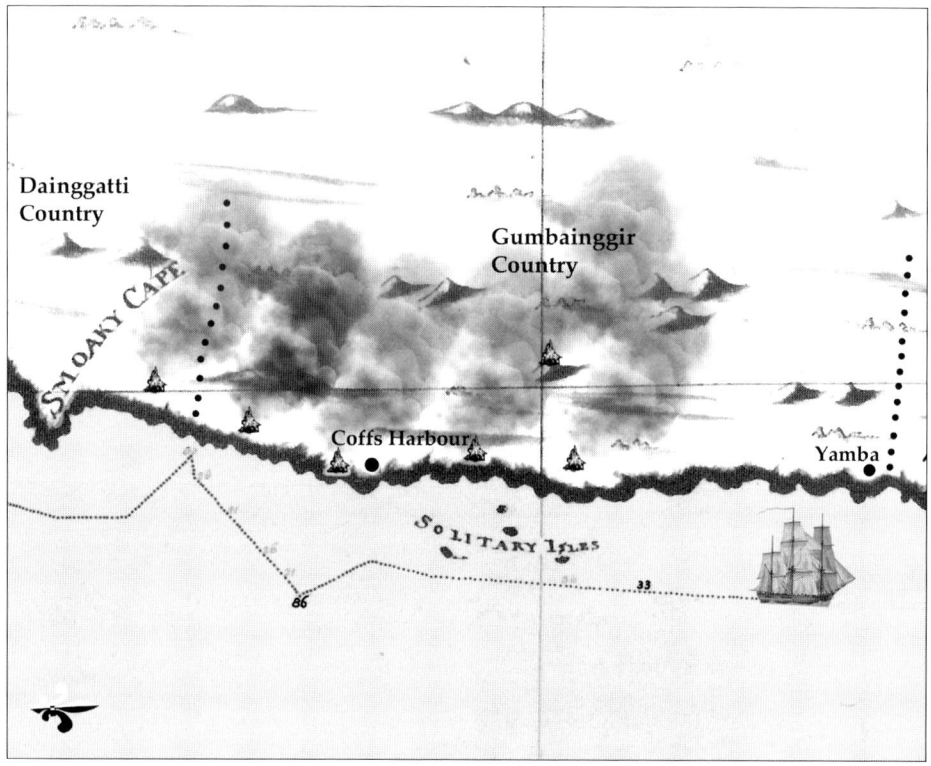

Portion of Cook's chart 'Smoaky Cape to Cape Townsend'.

COOK: At 8 o'clock [a.m.] it began to Thunder and rain which lasted about an hour and then fell calm which gave us an opportunity to sound and found 86 fathom water being about 4 or 5 Leagues from the Land. After this we got the wind southerly a fresh breeze and fair weather. We turned NBW for the northermost land in sight.

BANKS: The briskest breeze I think that the *Endeavour* has gone before during the voyage.

COOK: As we have advanced to the northward the land hath increased in height in so much that in this latitude it may be called a hilly country.

PARKINSON: We saw clouds of smoke arise from different distant parts of the country [Gumbainggir country].

15th May

BANKS: Several fires were seen and one high up on a hill side 6 or 7 miles at least from the beach.

COOK: In the p.m. some heavy squalls attended with rain and hail which obliged us to close reef our topsails. Between 2 and 4 we had some small rocky Islands between us and the land [Solitary Islands – marked on chart above]. At 5 p.m. we sounded and had 33 fathom water about 2 Miles without this last Island. At 8 o'clock we brought too until 10 at which time we made sail under our topsails having the advantage of the Moon.

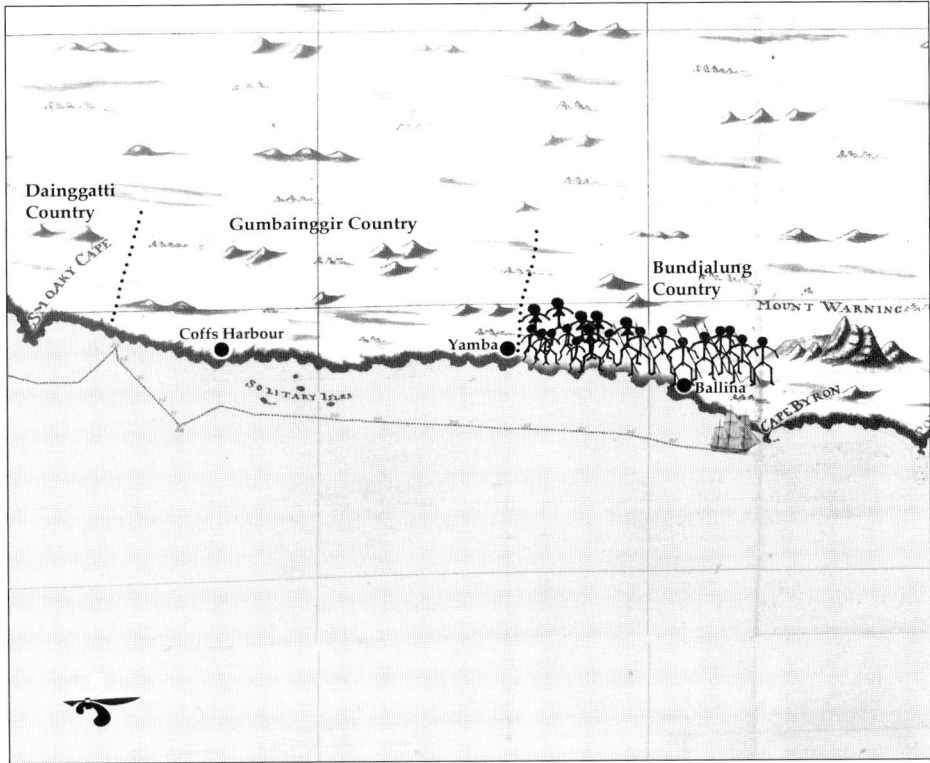

BANKS: In the Morning some people were seen, about 20, each of which carried upon his back a large bundle of something which we conjectured to be palm leaves for covering their houses; we observed them with glasses for near an hour during which time they walked upon the beach and then up a path up a gently sloping hill, behind which we lost sight of them. Not one was once observed to stop and look towards the ship; they pursued their way in all appearance entirely unmoved by the neighbourhood of so remarkable an object as the ship must necessarily be to people who have never seen one [Bundjalung country].

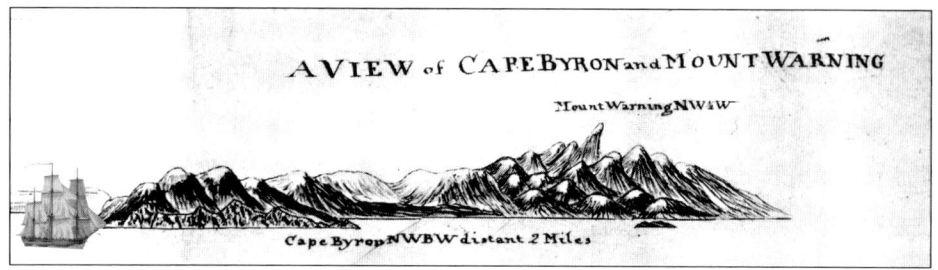

Charles Praval – copied from a lost original by Parkinson or Spöring.³

[Charles Praval joined the expedition later in Batavia (Jakarta) and was employed making copies of coastal views made earlier by Sydney Parkinson and Herman Spöring.]

COOK: At Noon a tolerable high point of land bore NWBW distant 3 Miles; this point I named <u>Cape Byron</u>. It may be known by a remarkable sharp peaked mountain lying inland from it [Mount Warning].

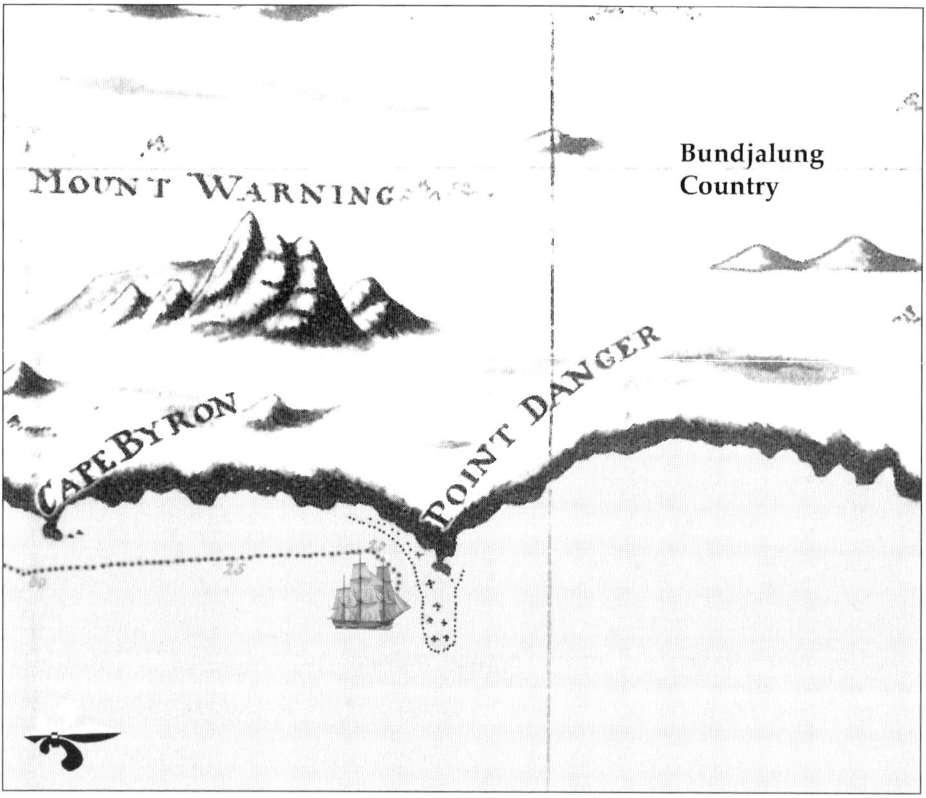

16th May

COOK: At sunset we discovered **breakers** ahead and on our larboard bow … [Off Point Danger on chart above].

BANKS: … in the very direction in which the ship sailed.

COOK: Hauled off east until 8 o'clock [p.m.]. We then brought too.

BANKS: It blowing rather fresh and a great sea running made the night rather uncomfortable. In the morn we saw the breakers which we last night escaped between us and the land.

Charles Praval – copied from a lost original by Parkinson or Spöring.[4]

COOK: There situation may always be found by the peaked mountain before mentioned which on this account I have named Mount Warning.

BANKS: About it was well wooded and looked beautiful as well as fertile.

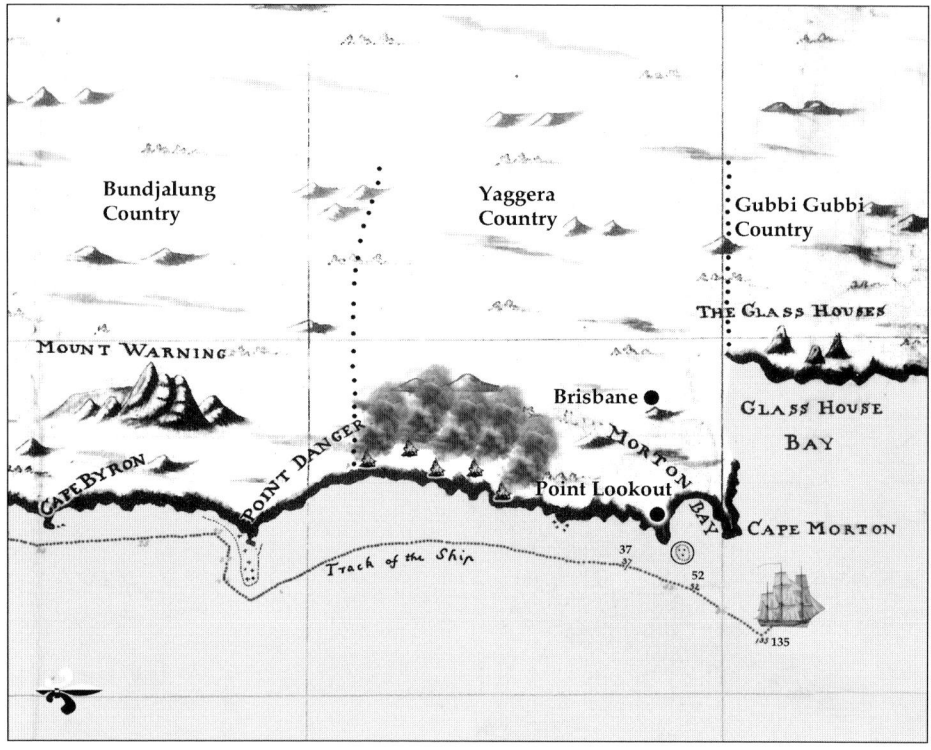

COOK: The point off which these shoals lay I have named <u>Point Danger</u>.

BANKS: At noon we were abreast of some very low land which looked like an extensive plain in which we supposed there to be a Lagoon, in the neighbourhood of which were many fires. [Estuary between North and South Stradbroke Islands. Midway between Point Danger and Morton Bay on chart above. Bordering Bundjalung & Yuggera country.]

17th May

COOK: Between 4 and 5 we discovered **breakers** on our larboard bow, our depth of water at this time was **37** fathom. At sunset the land which formed a point I named <u>Point Lookout</u>. The breakers lies about 3 or 4 Miles from Point Lookout. At this time we had a great sea from the south ward which broke prodigious high upon them. On the north side of this point the shore forms a wide-open Bay which I have named <u>Morton Bay</u> [named after Lord Morton, president of the Royal Society]. I named <u>Cape Morton</u> it being the north point of the Bay of the same name.

Stood on NNE until 8 o'clock [p.m.] when being past the breakers and having deepened our water to **52** fathom we brought too until 12 o'clock than made sail again to the NNE at 4 in the a.m. we sounded and had **135** fathom.

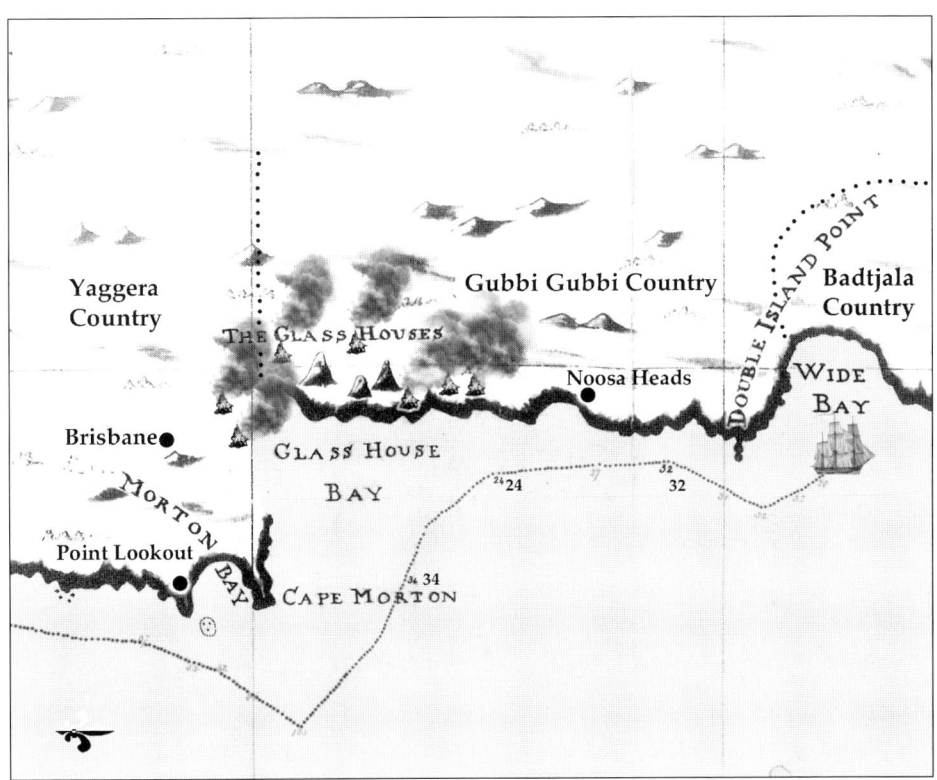

BANKS: About 10 [a.m.] we were abreast of a large bay the bottom of which was out of sight [Glass House Bay]. The sea in this place suddenly changed from its usual transparency to a dirty clay colour, appearing much as if charged with freshes [fresh water], from whence I was led to conclude that the bottom of the bay might open into a large river.

COOK: Some on board was of opinion that there is a River there because the Sea looked paler than usual, upon sounding we found 34 fathom water a fine white sandy bottom which a lone is sufficient to change the apparent colour of sea water without the assistance of Rivers. The land need only to be as low here as it is in a thousand other places upon the coast to have made it impossible for us to have seen it at the distance we were off. Be this as it may it was a point that could not be cleared up as we had the wind [present-day Brisbane River and Moreton Bay].

BANKS: About it were many smokes especially on the Northern side near some remarkable conical hills [Glass Houses – Yuggera & Gubbi Gubbi country].

COOK: These hills lay but a little way inland and not far from each other, they are very remarkable on account of their singular form of elevation which very

82   The Endeavour Journals

much resemble Glass Houses which occasioned my giving them that name. At noon we about 2 or 3 leagues from the land and in **24** fathom water. Several smokes seen today and some pretty far inland.

18th May

COOK: In steering along shore at the distance of two leagues off our soundings was from **24** to **32** fathom a sandy bottom. I named Double Island Point from its figure. It looks like two small islands laying under the land. It likewise may be known by the white cliffs on the north side of it, here the land trends to the NW and forms a large open bay [Wide Bay on Cook's chart].

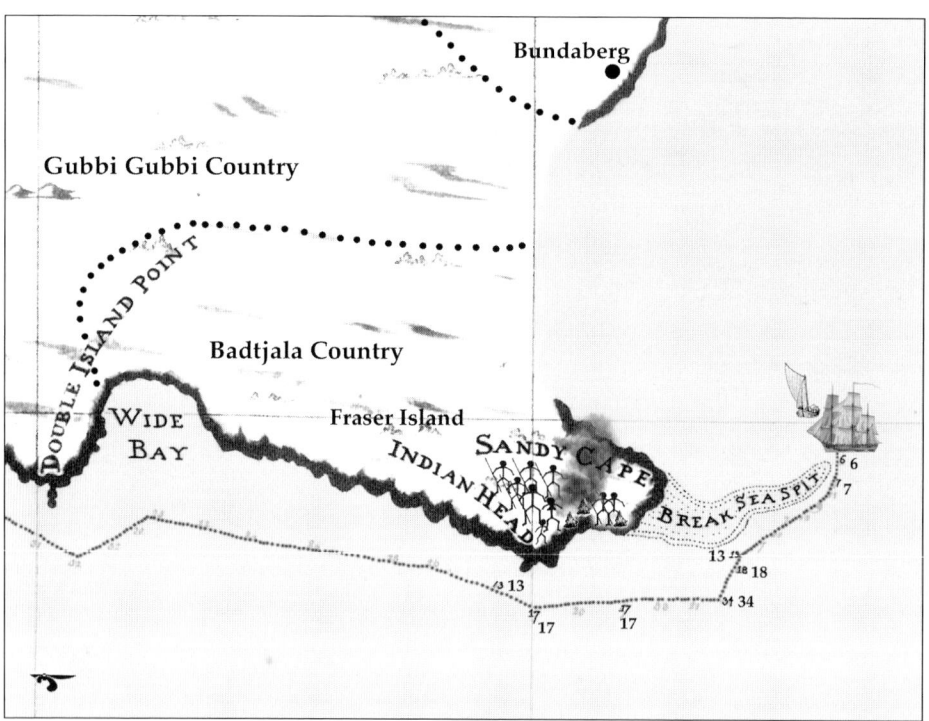

BANKS: Land this morn very sandy. We could see through our glasses that the sands which lay in great patches of many acres each were moveable: some of them had been lately moved, for trees which stood up in the middle of them were quite green, others of a longer standing had many stumps sticking out of them which had been trees killed by the sand heaping about their roots, Few fires were seen [present-day southern Fraser Island – part of the coast between Wide Bay and Indian Head on Cook's chart – Badtjala country].

COOK: At noon land here abouts appears more barren than any we have yet seen on this coast and the soil more sandy. Nor did we see many signs of inhabitants.

19th May

COOK: As we had but little wind we kept on to the northward all night. At Noon we were about 4 Miles from it and in this situation had but **13** fathom water.

20th May

COOK: At 1 o'clock in the p.m. we passed at a distant of 4 miles, having **17** fathom water a black bluff head or point of land on which a number of the natives were assembled, which occasioned my naming it Indian Head. We saw people [Badtjala people] in other places besides the one I have mentioned, some smokes in the day and fires in the night.

BANKS: At sun set the land appeared in a low bank to the sea over which nothing was seen.

COOK: Having but little wind all night we kept on to the northward – having from **17** to **34** fathom. At day light being at this time in **18** fathom water we discovered a reef stretching out to the northward as far as we could see.

BANKS: Our usual good fortune now again assisted us, for we discovered breakers which we had certainly ran upon had the ship in the night sailed 2 or 3 leagues farther than she did.

COOK: We then edged away along the East side of the shoal having regular even soundings from **13** to **7** fathom.

21st May

BANKS: This shoal extended a long way out from the land for we ran along it till 2 o'clock and then passed over the tail of it.

COOK: I sent a boat a head to sound and upon her making the signal for more than 5 fathom we hauled our wind and stood over the tail of it in **6** fathom. This Shoal I called Break Sea Spit, because now we had smooth water whereas upon the whole Coast to the South of it we had always a high sea or swell from ye SE.

BANKS: The Sea was so clear that we could distinctly see the bottom and indeed when it was 12 and 14 fathom deep the colour of the sand might be seen from the mast head at a large distance. While we were upon the shoal innumerable large fish, Sharks, Dolphins &c. and one large Turtle were seen. A grampus of the middle size leaped with his whole body out of water several times making a Splash and foam in the sea as if a mountain had fallen into it. [The term 'grampus' has been applied over the centuries to many large fish, including the Killer whale – *Orcinus orca*.]

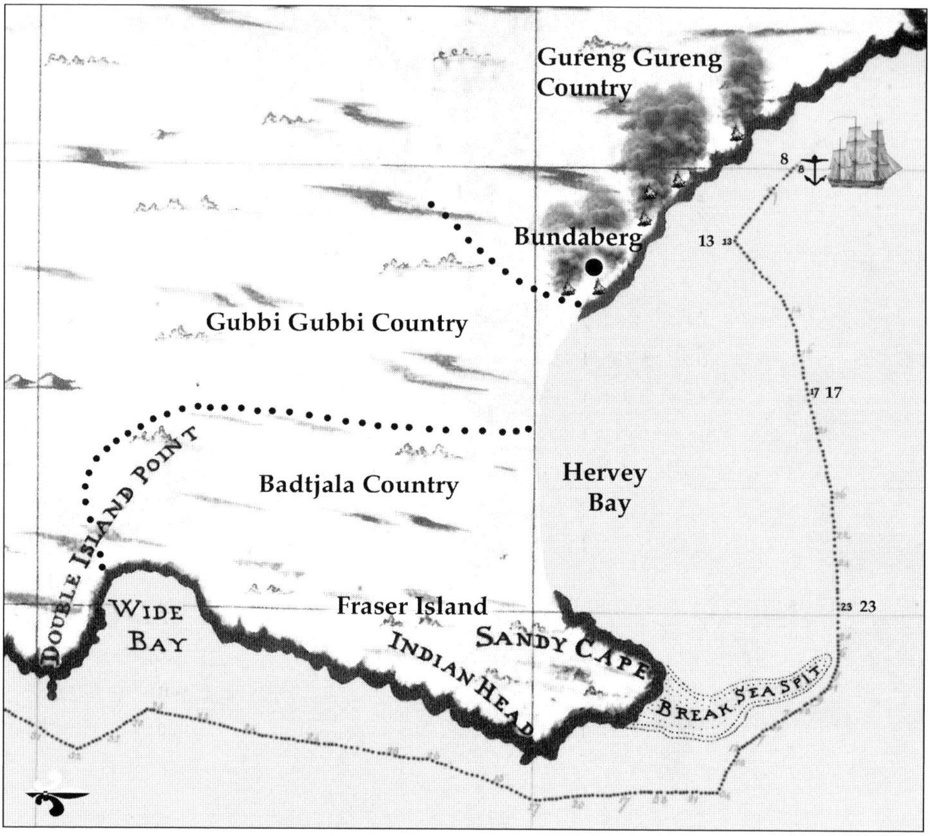

BANKS: The land in sight terminated in a sandy cape behind which a deep bay ran in [Hervey's Bay], across which we could not see.

'A VIEW of SANDY CAPE bearing S. distant 5 Leagues'.⁵ Charles Praval – copied from a lost original by Parkinson or Spöring.

COOK: This point I have named <u>Sandy Cape</u> on account of two very large white patches of sand upon it.

PARKINSON: The barren sandy land continued to this point, and was uninhabited.

COOK: At 6 o'clock [p.m.] the depth of water **23** fathom which depth we kept all night as we stood to the westward.

At 9 in the a.m. we discovered from ye mast head land to the westward and soon after saw smokes upon it [Gureng Gureng country]. Our depth of water was now decreased to **17** fathom and by Noon to **13**.

For these few days past we have seen at times a sort of Sea fowl we have nowhere seen before that I remember, they are of that sort called Boobies [Brown Boobies – *Sula leucogaster plotus*]. (Colour Plate No. 11)

Before today we seldom saw more than 2 or 3 at a time and only when we were near the land. Last night a small flock of these birds passed the Ship and went away to the NW and this morning from half an hour before sun rise to half an hour after flights of them were continually coming from the NNW and flying to the SSE and not one was seen to fly in any other direction. From this we did suppose that there was a Lagoon, River or Inlet of shallow water to the southward of us in the bottom of the deep Bay, which I named Hervey's Bay, where these birds resorted to in the day to feed, and that not very far to the northward lay some Island where they retired to in the night.

[The Capricorn and Bunker Group of Islands which lie NNW of Hervey's Bay, contain 73–75 per cent of all seabird biomass in the Great Barrier Reef World Heritage Area. The shallows of Great Sandy Strait, at the bottom of Hervey's Bay running between the southern half of Fraser Island and the mainland, would seem ideal for these birds to feed – Ray Parkin 'HM Bark *Endeavour*' – p. 238.]

22nd May

BANKS: At 4 in the evening the land appeared very low but covered with fine wood; on it were many very large Smoaks several of which were seen before we could see the land itself.

COOK: In running along shore we shoaled in our water from 9 to 7 fathom and at one time had but 6 fathom which determined me to **anchor** for the night [anchorage marked on chart] and accordingly at 8 o'clock we came too in **8** fathom water a fine gravely bottom about 5 Miles from the land.

Chapter 6

Mister Orton He Is A Man Not Without Faults

James Cook.

22nd May (cont'd)

COOK: Last night sometime in the Middle Watch [between midnight and 4 a.m.] a very extraordinary affair happened to Mr Orton my Clerk, he having been drinking in the Evening, some Malicious person or persons in the Ship took the advantage of his being drunk and cut off all the clothes from off his back, not being satisfied with this they sometime after went into his Cabin and cut off a part of both his Ears as he lay asleep in his bed, the person whom he suspected to have done this was Mr Magra one of the Midshipmen, but this did not appear to me upon inquiry, however as I know'd Magra had once or twice before this in their drunken frolics cut off his Clothes and had been heard to say / as I was told / that if it was not for the Law he would Murder him, these things considered induced me to think that Magra was not altogether innocent, I therefore, for the present dismissed him the quarter deck and suspended him from doing any duty in the Ship, he being one of those gentlemen, frequently found on board Kings Ships, that can very well be spared, or to speak more planer 'good for nothing'.

Besides it was necessary in me to show my immediate resentment against the person on whom the suspicion fell least they should not have stopped here.

With respect to Mr Orton he is a man not without faults, yet from all the enquiry I could make, it evidently appeared to me that so far from deserving such treatment he had not designedly injured any person in the Ship, so that I do and shall all ways look upon him as an injured man. Some reasons might however be given why this misfortune came upon him in which he himself was in some measure to blame, but as this is only conjecture and would tend to fix it up some people in the Ship whom I would fain believe would hardly be guilty of such an action, I shall say nothing about it unless I shall hereafter discover the Offenders which I shall take every method in my power to do, for I look upon such proceedings as highly dangerous in such Voyages as this and the greatest insult that could be offered to my authority in this Ship, as I have always been ready to hear and redress every complaint that have been made against any Person in the Ship.

Sydney Parkinson.

PARKINSON: The Captain and Officers offered, sometime after, at Batavia, a reward of fifteen guineas, to any one that should discover the person or persons who cut off his ears, and fifteen gallons of arrack, to any one that should discover him or them who had cut off his clothes. One of our midshipmen, Mister Saunders, ran away from us in Batavia, and it was suspected that he was the person who cut off Orton's ears.

[Patrick Saunders was disrated to Able Seaman on the following day, 23rd May 1770.]

Isaac Smith (1752–1832).[1]

[On the same day (23rd May) Isaac Smith, Able Seaman, was made a midshipman. Isaac Smith and Elizabeth Cook (Cook's wife) were first cousins once removed. Isaac played an important part in drawing up the charts Cook surveyed].

Portion of Cook's chart 'Smoaky Cape to Cape Townsend'.[2]

BANKS: In the morn we got under sail again. The land as last night fertile and well wooded. At noon the land appeared much less fertile, near the beach it was sandy and we plainly saw with our glasses that it was covered with Palm nut trees, *Pandanus tectorius* which we had not seen since we left the Islands within the tropicks. (Colour Plate No. 12)

Along shore we saw 2 men walking along who took no kind of notice of us [Gureng Gureng people].

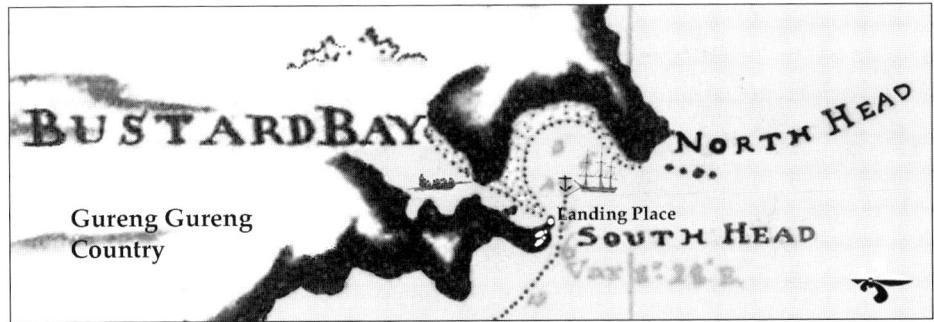

23rd May

BANKS: At night we were working into a bay in which seemed to be good anchorage, where we came to an **anchor** resolved to go ashore tomorrow and examine a little the produce of the country [Bustard Bay – anchorage marked on the chart above].

COOK: In the a.m. I went ashore with a party of men accompanied by Mr Banks and the other gentlemen [Gureng Gureng country].

BANKS: Wind blew fresh off the land so cold that our cloaks were very necessary in going ashore; as the ship lay a good way from the land we were some time before we got there; when landed however the sun recovered its influence and made it sufficiently hot, in the afternoon almost intolerably so. We landed near the mouth of a large lagoon which ran a good way into the country and sent out a strong tide. Here we found a great variety of Plants [one of which was *Dendrobium discolor*]. (Colour Plate No. 13)

COOK: I made a little excursion into the woods while some hands made 3 or 4 hauls with the seine but caught not above a dozen very small fish.

BANKS: The sea seemed to abound in fish but unfortunately at the first haul we tore our seine to pieces.

COOK: As yet we had seen no people but saw a great deal of smoke up and on the west side of the lagoon which was all too far off for us to go by land.

BANKS: One small one was in our neighbourhood, to this we went.

BANKS: It was burning when we came to it, but the people were gone; near it was left several vessels of bark which we conceived were intended for water buckets, several shells and fish bones, the remainder I suppose of their last meal. Near the fires, for there were 6 or 7 small ones, were as many pieces of soft bark of about the length and breadth of a man: these we supposed to be their beds.

COOK: On the windward of one fire was stuck up a little bark about a foot and a half high, these we concluded were all the covering they had in the night, and many of them I firmly believe have not this but naked as they are sleep in the open air. Tupia who was with us observed that they were Taata Eno's, that is bad or poor people.

BANKS: The whole was in a thicket of close trees, defended by them from the wind; whether it was really or not the place of their abode we can only guess. We saw no signs of a house or anything like the ruins of an old one, and from the ground being much trod we concluded that they had for some time remained in that place.

PARKINSON: We saw, too, about twenty of them from the ship, who stood gazing at us upon the beach; also smoke arising out of the woods, which, perhaps, was only an artifice of theirs, to make us think they were numerous [Gureng Gureng people].

BANKS: Fresh water we saw none, but several swamps and bogs of salt water; in these and upon the sides of the lagoon grew many mangrove trees [Large-Leafed Orange Mangrove – *Bruguiera gymnorhiza*] (Colour Plate No. 14) in the branches of which were many nests of ants, one sort of which were quite green [Weaver ant – *Oecophylla smaragdina virescens*]. These when the branches were disturbed came out in large numbers and revenged themselves very sufficiently upon their disturbers, biting sharper than any I have felt in Europe. (Colour Plate No. 15)

The mangroves had also another trap which most of us fell into, a small kind of caterpillar [Cup moth caterpillars – *Doratifera – Limacodidae*], green and beset with many hairs: these sat upon the leaves many together ranged by the side of each other like soldiers drawn up, 20 or 30 perhaps upon one leaf; if these wrathful militia were touched but ever so gently they did not fail to make the person offending them sensible of their anger, every hair in them stinging much as nettles do but with a more acute though less lasting smart. (Colour Plate No. 16)

COOK: The Country is visibly worse than at the last place we were at, the soil is dry and sandy and the woods are free from under wood of every kind.

BANKS: On the shoals and sand banks near the shore of the bay were many large birds far larger than swans which we judged to be Pelicans [*Pelecanus conspicillatus temminck*] but they were so shy that we could not get within gunshot of them. (Colour Plate No. 17)

PARKINSON: We saw some which were near five feet high.

We shot a duck of a beautiful plumage, with a white beak, black body, and white and green on the wings [White-headed Shelduck – *Tadorna radjah rufitergum*]. (Colour Plate No. 18)

COOK: We saw here Bustards [Australian bustard or Plains turkey – *Ardeotis australis – Otididae*] (Colour Plate No. 19) such as we have in England one of which we killed that weighed 17½ pounds which occasioned my giving this place the name of <u>Bustard Bay</u> [present Township of Seventeen Seventy].

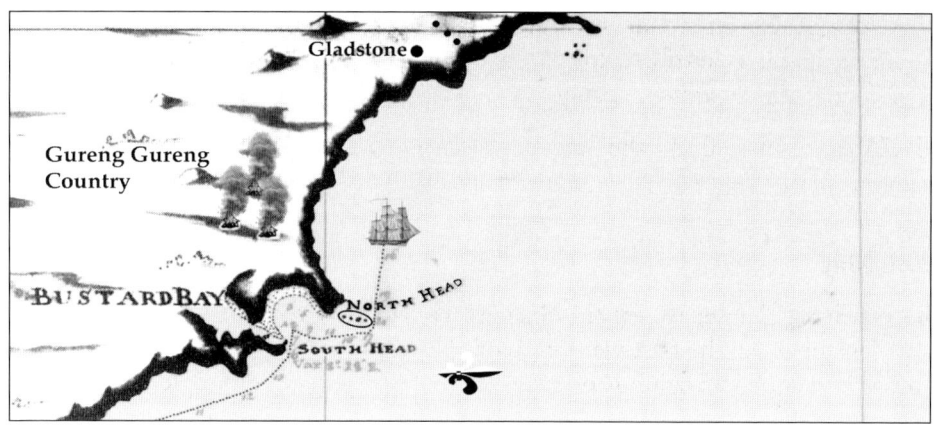

24th May

COOK: At 4 in the a.m. we weighed with a gentle breeze at south and made sail out of the Bay. In standing out we were abreast of the north point and being daylight discovered **breakers** stretching out from it about 2 or 3 miles.

BANKS: The weather was fine; we however were too far from the land to distinguish anything but that there were some fires upon it tho not many [Gureng Gureng country].

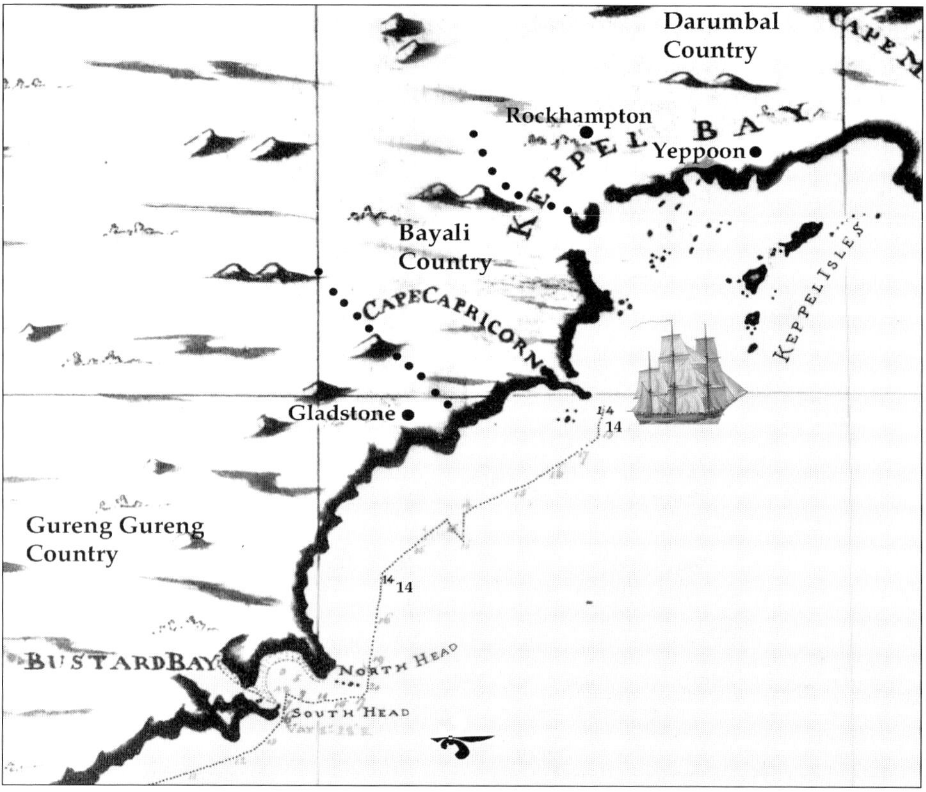

PARKINSON: At noon we were becalmed and caught with hook and line several sorts of beautiful-coloured fish of the snapper kind [including Sucker shark – *Echeneis neucrates* – caught 24th May]. (Colour Plate No. 20)

COOK: Distant from the nearest shore 6 Miles in this situation had **14** fathom water.

25th May

BANKS: At Dinner we eat the Bustard we had shot yesterday, it turned out an excellent bird, far the best we all agreed that we have eat since we left England, and as it weighed 15 pounds our Dinner was not only good but plentiful.

COOK: At 5 in the a.m. we made sail, at day light we saw land making like Islands. At 9 o'clock we were abreast of a point distant from it 1-mile depth of water **14** fathom. I found to lay directly under the Tropic of Capricorn and for that reason called it by that name [Cape Capricorn – NE point of Curtis Island – Bayali country].

'A VIEW of CAPE CAPRICORN…'[3] Charles Praval – copied from a lost original by Parkinson or Spöring.

PARKINSON: The land appeared very desolate, being little else than sand and rocks, parcelled out into several islands and ragged points.

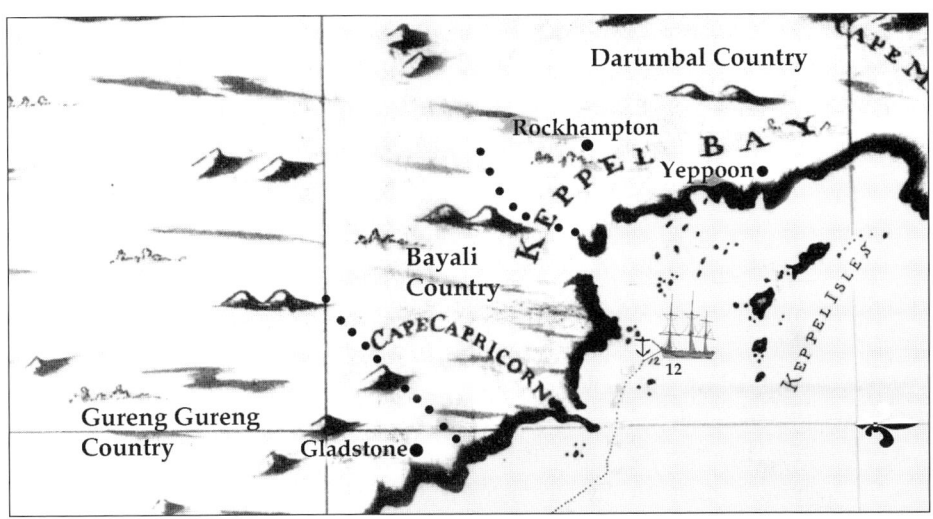

COOK: On the west side of the Cape there appeared to be a Lagoon.

26th May

COOK: In the p.m. we stood NW until 4 o'clock when it fell calm and soon after we **anchored** in **12** fathom water, having the Main land and Islands in a manner all round us.

BANKS: We examined the orange juice and brandy which had been sent on board as prepared by Dr Hulmes directions. It had never been moved from the cag in which it came on board. About half of it had been used or leaked out; the remainder was covered with a whitish mother [yeast] but otherwise was not at all damaged either to taste or sight when it came out of the cag, but when put into a bottle in 3 or 4 days it became ropey and good for nothing. On this we resolved to have it evaporated immediately to a strong essence and put up in Bottles immediately.

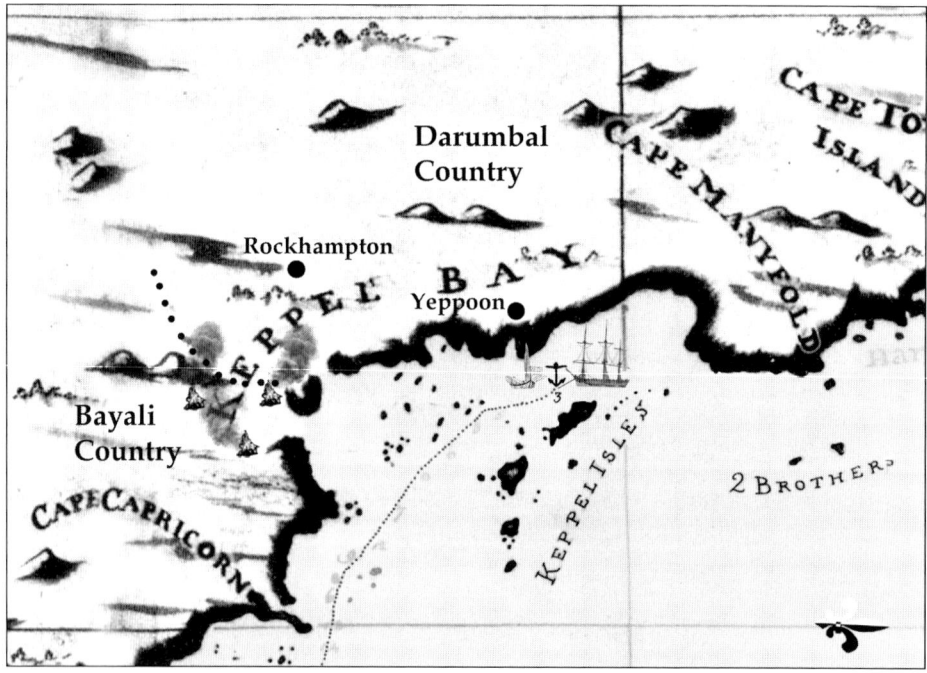

COOK: At 6 in the a.m. we weighed and stood away to the NW between the outermost range of Islands and the Main land, leaving several small Island between us and the latter which we passed close by. Our soundings was a little irregular from which caused me to send a boat a head to sound.

We saw smokes a good way inland which makes me think there must be a River Lagoon or Inlet into the Country and we passed two places that had the appearance of such this morning [present-day Fitzroy River & Cawarral Creek. Darumbal country]. At Noon we were about 3 Miles from the Main, about the same distance from the Islands without us.

27th May

COOK: We had not stood on quite an hour before we fell into **3** fathom water upon which **I anchored** and sent away the Master with two boats to sound the Channel.

BANKS: While the ships boats were employed in sounding round about her myself in my small boat went a shooting and killed several bobies and a kind of white bird called by the seamen Egg bird [Crested tern – *Sterna bergii*]. (Colour Plate No. 21)

Before I went out we tried in the cabin to fish with hook and line but the water was too shoal for any fish. This want was however in some degree supplied by Crabs of which vast numbers were on the ground who readily took our baits, and sometimes held them so fast with their claws that they suffered themselves to be hauled into the ship. They were of 2 sorts.

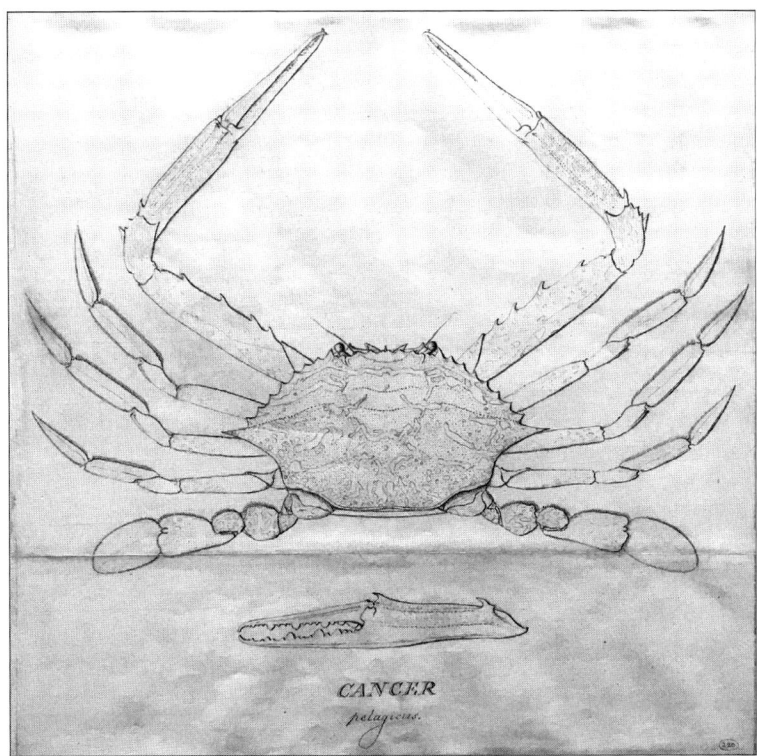

Blue swimmer crab – *Portunus pelagicus*.[4] Pencil sketch by Herman Spöring.

BANKS: The first was ornamented with the finest ultramarine blew conceivable with which all his claws and every Joint was deeply tinged; the under part of him was a lovely white, shining as if glazed and perfectly resembling the white of old China.

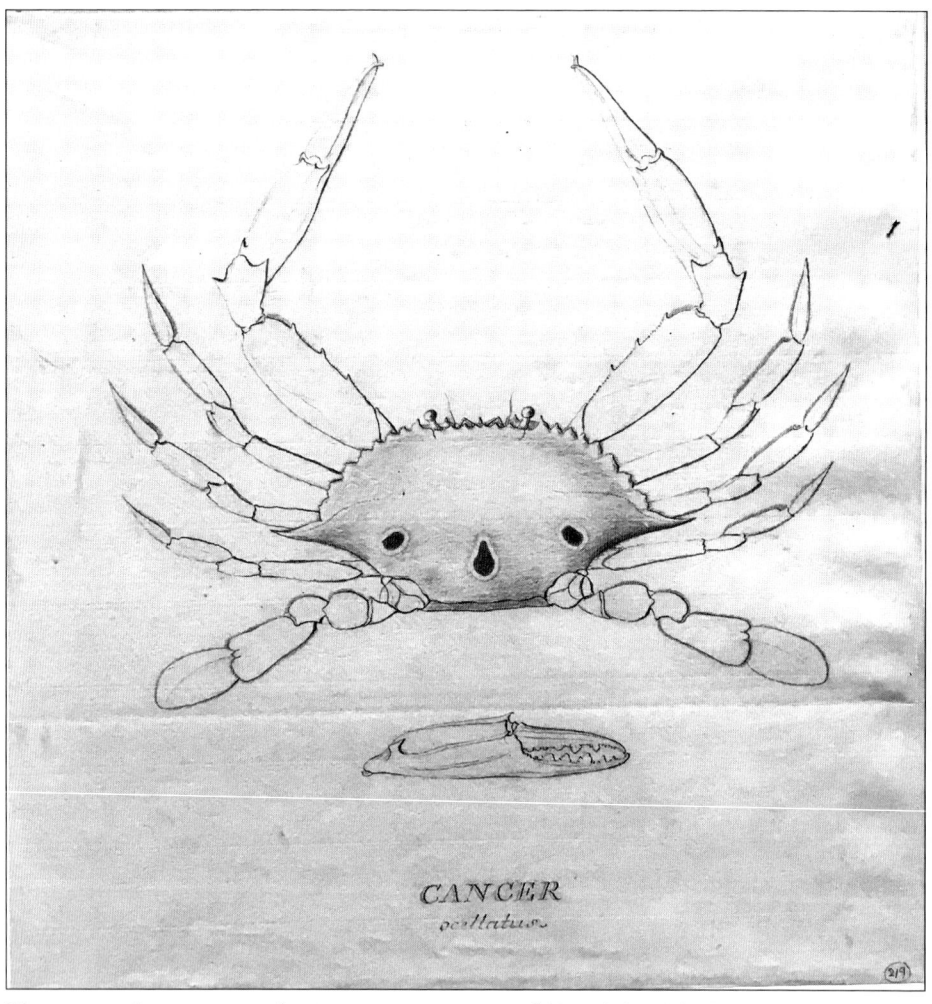

Three-spotted swimming crab – *Portunus sanguinolentus*[5]. Pencil sketch by Herman Spöring.

BANKS: The other had a little of the ultramarine on his Joints and toes and on his back 3 very remarkable brown spots.

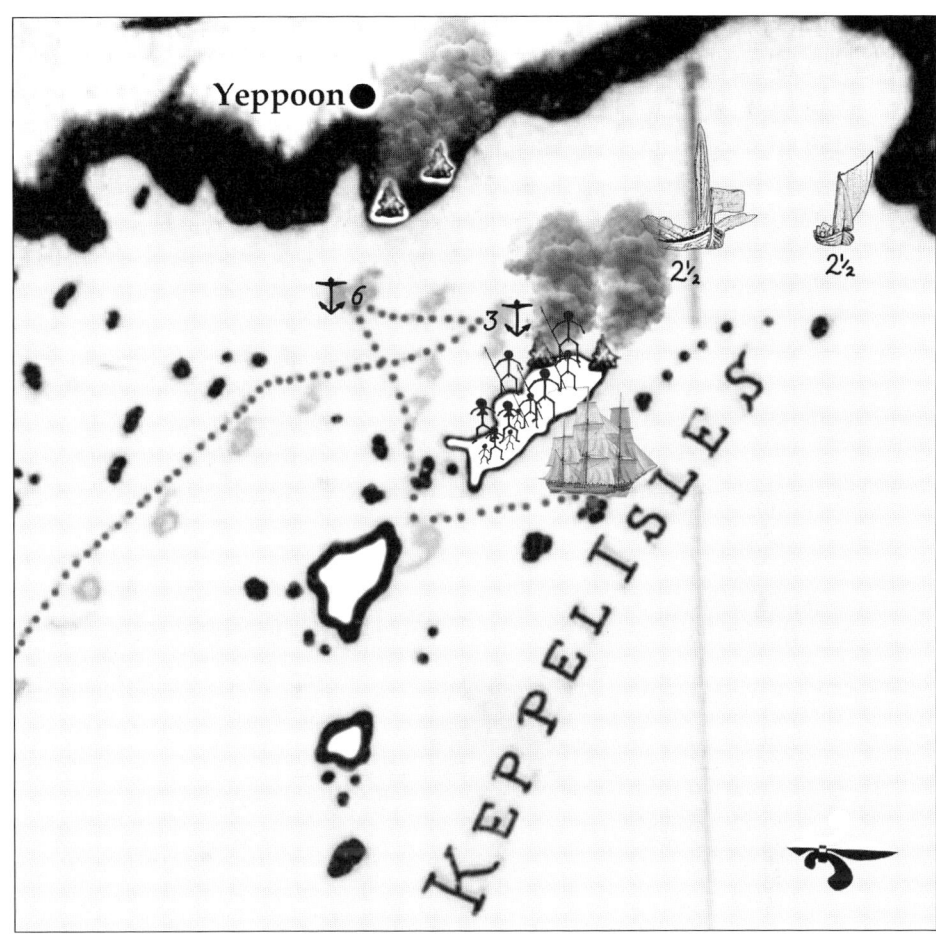

BANKS: Two fires were seen upon an Island, and those who went to sound in the boats saw people [Woppaburra people] upon an Island also who called to them and seemed very desirous that they should land.

PARKINSON: They were of the same sort as those we had seen before.

COOK: The Master reported to me upon his return that he found in many places only 2½ fathom and where we lay at **anchor** we had only 16 feet which was not 2 feet more than the Ship draw'd. In the evening the wind veered to ENE which gave us an opportunity to stretch 3 or 4 miles back the way we came before the wind shifted to South and obliged us again to **anchor** in **6** fathom.

BANKS: The main land was barren to appearance; on it were some smoaks.

PARKINSON: The next morning, we had a fine breeze, and went through a passage to the north-east, between two islands.

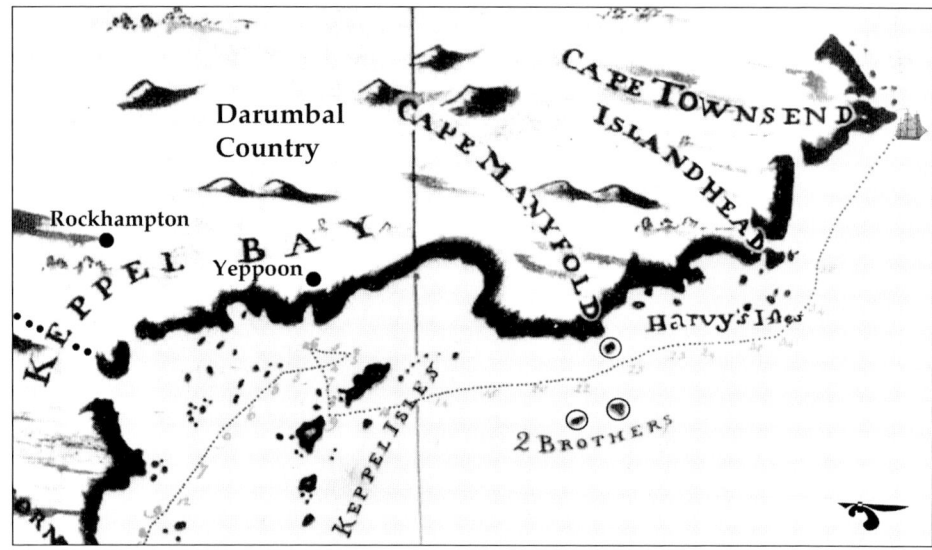

COOK: After we had got out and into deep water we hoisted in the boats and made Sail to the Northward as the land lay. I named Cape Manyfold from the number of high hills over it. It may be known by **three Islands** laying off it, one near the shore and the other two 8 Miles out at sea [2 Brothers on chart] the one of these is low and flat [now Flat Island] and the other high and round [now Peak Island] – The shore forms a large bay which I called Keppel Bay and the Islands which lay in and off it are known by the same name [Keppel Isles on chart].

28th May

COOK: At 6 o'clock [p.m.] we shortened sail and brought too. At day light in the Morning we made sail. At 9 o'clock in the a.m. we were abreast of a point which I named Cape Townsend.

1a. Holothuria obtusata. **1b.** Mimus volutator.

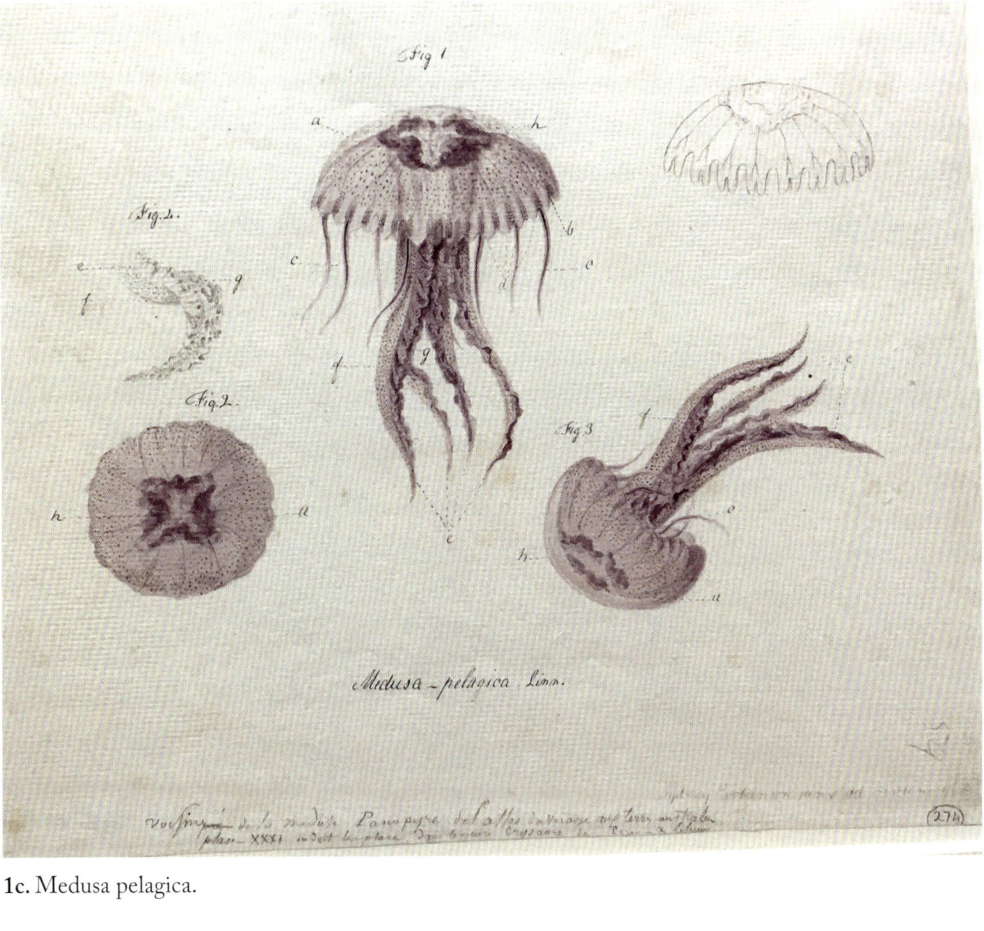

1c. Medusa pelagica.

2. Giant cuttlefish.

3. Cabbage tree.

4. Epacris longiflora – outline drawing.

Epicaris longiflora – finished drawing.

5. Lambertia formosa – outline drawing.

Lambertia formosa – finished drawing.

6. Rainbow lorikeet.

7. Magenta lilly pilly.

8. Brown quail.

9. Warrigal greens.

10. Wedge-tailed shearwater.

11. Brown booby.

12. Pandanus.

13. Dendrobium discolor – outline drawing.

Dendrobium discolor – finished drawing.

14. Large-leafed orange mangrove.

15. Weaver ant.

16. Cup moth caterpillars.

18. White-headed shelduck.

17. Australian pelicans.

19. Australian bustard.

20. Sucker shark.

23. Nest of tree termite.

21. Crested tern.

22. Spear grass.

24. Blue tiger butterfly.

25. Silver pupa and butterfly of two-brand crow.

26. Marine snail.

27. Silver lined mud-skipper.

28. Spurred mangrove – outline drawing.

Spurred mangrove – finished drawing.

29. Audubon's shearwater.

30. Hibiscus meraukensis – outline drawing.

Hibiscus meraukensis – finished drawing.

31. Planchonia careya – outline drawing.

Planchonia careya – finished drawing.

32. Endeavour River.

33. Topknot pigeon.

34. Flying fox.

35. Kale/taro.

36. Wild plantain.

37. Burdekin plum.

38. Threadfin salmon.

39. Giant clam.

40. Blue-black urchin.

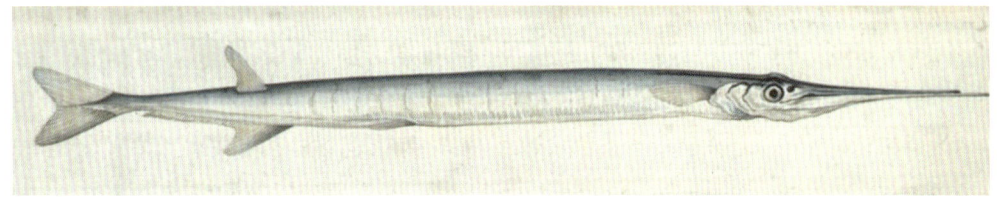

41. Garfish.

42. Coconut-opening crab.

43. Encrusted coconut.

44. Endeavour River.

45. Sea hibiscus.

46. Dingo.

47. Fresh water mussels.

48. Whistling tree duck.

49. Estuarine crocodile.

50. Cochlospermum gillivraei – outline drawing.   Cochlospermum gillivraei – finished drawing.

51. Common baler shell.

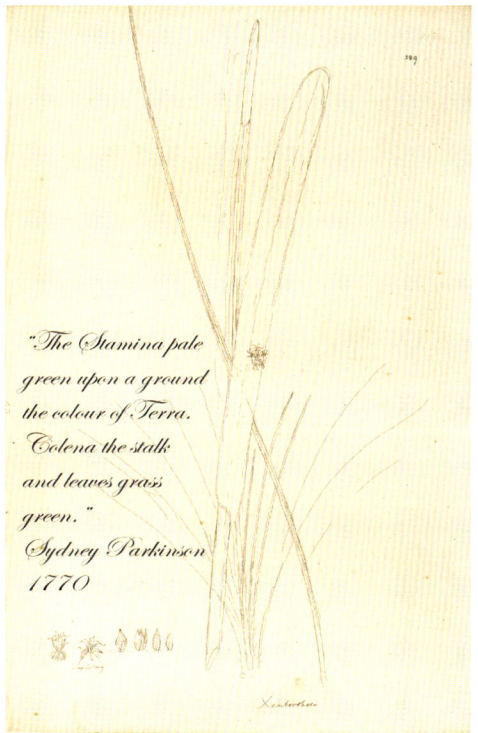

52. Xanthorrhoea resinosa – outline drawing.

Xanthorrhoea resinosa – finished drawing.

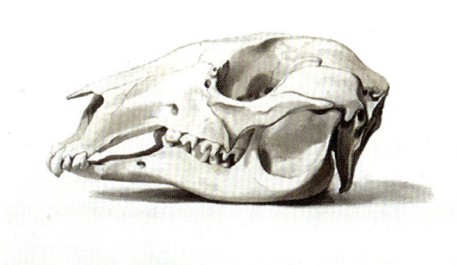

53. Macropus robustus – skull.

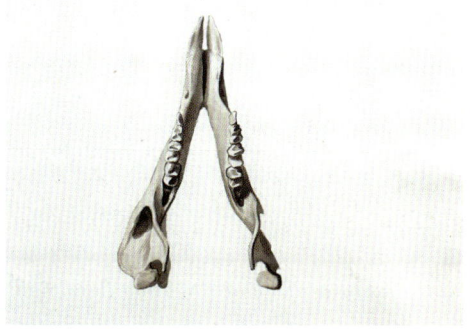

Macropus robustus – lower jaw.

54. Green turtle.

55. Turtle grass.

56. Loggerhead turtle.

57. Native yam.

58. Endeavour bushland.

59. Native cashew.

60. Kongouro of New Holland.

61. Dillenia alata – outline drawing.

Dillenia alata – finished drawing.

62. Josephinia imperatricis – outline drawing.

Josephinia imperatricis – finished drawing.

63. Yellow-spotted monitor.

64. Osprey.

65. Orange-footed scrub fowl.

66. Sea snake.

67. Organ pipe coral.

69. King George III.

68. Common noddy.

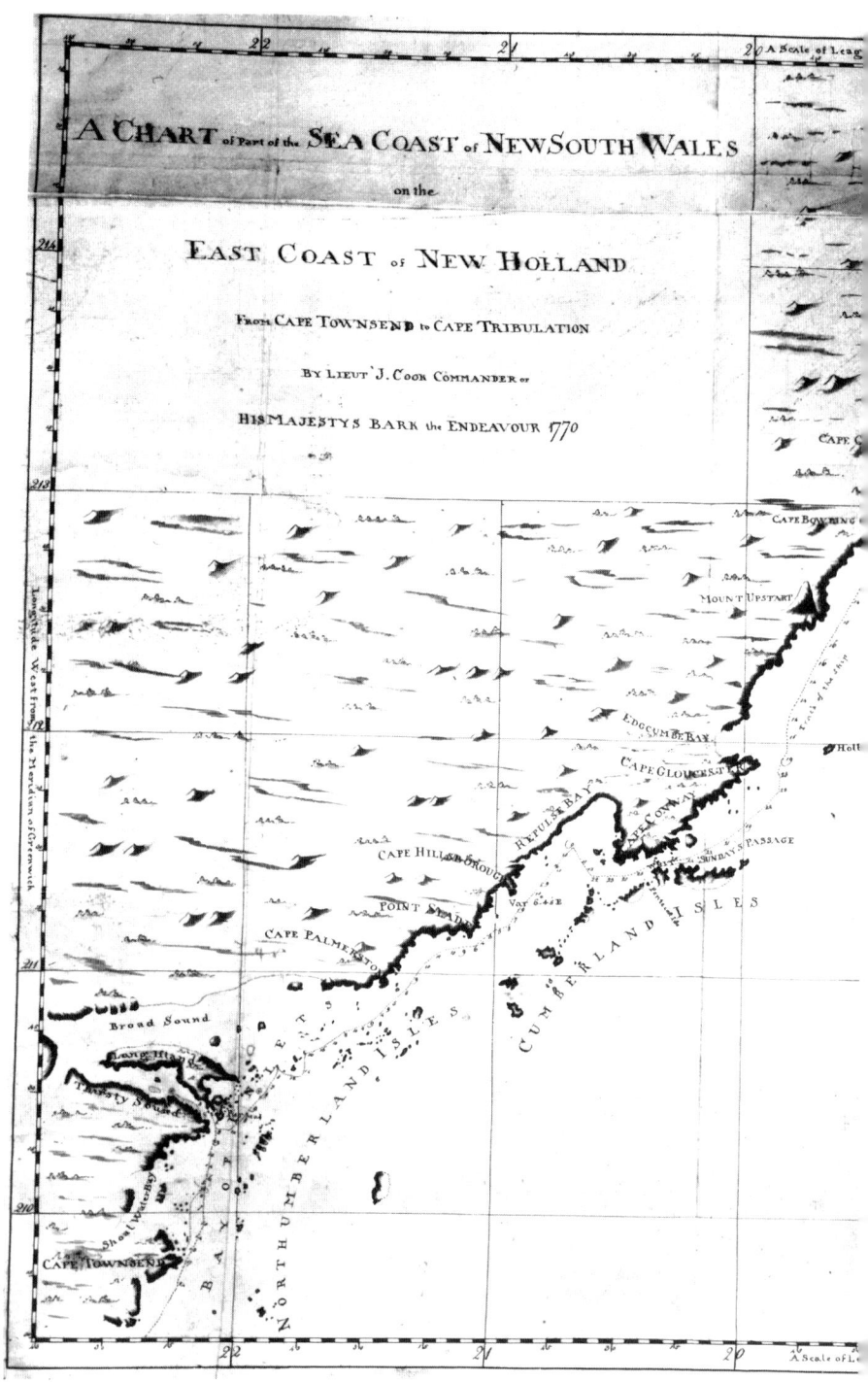

'A CHART of the SEA COAST of NEW SOUTH WALES on the EAST COAST of NEW HOLLAND from Cape Townsend to Cape Tribulation by Lieut. J. Cook Commander of His Majesty's Bark the *Endeavour* 1770'.[6] (The following charts in this chapter and chapter 7 are enlargements taken from this chart.)

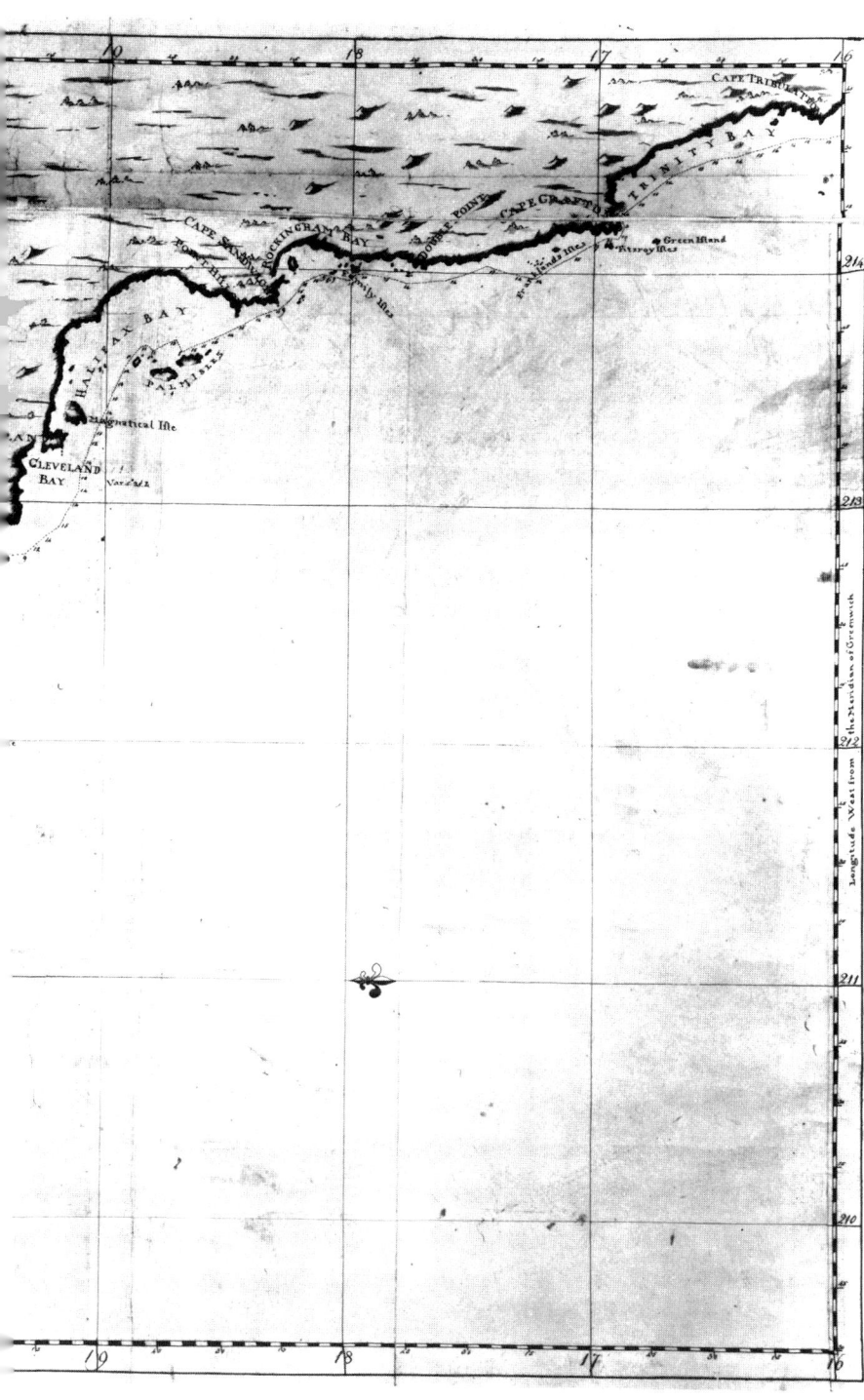

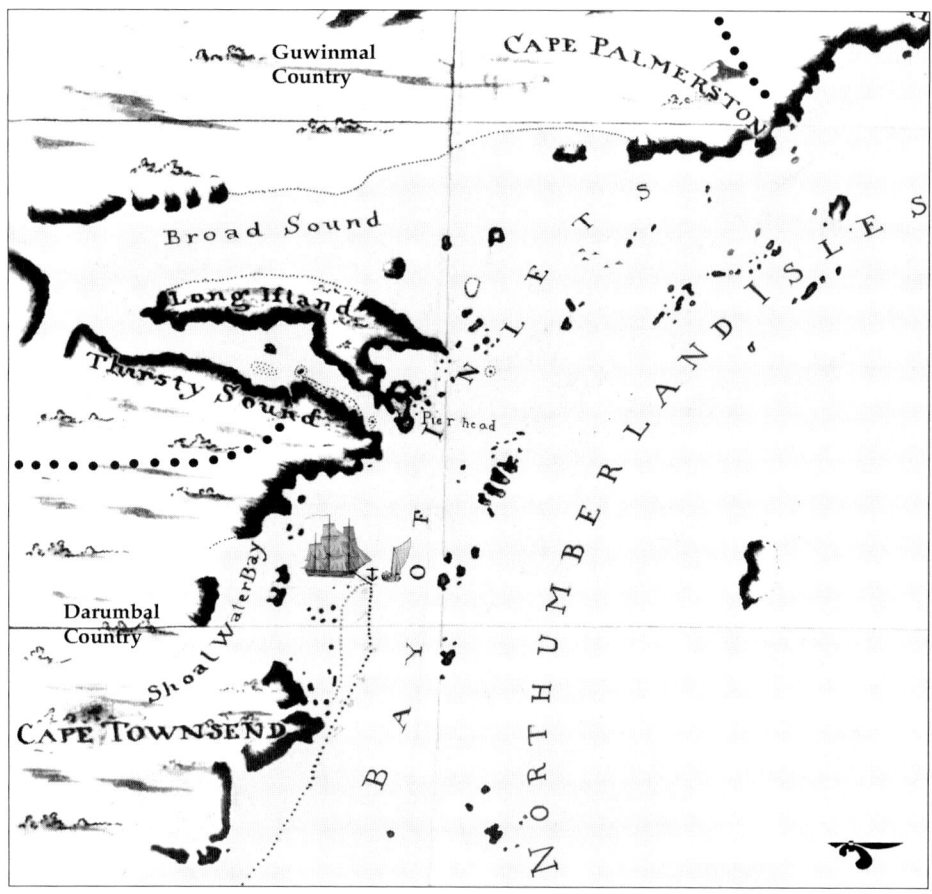

Portion of Cook's chart 'Cape Townsend to Cape Tribulation'.[7]

COOK: To the westward of the Cape the land forms a very large bay [Shoalwater Bay on chart]. As soon as we got round this Cape we hauled our wind to the westward in order to get within the Islands which lay scattered upon and down in this bay in great number and extend out to sea as far as we could see from the Mast-head; how much farther will hardly be in my power to determine. They are as various both in their height and circuit as they are numerous. We had not stood long upon a wind before we met with shoal water and was obliged to tack at once to avoid it, after which I sent a boat ahead and we bore away WBN leaving many small Islands, Rocks & Shoals between us and the Main, and a number of larger Islands without us. A little before noon the boat made the Signal for meeting with Shoal water upon this we hauled close upon a wind to the Eastward but suddenly fell into 3½ fathom water upon which we immediately let go an **Anchor** and brought the Ship up with all sails standing [anchorage marked on chart above].

COOK: We found here a Strong tide setting at the rate of between 2 and 3 Miles an hour which was what carried us so quickly upon the Shoal.

BANKS: The boats now sounded all round her and found that she was upon the shoalest part, on which the anchor was got up and we stood on.

29th May

COOK: At 6 o'clock in the p.m. we **anchored** in 10 fathom a sandy bottom about 2 Miles from the Main land [anchorage marked on chart above].

Chapter 7

Not One Drop of Which We Could Find

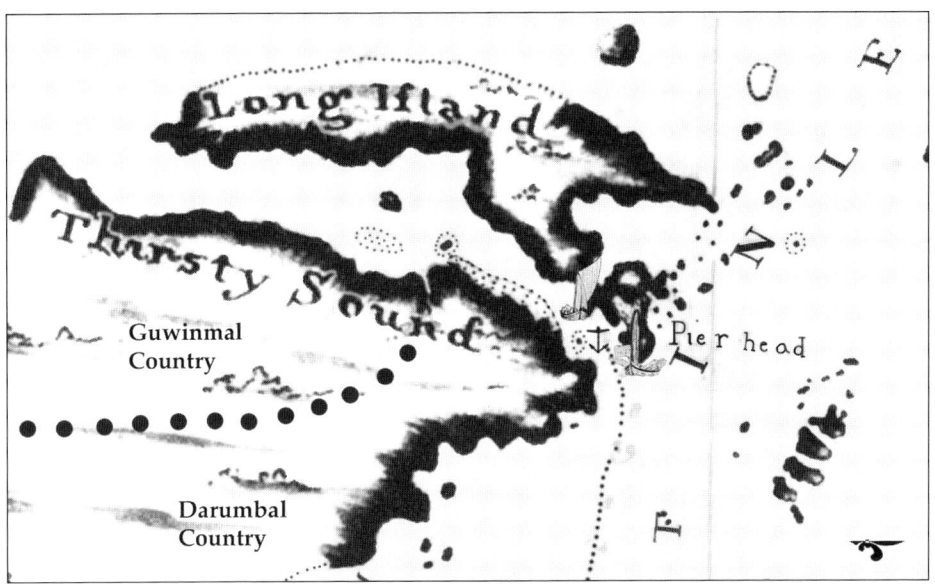

Portion of Cook's chart 'Cape Townsend to Cape Tribulation'.[1]

29th May (cont'd)

COOK: At 5 o'clock in the a.m. I sent away the Master with two boats to sound the entrance of an inlet.

BANKS: We got up our anchor and stood in for an opening in which by 9 o'clock we came to an **anchor**.

COOK: We anchored in 5 fathom water about a League within the entrance of the inlet which we judged to be a River running a good way in land.

BANKS: We saw in coming in no signs of people.

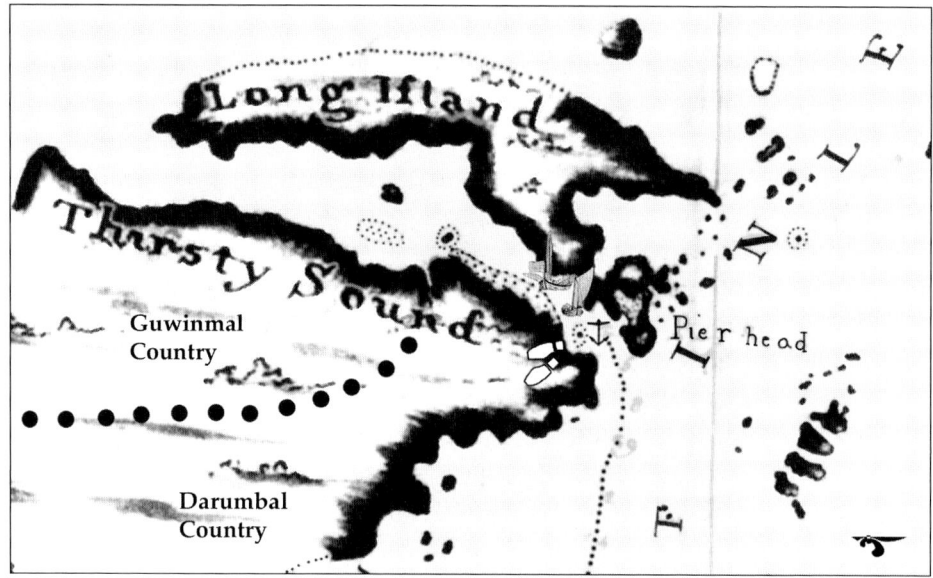

COOK: I intended to wait a few days until the Moon increased and in the meantime to examine the Country. I had some thoughts of laying the Ship a shore to clean her bottom. With this view both the Master and I went to look for a convenient place for that purpose, and at the same time to look for fresh water, not one drop of which we could find, but met with several places where a Ship might be laid a shore with safety.

BANKS: After breakfast we went ashore and found several plants we had not seen before. One kind of Grass was very troublesome to us [Spear grass – *Heteropogon contortus*] (Colour Plate No. 22) its sharp seeds were bearded backwards and whenever they stuck into our cloths were by these beards pushed forward till they got into the flesh: this grass was so plentiful that it was hardly possible to avoid it and with the Musketos [mosquitoes] that were likewise innumerable made walking almost intolerable. We were not however to be repulsed but proceeded into the country.

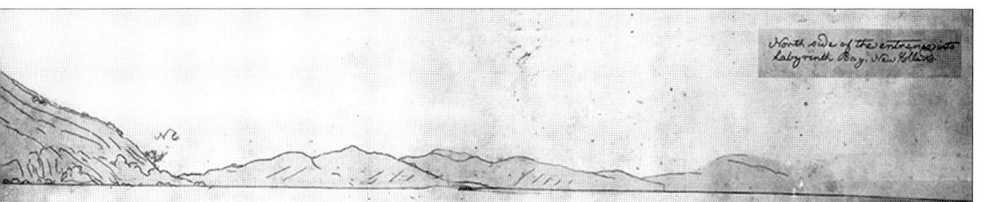

'North side of the entrance into Labyrinth Bay. New Holland.'[2] (One of several first ever depictions of an Eastern Australian landscape by Sydney Parkinson.)

The gum trees were like those in the last bay [Bustard Bay]. On the branches of them and other trees were large ants nests made of Clay as big as a bushel [nest of Tree termite – *Nasutitermes walker*]. (Colour Plate No. 23)

Insects in general were plentiful, Butterflies especially: the air was for the space of 3 or 4 acres crowded with them to a wonderful degree: the eye could not be turned in any direction without seeing millions and yet every branch and twig was almost covered with those that sat still: of these we took as many as we chose, knocking them down with our caps or any thing that came to hand. [Banks is thought to have encountered the Blue tiger – the subspecies *Tirumala hamata hamata*.] (Colour Plate No. 24)

On the leaves of the gum tree we found a Pupa or Chrysalis which shone almost all over as bright as if it had been silverd over with the most burnished silver and perfectly resembled silver; [pupa of the Two-brand Crow, *Euploea sylvester sylvester*, is commonly found in this area of the east coast]. It was brought on board and the next day came out into a butterfly of a velvet black changeable to blue, his wings both upper and under marked near the edges with many light brimstone-coloured spots, those of his under wings being indented deeply at each end. (Colour Plate No. 25)

We saw several swamps of salt overgrown with mangroves; in these we found some species of shells, among them the *Trochus perspectivus* Linn [now *Architectonica perspectiva*]. (Colour Plate No. 26)

Here was a very singular Phaenomenon in a small fish of which there were great abundance. It was about the size of a minnow in England and had two breast fins very strong. We often found him in places quite dry where may be he had been left by the tide: upon seeing us he immediately fled from us leaping as nimbly as a frog by the help of his breast fins: nor did he seem to prefer water to land for if seen in the water he often leaped out and proceeded upon dry land, and where the water was filled with small stones standing above its surface would leap from stone to stone rather than go into the water: in this manner I observed several pass over puddles of water and proceed on the other side leaping as before. [There are multiple species of mudskippers in Queensland one of the common species that Banks might have seen is the Silver lined mudskipper *Gobiidae, Periophthalmus argentilineatus*.] (Colour Plate No. 27)

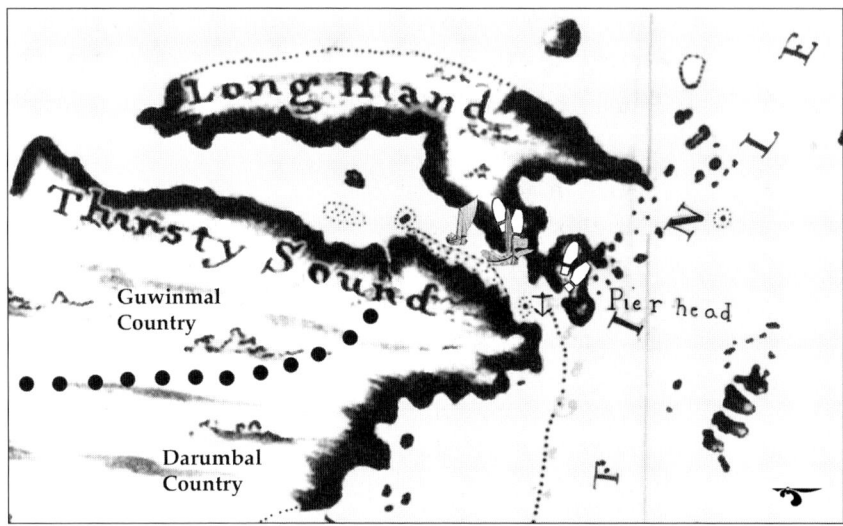

30th May

BANKS: In the afternoon we went ashore on the opposite side of the bay. In neither morning nor evening were there any traces of inhabitants ever having been where we were, except that here and there trees had been burnt down.

COOK: In the p.m. I went again in search of fresh water but had no better success than before wherefore I gave over all thoughts of laying the Ship a shore being resolved on spending as little time as possible in a place that was likely to afford us no sort of refreshment, but as I had observed from the hills the Inlet to run a good way in land I thought this a good time to penetrate into the Country to see a little of the inland parts. Accordingly I prepared for making that excursion in the morning.

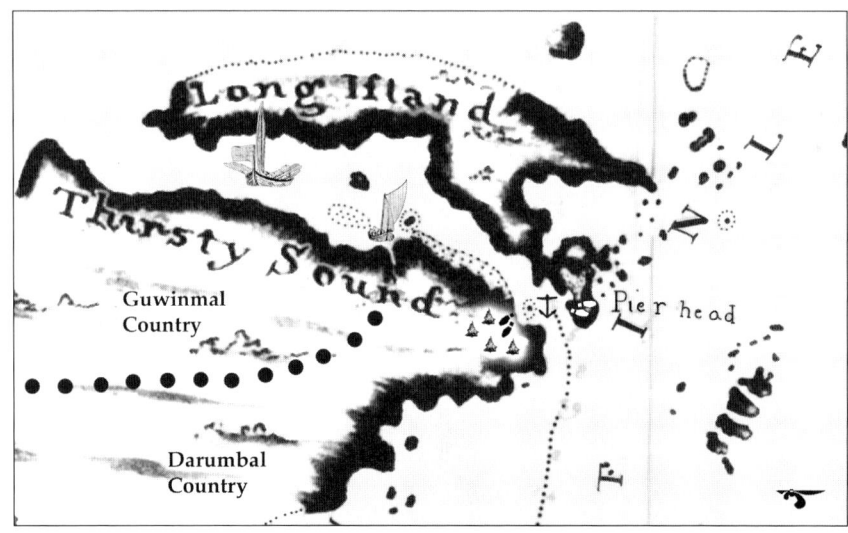

COOK: The first thing I did was to get upon a pretty high hill [Pier Head] which is at the NW entrance of the inlet before sunrise in order to take a view of the sea coast and the Islands that lay off it. As soon as I had done this I proceeded up the inlet.

BANKS: We went again ashore in the same place as yesterday. In attempting to penetrate farther into the country it was necessary to pass a swamp covered with mangrove trees [Spurred mangrove – *Ceriops tagal*]. (Colour Plate No. 28)

This we attempted cheerfully tho the mud under them was midleg deep, yet before we had got half way over we heartily repented of our undertaking: so entangled were the arched branches of those trees that we were continually stooping and often slipping off from their slimy roots on which we stepped; we resolved however not to retreat and in about an hour accomplish our walk of about quarter of a mile. Beyond this we found a place where had been 4 small fires; near them were fish bones, shells &c. that had there been roasted, and grass layd together upon which 4 or 5 people had slept as I guessed about a fortnight before.

The 2nd Lieutenant [John Gore] and one more man who were in very different places declared that they heard the voices of Indians near them, but neither saw the People. It was most destitute of fresh water, probably that was the reason why so few inhabitants were seen [Guwinmal country].

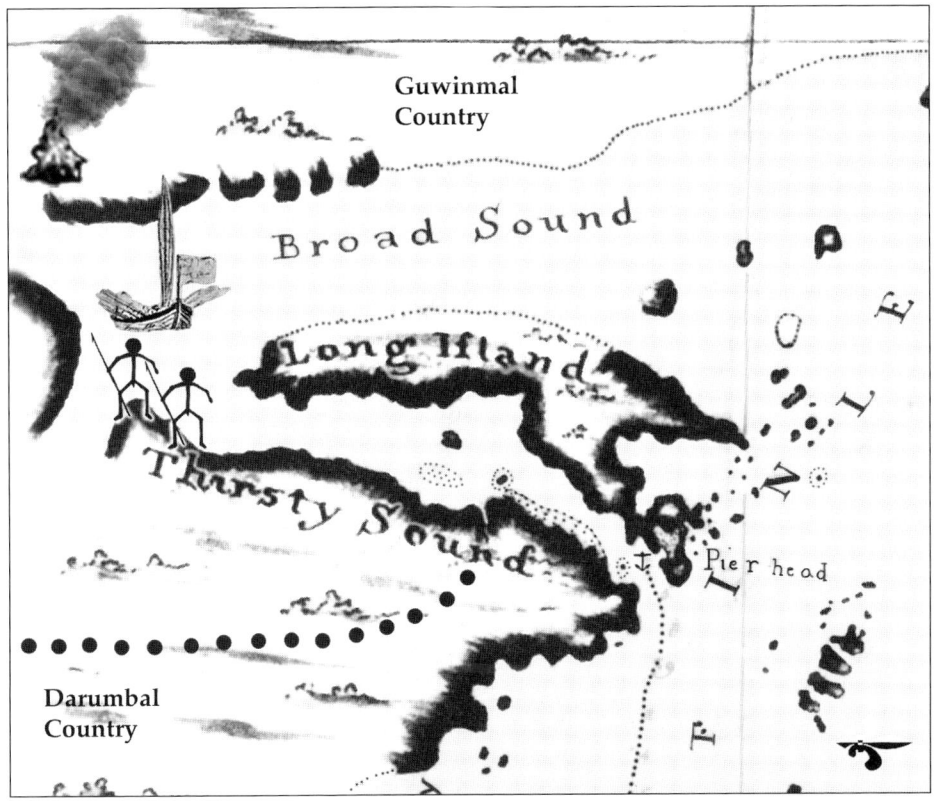

BANKS: The Captn and Dr Solander went to examine the bottom of the inlet which appeared to go very far inland; They saw two men [Guwinmal people] who followed the boat along shore a good way but the tide running briskly in their favour they did not choose to stop for them; at a distance from them far up the inlet they saw a large smoke.

31st May

COOK: Finding no one inducement to stay longer in this place we at 6 in the a.m. weighed and put to Sea. This inlet which I have named Thirsty Sound by reason we could find no fresh water.

PARKINSON: On our first view of this coast, we conceived the most pleasing hopes, but were unhappily disappointed.

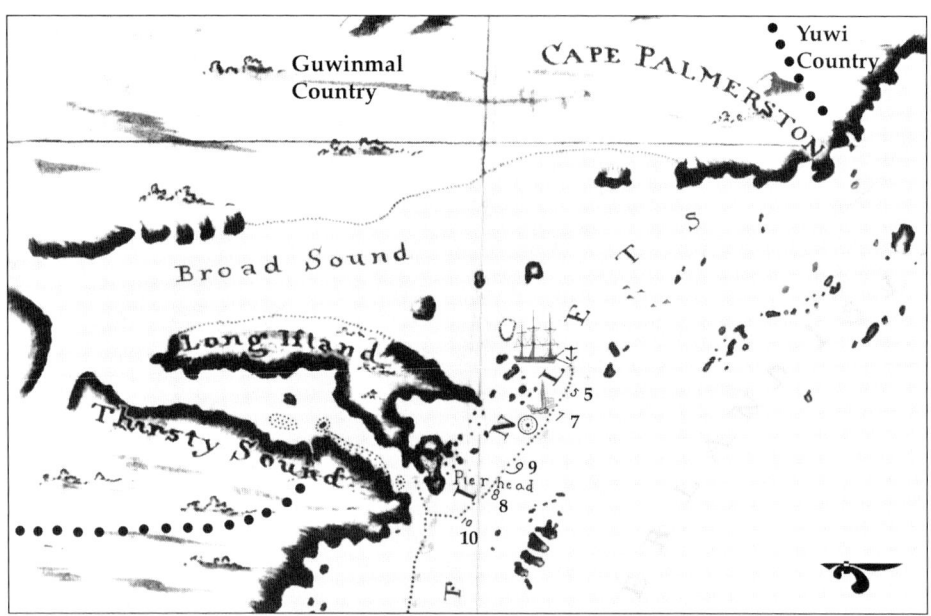

Portion of Cook's chart 'Cape Townsend to Cape Tribulation'.

COOK: The north point of Thirsty Sound I have named Pier Head.
 We kept without the group of islands which lay in shore as there appeared to be no safe passage between them and the main; at the same time we had a number of islands without us extending out to sea as far as we could see.

PARKINSON: From a hill, at the entrance into the bay, we had thirty islands in view.

COOK: As we run in this direction our depth of water was **10, 8** and **9** fathom. At 11 passed by some **rocks**. Sent the pinnace ahead to sound a passage between some islands.

1st June

COOK: At half an hour after noon upon the boat, which we had ahead sounding, making the signal for shoal water, having at that time **7** fathom, the next cast **5**, then 3, upon which we let go the **anchor** and brought the ship up.

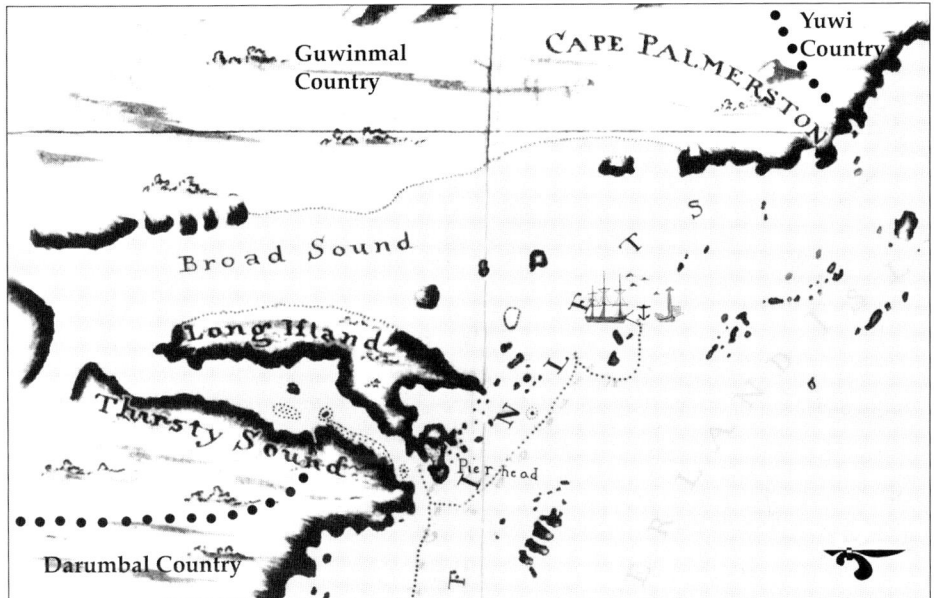

BANKS: The boats sounded and found a passage on which we proceeded and at night came to an **anchor** under the shelter of an island [Bedwell group], in the midst of innumerable islands, rocks and shoals.

In the night it rained and at times blew strong. If our anchor should come home or cable break we had nothing to expect but going ashore on some one or other of the shoals which lay round us. The night passed however without the least accident, and at daylight in the morn the anchor was got up and we proceeded, in hopes of getting out of this archipelago.

COOK: We got again under sail having the Main land in sight and a number of Islands all round us some of which lay out at Sea as far as we could see.

PARKINSON: Through this labyrinth of islands we passed with some difficulty, on account of the number of shoals which we met with; one of which we should have been upon, had not the men in the boat given us timely notice.

COOK: The western inlet known in the chart by the name <u>Broad Sound</u> we now had all open. The point of land which forms the NW entrance into Broad Sound I have named <u>Cape Palmerston</u>. Between this Cape and <u>C. Townshend</u> lies the Bay of Inlets, so named from the Number of Inlets Creeks &c. in it.

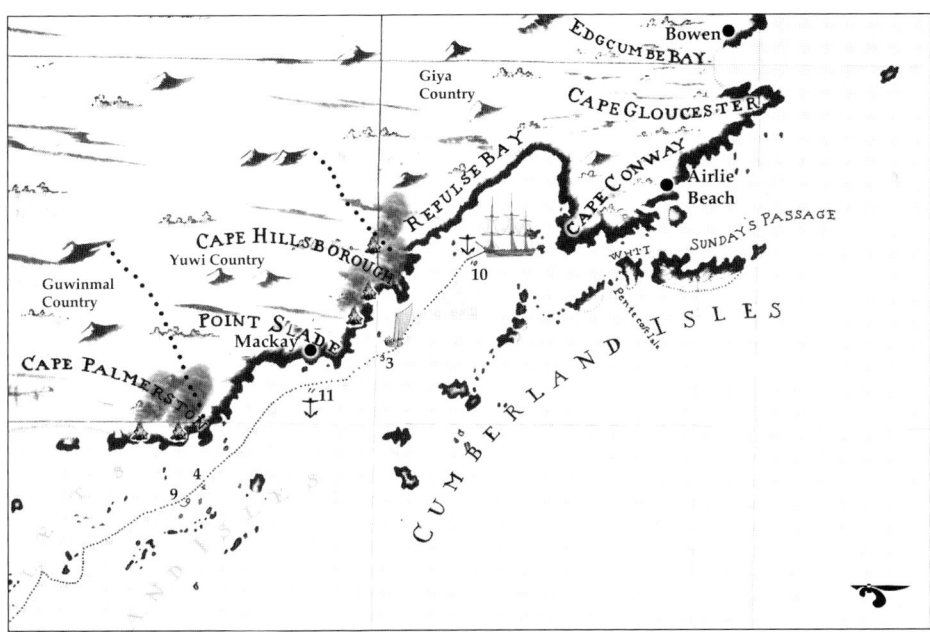

BANKS: By noon we got in with the mainland which made hilly and barren; on it were some smokes [on Cape Palmerston – Guwinmal country].

2nd June

COOK: Our soundings at first were very irregular from **9** to **4** fathom. At 8 [p.m.] o'clock being about two Leagues from the Main Land we **anchored** in **11** fathom.

BANKS: Tupia complained this evening of swelled gums; he had it seems had his mouth sore for near a fortnight, but not knowing what cause it proceeded from did not complain. The Surgeon immediately put him upon taking extract of Lemons in all his drink.

COOK: At 6 [a.m.] we got under sail.

PARKINSON: Still among the islands, through which we were obliged to steer with great caution.

COOK: About 11 o'clock we were again embarrassed by shoal water but got clear without letting go the anchor, we had at one time not quite **3** fathom.

PARKINSON: We so narrowly escaped a bank, the soundings were so unequal.

BANKS: The irregularity of soundings made it necessary to send a boat ahead.

COOK: At noon a pretty high promontory which I named <u>Cape Hillsborough</u>. Some few smokes were seen on the mainland [Yuwi country].

PARKINSON: The land seemed to be thinly inhabited.

3rd June

BANKS: In the evening the country was moderately hilly and seemed green and pleasant; one smoke was seen upon it [just west of Cape Hillsbrough – Yuwi country].

COOK: As I was not sure that there was a passage this way we at 8 o'clock [p.m.] came too an **Anchor** in **10** fathom a muddy bottom.

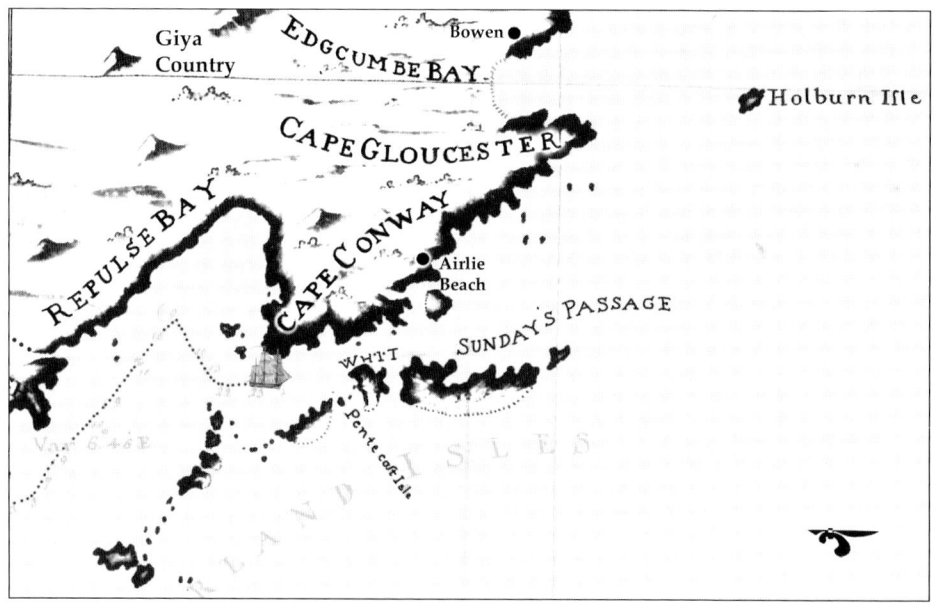

COOK: The flood came from the northward which was from the Islands out at Sea and plainly indicated that there was no passage to the NW. But as this did not appear at day light when we got under sail we stood away to the NW until

'Cumberland Isles, Islands at the South Entrance of the Streights. N. Holland.'[3] (A panorama by Sydney Parkinson. One of several first ever depictions of an Eastern Australian landscape.)

8 o'clock [a.m.], at this time we discoverd low land quite across what we took for an opening between the Main and the Islands which proved to be a Bay about 5 or 6 Leagues deep [Repulse Bay].

BANKS: The country within the bay, especially on the innermost side, was well wooded, looked fertile and pleasant [Giya country].

COOK: Upon this we hauled our wind to the Eastward round the north point of the Bay. This point I have named <u>Cape Conway</u> and the Bay <u>Repulse Bay</u>.

PARKINSON: There appeared to be an opening, which carried us into a strait, in which we found deep water [Whitsunday's Passage].

COOK: At Noon we were just within the entrance.

'One of the Cumberland Isles New Holland'.[4] (Two views of Pentecost Island, NE and E, by Sydney Parkinson.)

COOK: Among the many Islands that lay upon this coast there is one more remarkable than the rest being small very high and peaked [Pentecost Island – Ngaro country].

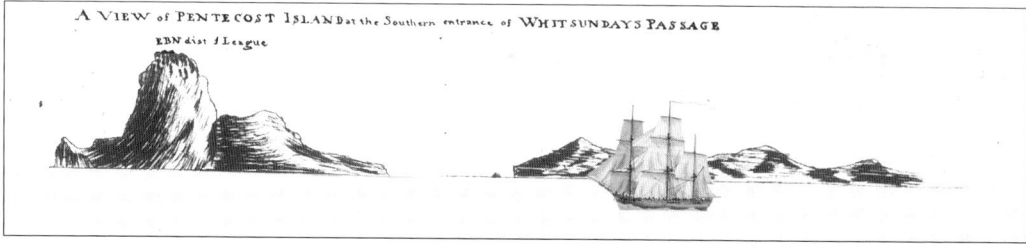

'A VIEW of PENTECOST ISLAND at the Southern entrance of WHITSUNDAYS PASSAGE'.[5] Charles Praval – copied from a lost original by Parkinson or Spöring.

'Whit Sunday's Passage'. (Ngaro people. On Pine Island or Dent Island or South Mole.)

4th June

COOK: On a sandy beach upon one of the Islands we saw two people …

PARKINSON: … three persons …

BANKS: … two men a woman … we saw with our glasses …

COOK: … and a Canoe with an outrigger that appeared to be both larger and differently built to any we have seen upon the Coast.

PARKINSON: … with outriggers, like those of Otaheite [Tahiti].

BANKS: … which made us hope that the people were something improved as their boat was far preferable to the bark Canoes of Stingrays Bay [Botany Bay].

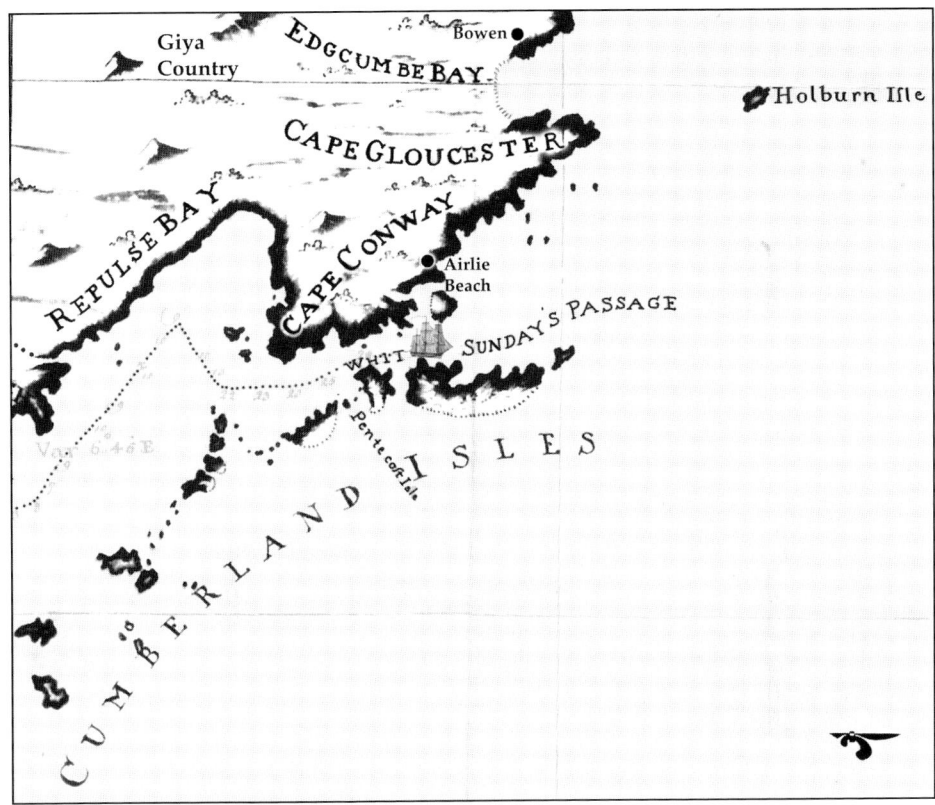

PARKINSON: This strait lies almost north and south; is about seven leagues long, and one and a half broad. On the west of it lies the main [Gyia country] and, on the east, a row of islands which extend a considerable way to the south [Ngaro country].

COOK: The whole passage is one continued safe harbour, besides a number of small Bays and Coves on each side where Ships might lay as it were in a Bason.

PARKINSON: The land on both sides looked much better than that which we had seen before; being high, abounding in trees, and not sandy.

COOK: This passage I have named <u>Whitsunday's Passage</u>, as it was discovered on the day the Church commemorates that Festival and the Isles which form it <u>Cumberland Isles</u> — in honour of His Royal Highness the Duke of Cumberland.

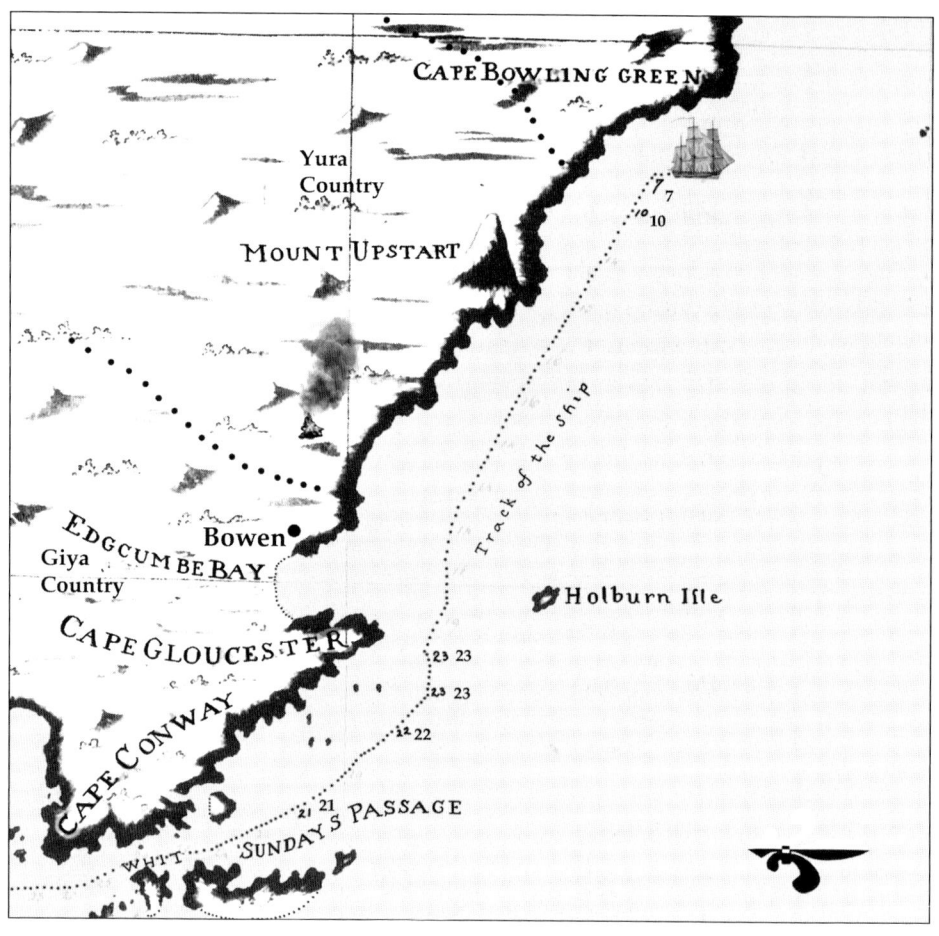

COOK: We kept under an easy Sail and the lead going all night having **21, 22** and **23** fathom at the distance of 3 Leagues from the land.

PARKINSON: On the 4th, we cleared the straits and islands, and got into an open sea.

COOK: At day light in the Morning we were abreast of a Lofty promontory that I named <u>Cape Gloucester</u>. It may be known by an Island which lies out at Sea from it this I called <u>Holburn Isle</u>. There are also Islands Laying under the Land between it and Whitsunday's Passage. On the west side of the Cape the land forms a deep bay which I called <u>Edgecumbe Bay</u>. We continued our Course to the westward.

PARKINSON: The land upon the coast was full of very high hills, whose bowels are probably rich in ore; but their surface is poor indeed, being more barren, and fuller of stones, than any land we had seen.

BANKS: At noon one smoke was seen behind some hills inland.

5th June

COOK: At 6 o'clock [p.m.] we were abreast of the western point of land which I have named Cape Upstart [similar location to Mount Upstart] because being surrounded with low land it starts or riseth up singly at the first making of it.

BANKS: A headland miserably rocky and barren.

COOK: Having past this Cape we continued under an easy sail having from **16** to **10** fathom water until 2 o'clock in the A.M. when we fell into **7** fathom upon which we hauled our wind to the northward judging ourselves to be very near the land as so we found for at day light we were little more than 2 Leagues off. What deceived us was the lowness of the land which is but very little higher than the surface of the Sea, but in this low Country were some hills.

BANKS: Very little appearance of fertility however, either on one or the other.

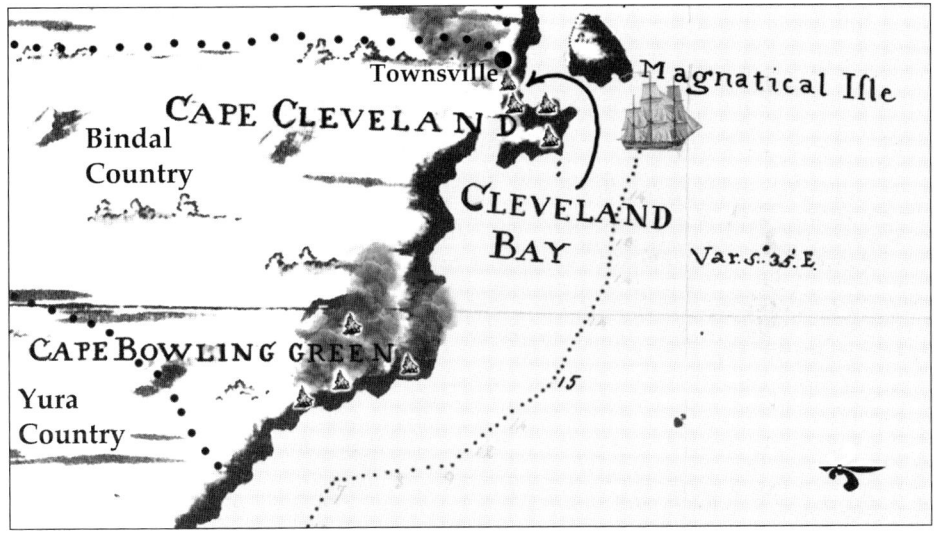

COOK: At Noon we were in **15** fathom water [off Cape Bowling Green] and about 4 Leagues from the land. At and before noon some very large smokes were seen rise up out of the low lands [a little south and north of Cape Bowling Green – Bindal country].

At sun rise I found the **Variation to be 5..35' Easterly** – at sun set last night the same needle gave near 9°, this being close under Cape Upstart I judged that it was owing to Iron ore or other Magnetical matter lodged in the earth.

6th June

BANKS: In the Even the Thermometer was at 74 [23°C] and the air felt to us hotter than we have felt it on the coast before. Many Clouds of a thin scum lay floating upon the water the same as we have before seen off Rio de Janiero; some few flying fish also.

Land made in Barren rocky capes; one in particular [Cape Cleveland] which we were abreast of in the morn appeared much like Cape Roxent [Cabo da Roca – Portugal] at noon 3 fires upon it [Bindal people].

COOK: At noon we had the mouth of a bay all open. This bay which I named Cleveland Bay appeared to be about 5 or 6 Miles in extent every way. The East point I named Cape Cleveland and the West Magnetical Head or Isle as it had much the appearance of an Island and the compass did not travis well when near it. They are both tolerable high and so is the Mainland within them and the whole appeared to have the most rugged, rocky and barren Surface of any we have yet seen. However it is not without inhabitants as we saw smooks in several place in the bottom of the Bay.

7th June

BANKS: In the evening it fell quite calm and I went out in the small Boat and shot *Nectris nugax* [now *Puffinus l'herminieri* – Audubon's shearwater] (Colour Plate No. 29) but saw nothing remarkable on the water; the weather most sultry hot in an open boat.

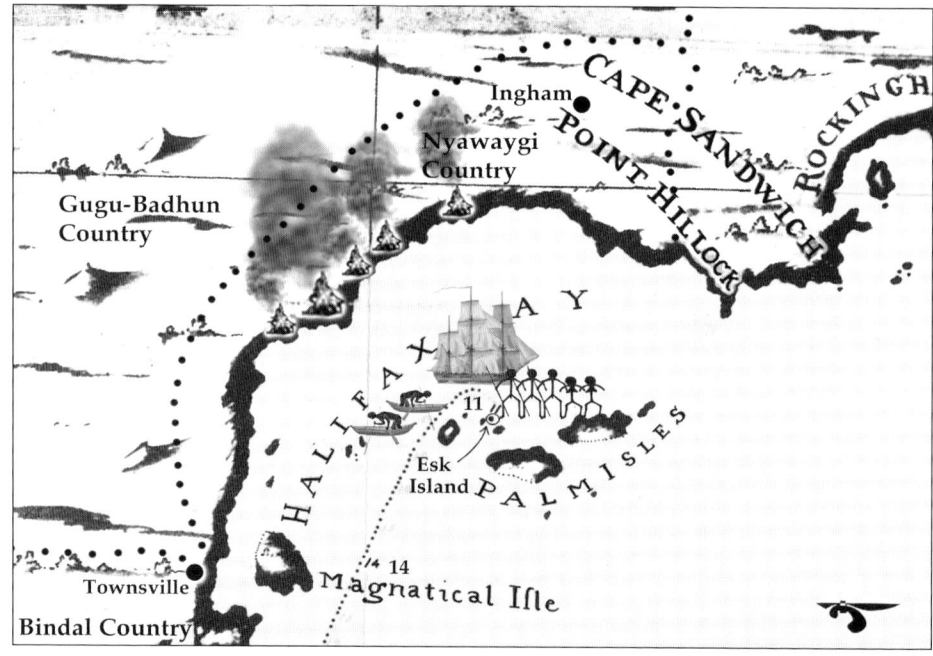

COOK: Punished Thomas Dunster, marine, with a dozen lashes for theft [see glossary 'punishment']. Our soundings in the course of this days sail were from 14 to 11 fathom.

PARKINSON: At noon we were between a parcel of islands and the main [between Palm Isles and the mainland on the chart]. The main looked very barren and dreary: the hills upon it looked like a heap of rubbish on which nothing was to be seen, excepting a few low bushes.

BANKS: As barren as ever with several fires upon it, one vastly large [Nyawaygi country].

PARKINSON: But the islands made a better appearance.

BANKS: People were seen on them [on Esk Island – Nyawaygi people].

PARKINSON: We saw a few people in canoes striking fish.

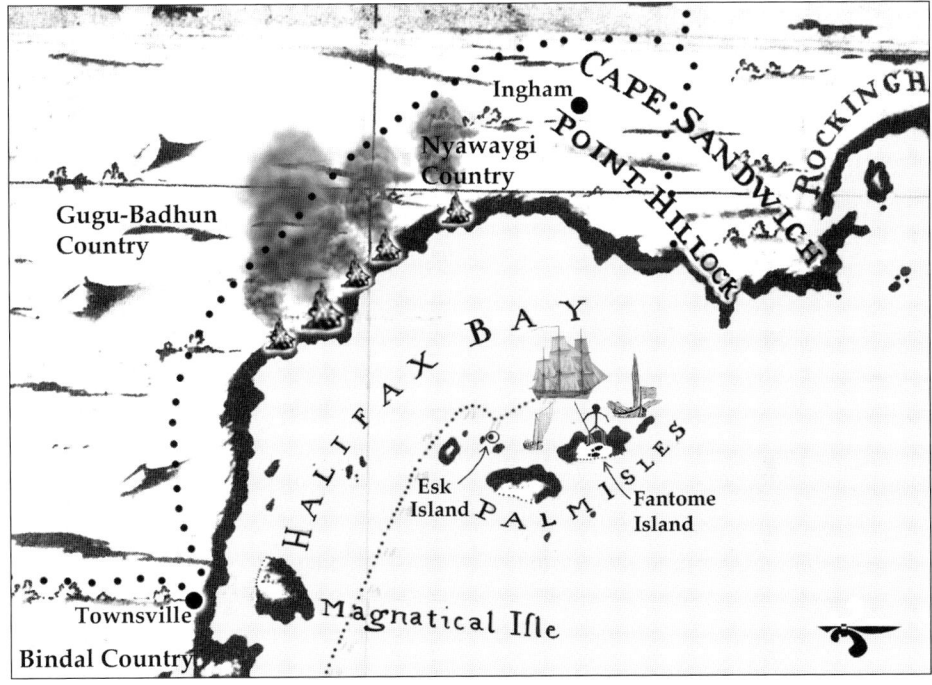

8th June

COOK: In the p.m. we saw as we thought Coconut trees upon one of the Islands, and as a few of these nuts would have been very acceptable to us at this Time I sent Lieut Hicks ashore with whom went Mr Banks and Dr Solander to see what was to be got, in the mean time we kept standing in for the Island with the Ship [Fantome Island, one of the Palm Islands – Nyawaygi people].

BANKS: We found our supposed Cocoanut trees to be no more than bad Cabbage trees [Cabbage palm – *Livistona australis*]. The country about them was very stony and barren and it was almost dark when we got ashore; we made a shift however to gather 14 or 15 new plants [one of which was *Hibiscus meraukensis*]. (Colour Plate No. 30)

We repaired to our boats, but scarce were they put off from the shore when an Indian came very near it and shouted to us very loud; it was so dark that we could not see him, we however turned towards the shore by way of seeing what he wanted with us, but he I suppose ran away or hid himself immediately for we could not get a sight of him [Nyawaygi people].

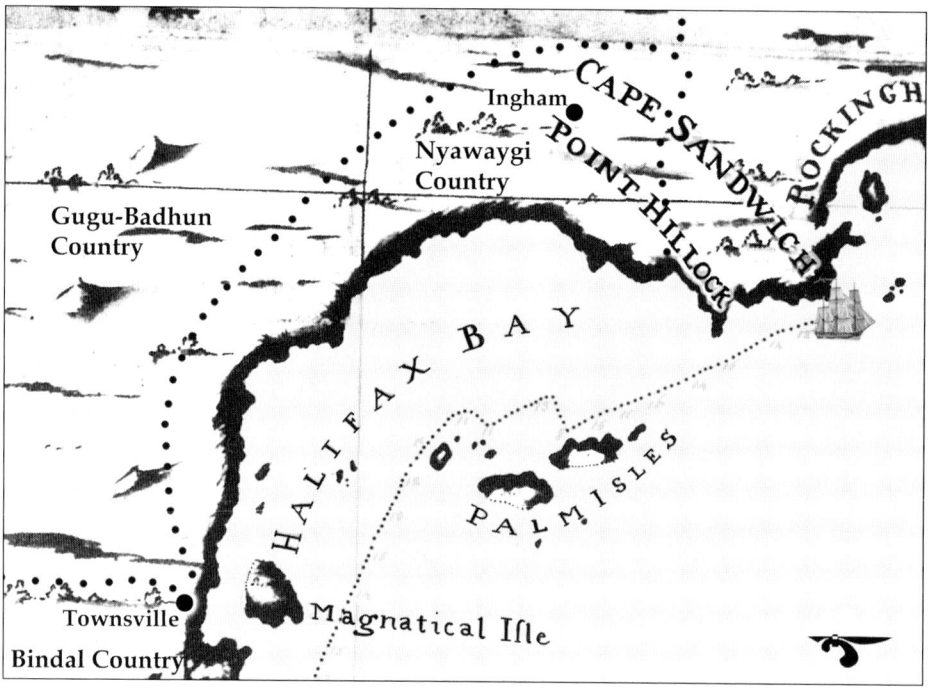

COOK: After the boats were hoisted in we stood away NbW. I named Point Hillock on account of its figure. Between this cape and Cape Cleveland the shore forms a large bay which I named Halifax Bay. At 6 o'clock in the a.m. we were abreast of a point of land which I named Cape Sandwich.

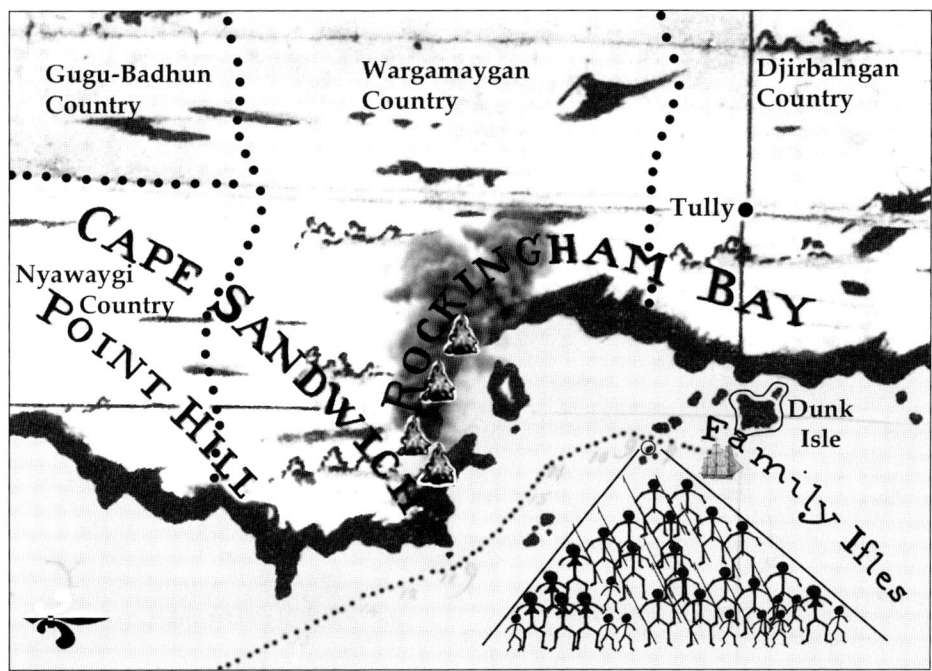

COOK: From Cape Sandwich the land forms a fine large bay which I called Rockingham Bay.

BANKS: Still sailing between the Main and Islands; the former rocky and high looked rather less barren than usual and by the number of fires seemed to be better peopled.

PARKINSON: We discovered several islands [Family Isles on chart] that looked like so many heaps of rubbish, which had lain long enough to have a few weeds and bushes grow on them.

COOK: Finding a channel of a mile broad between the three outermost and those nearer the shore we pushed through.

PARKINSON: On one of them [possibly Bowden Island], which is not more than two miles in circumference, we saw a company of the natives [Bandjin and Djiru people].

BANKS: We passed within quarter of a mile. We saw with our glasses about 30 men women and children standing all together and looking attentively at us.

PARKINSON: Standing quite still and beholding the ship with astonishment.

BANKS: The first people we have seen shew any signs of curiosity at the sight of the ship.

COOK: They were quite naked and of a very dark Colour with short hair.

At noon we were abreast the north point of Rockingham Bay. This boundary of the Bay is formed by a tolerable high Island known on the chart as Dunk Isle. [The island appears on the chart but is not named there.]

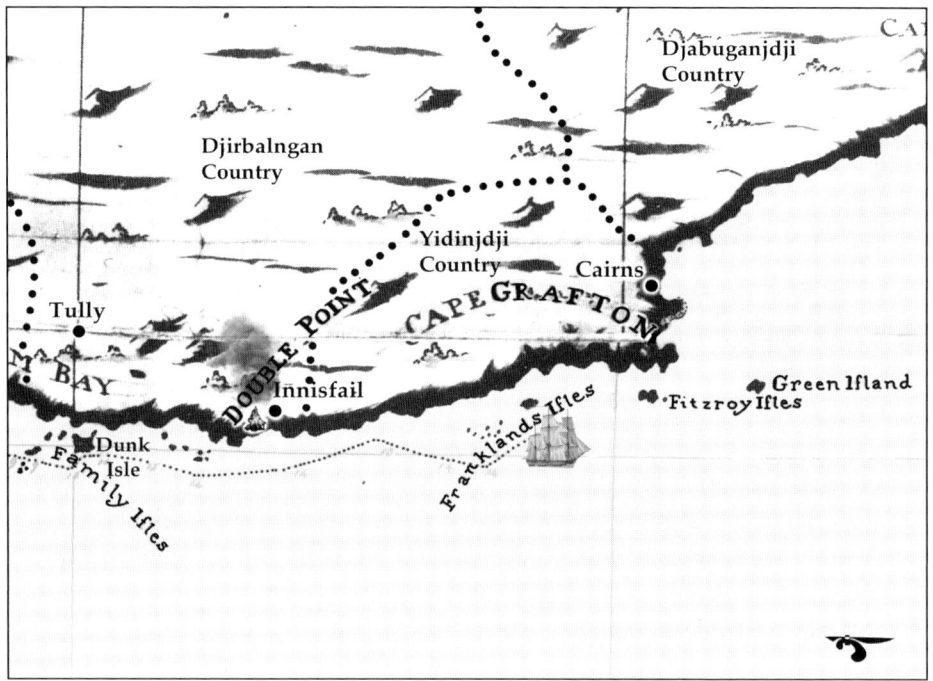

9th June

COOK: We kept our course under an easy sail all night.

PARKINSON: At night we saw a fire, which yielded a very grateful odour, not unlike that produced by burning the wood of gum benjamin. [Benjamin is a balsamic resin obtained from the bark of several species of trees in the genus *styrax*.]

COOK: At 6 o'clock in the a.m. we were abreast of some small islands which we called Frankland Isles.

BANKS: At noon a fire and some people were seen [Yidinjdji people].

COOK: An island tolerable high [Fitzroy Isles/Fitzroy Island – Gungandji country] lies about 2 miles from the point of the main between which we went with the ship and were in the middle of the channel at noon, where we had **20 fathom water**.

The point of land we were now abreast off I called Cape Grafton.

10th June

COOK: Sent a Mate in the yawl to sound ahead. Three Miles to the Westward of the Cape is a Bay [Mission Bay] wherein we **anchored** about 2 Miles from the shore.

BANKS: Just without us as we lay at an anchor was a small sandy Island laying upon a large Coral shoal, much resembling the low Islands to the eastward of us but the first of the kind we had met with in this part of the South Sea.

COOK: It is known in the chart by the name Green Island [Gungandji and Mandingalbay country]. As soon as the Ship was brought to an Anchor I went ashore accompanied by Mr Banks and Dr Solander.

BANKS: We saw no people. The country was hilly and very stony affording nothing but fresh water, at least that we found, except a few plants that we had not before met with [one of which was *Planchonia careya*]. (Colour Plate No. 31)

COOK: My intention was to have stayed here at least one day to have looked into the Country had we met with fresh water convenient or any other refreshment but as we did not I thought it would be only spending time and losing so much of a light moon to little purpose and therefore at 12 o'clock at night we weighed and stood away to the NW having at this time but little wind attended with showers of rain.

Chapter 8

Fear of Death Now Stared Us in the Face

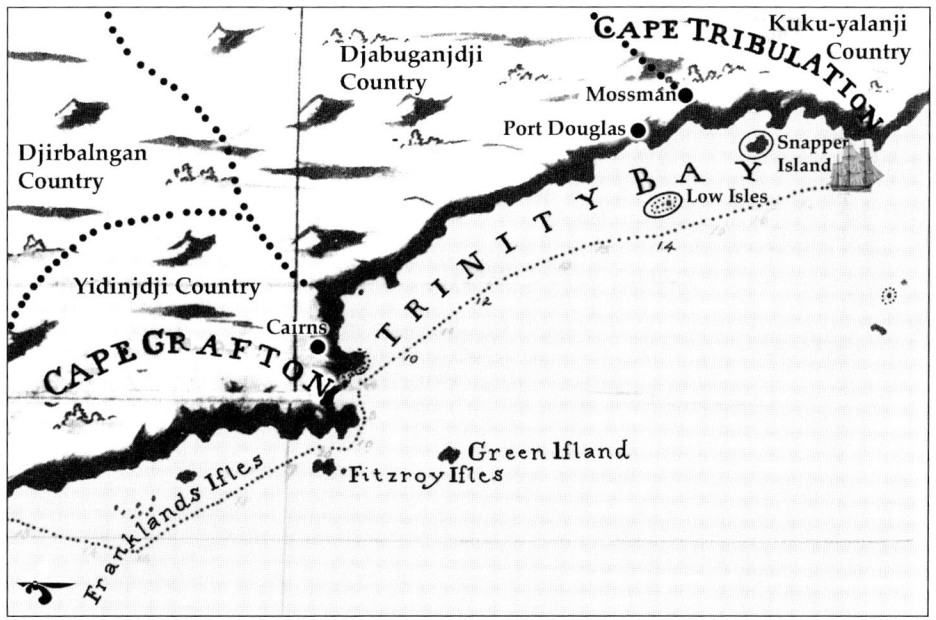

Portion of Cook's chart 'Cape Townsend to Cape Tribulation'.[1]

10th June (cont'd)

COOK: We continued steering as the land lay having **10, 12** and **14** fathom.

At 10 o'clock [a.m.] we hauled off north in order to get without a small low Island which lay about 2 Leagues from the Main [now <u>Low Isles</u>]; it being about high water at the time we passed it the greater part of it lay under water. About 3 Leagues to the north-westward of this Island close under the main land is another Island tolerable high which bore from us at Noon N. 55° west distant 7 or 8 Miles [now <u>Snapper Island</u>].

The shore between Cape Grafton and the northern point forms a large but not very deep Bay which I named <u>Trinity Bay</u> after the day on which it was discovered – the north point <u>Cape Tribulation</u>, because here begun all our troubles.

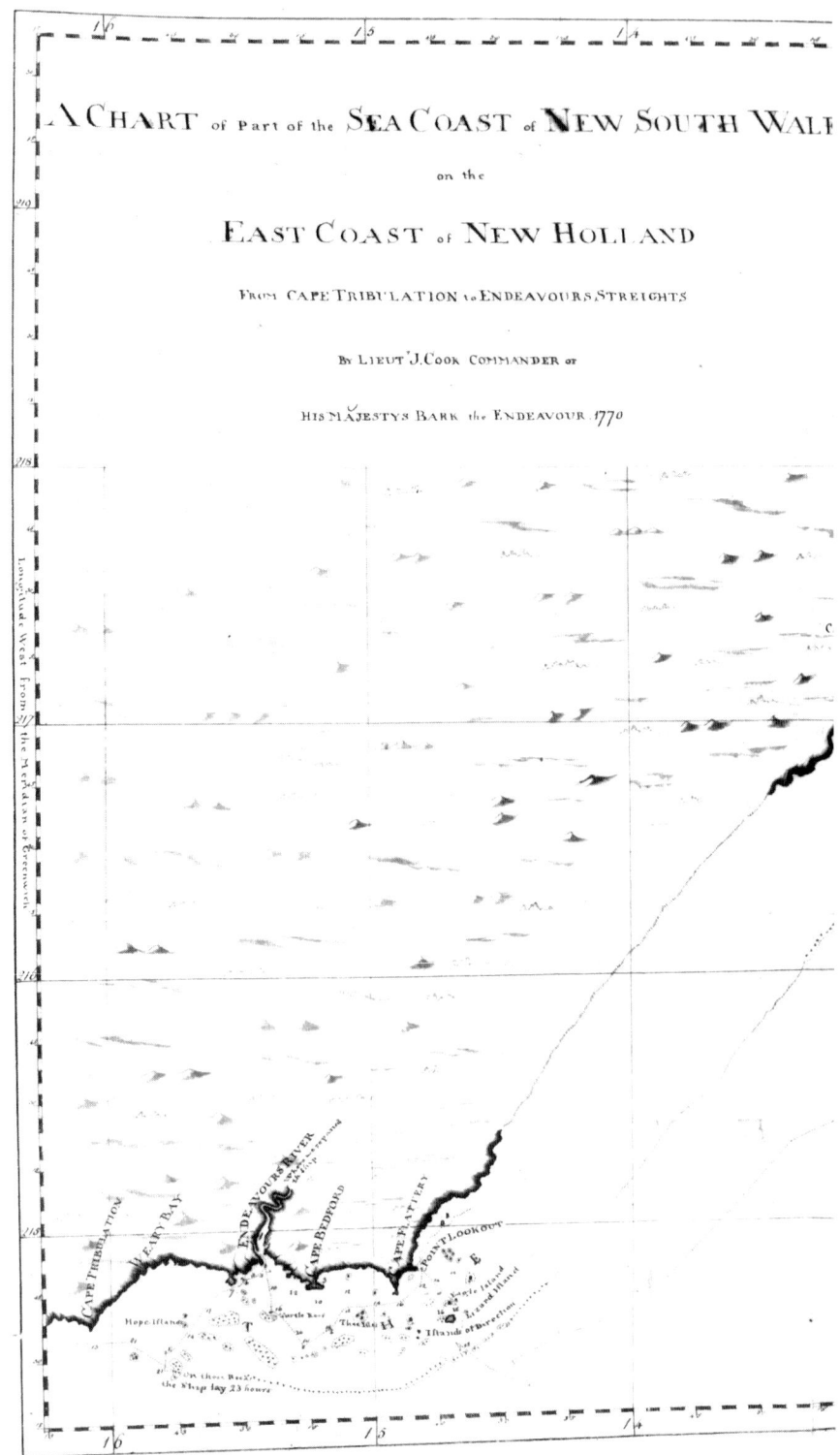

'A CHART of the SEA COAST of NEW SOUTH WALES on the EAST COAST of NEW HOLLAND from Cape Tribulation to Endeavours Streights by Lieut. J. Cook Commander of His Majesty's Bark the *Endeavour* 1770'.[2]

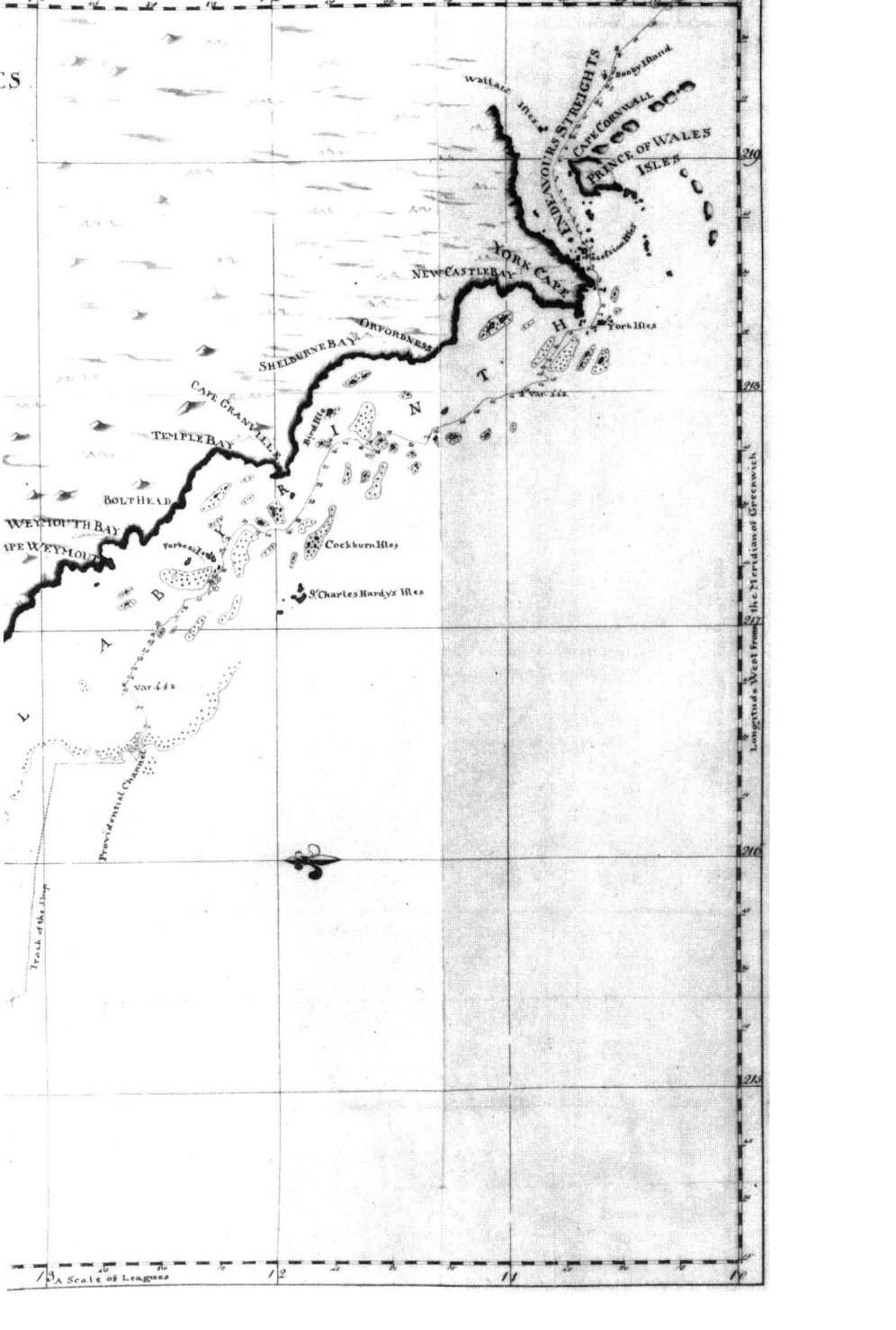

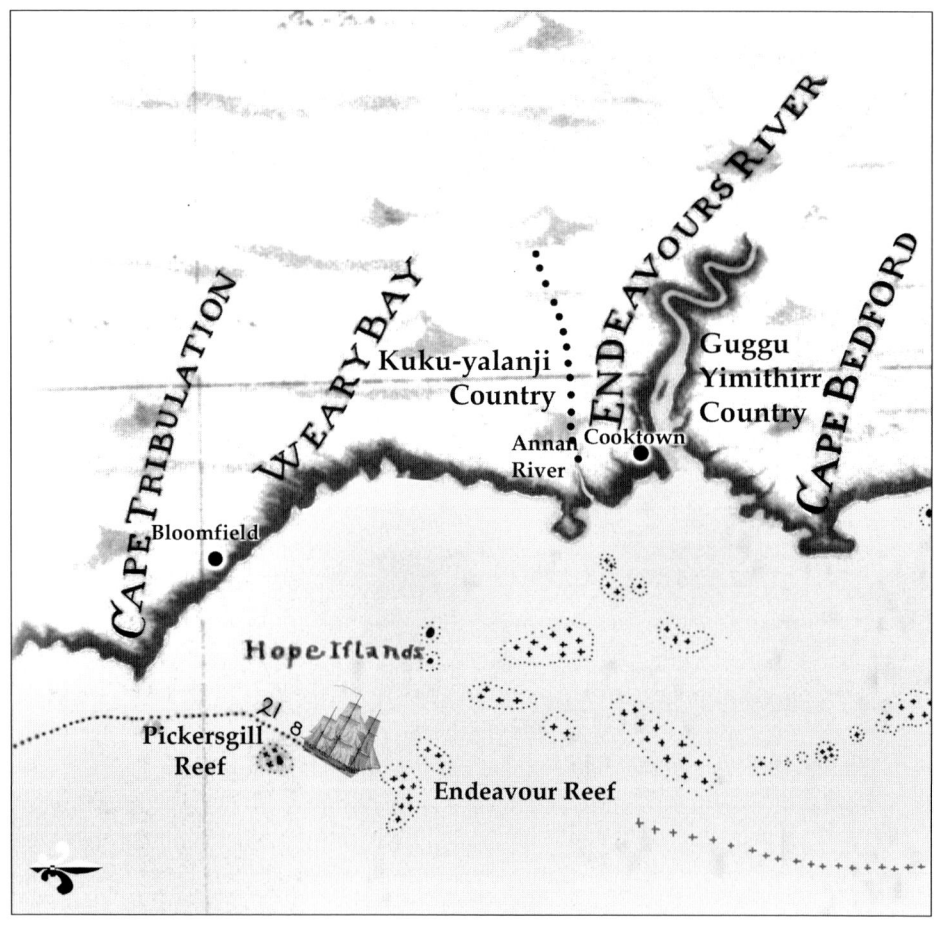

Portion of Cook's chart above 'Cape Tribulation to Endeavours Streights'.

11th June

BANKS: At night fall rocks and shoals were seen ahead [Hope Islands on chart] on which the ship was put upon a wind off shore.

COOK: My intention was to stretch off all night as well to avoid the dangers we saw ahead as to see if any Islands lay in the offing. Having the advantage of a fine breeze of wind and a clear moon light night in standing off from 6 until near 9 o'clock we deepened our water from 14 to **21** fathom.

BANKS: While we were at supper she went over a bank of 7 or **8** fathom water which she came upon very suddenly [now Pickersgill Reef].

PARKINSON: … which alarmed us very much: every countenance expressed surprise, and every heart felt some trepidation.

BANKS: This we concluded to be the tail of the Sholes we had seen at sunset [Hope Islands] and therefore went to bed in perfect security.

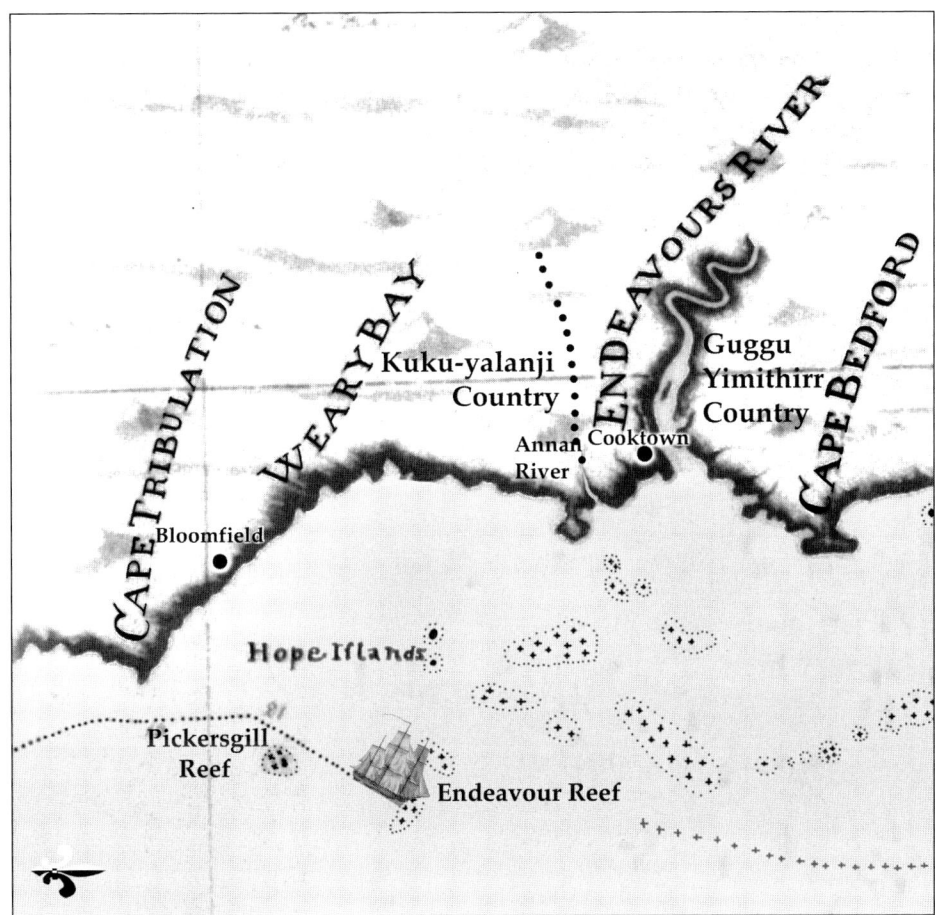

COOK: At this time I had everybody at their stations to put about and come too an anchor but in this I was not so fortunate for meeting again with deep water I thought there could be no danger in standing on. Before 10 o'clock we had 20 and 21 fathom and continued in that depth until a few Minutes before a 11 when we had 17 and before the Man at the lead could heave another cast the Ship Struck and stuck fast.

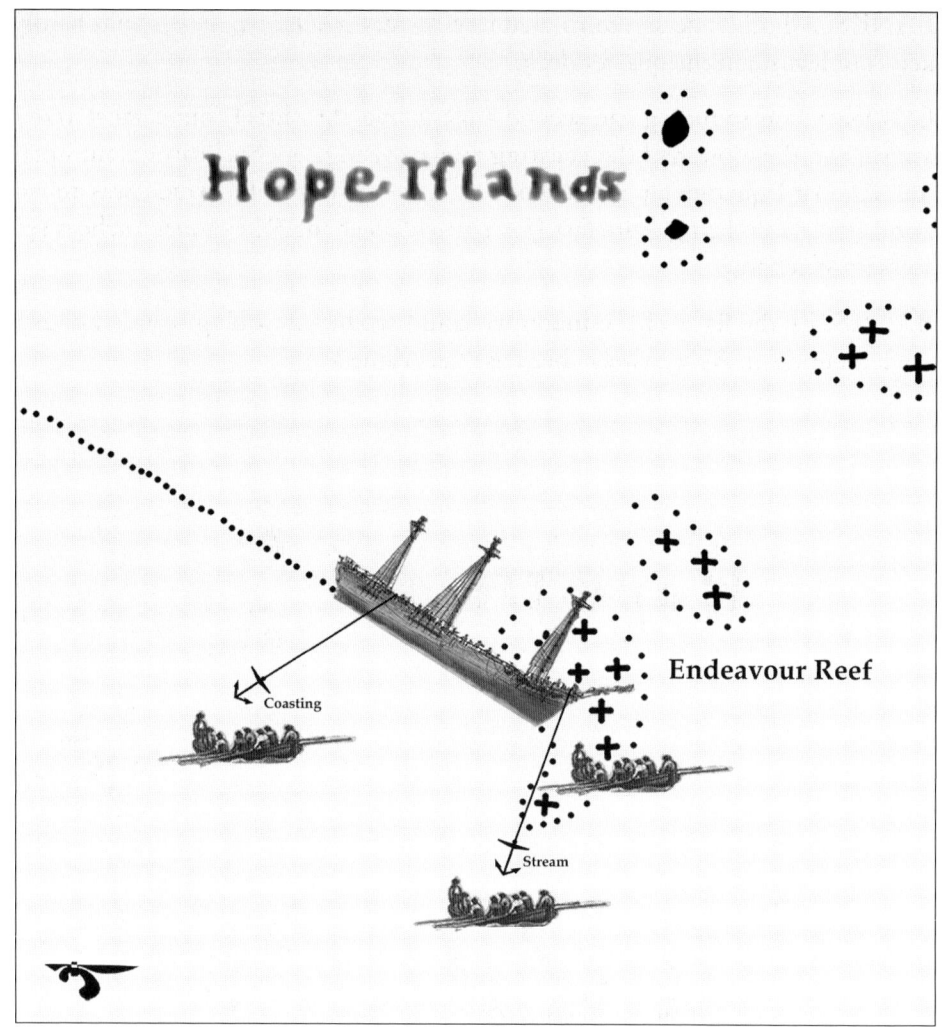

Detail of Cook's chart 'Cape Tribulation to Endeavours Streights'.

BANKS: Our situation became now greatly alarming. We were little less than certain that we were upon sunken coral rocks, the most dreadful of all others on account of their sharp points and grinding quality which cut through a ships bottom almost immediately. The officers however behaved with inimitable coolness void of all hurry and confusion.

COOK: Immediately we took in all our sails and hoisted out our boats and sounded round the ship. As soon as the long boat was out we struck yards and topmasts and carried out the **stream anchor** upon the starboard bow [see glossary 'relative bearings' illustration number 3].

BANKS: All this time she continued to beat very much so that we could hardly keep our legs upon the quarter deck.

COOK: Upon sounding the second time round the ship I found most water astern, and therefore had the **coasting anchor** [kedge anchor] carried out upon the starboard quarter and hove upon it a very great strain which was to no purpose the ship being quite fast [for particulars about *Endeavour*'s anchors see glossary 'Anchors'].

BANKS: By the light of the moon we could see her sheathing boards floating thick around her. About twelve her false keel came away.

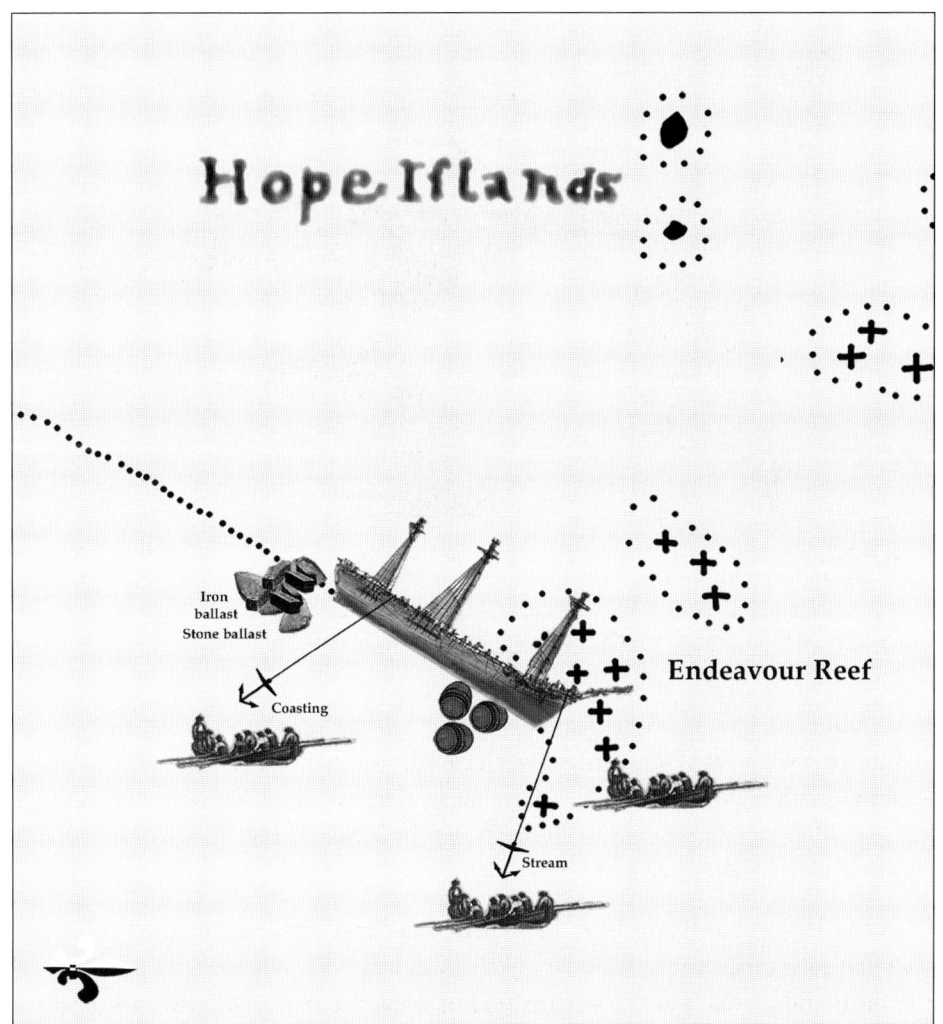

COOK: We went to work to lighten her as fast as possible which seemed to be the only means we had left to get her off as we went ashore about the top of high water. We hove hoops, starves, empty casks, and oil jars overboard, as well to lighten her as to clear away for heavier articles. All this time the ship made little or no water.

BANKS: A rock however under her starboard bow kept grating her bottom and making a noise very plainly to be heard in the fore store rooms [see Prologue illustration 'The Plan of His Majesty's Bark *Endeavour* as fitted at Deptford in July 1776' page XX]; this we doubted not would make a hole in her bottom, we only hoped that it might not let in more water than we could clear with our pumps.

[The four pump shafts were made of a single elm tree, bored through the centre of its whole 23-foot length by hand augers. Each went right to the bottom of the ship and sucked from the pump well by the keel – Ray Parkin.]

BANKS: In this situation the day broke upon us and showed us the land about 8 leagues off as we judged; nearer than that was no island or place on which we could set foot. It however brought with it a decrease of wind and soon after that a flat calm, the most fortunate circumstance that could possibly attend people in our circumstances.

COOK: Started 30 tons of water [their fresh water was pumped or siphoned out of the water casks into the bilges from whence it was pumped over the side], hove some of the boatswain's and carpenter's condemned stores overboard, got the stone and iron ballast out of the hold, with a large quantity of firewood, and hove them all overboard.

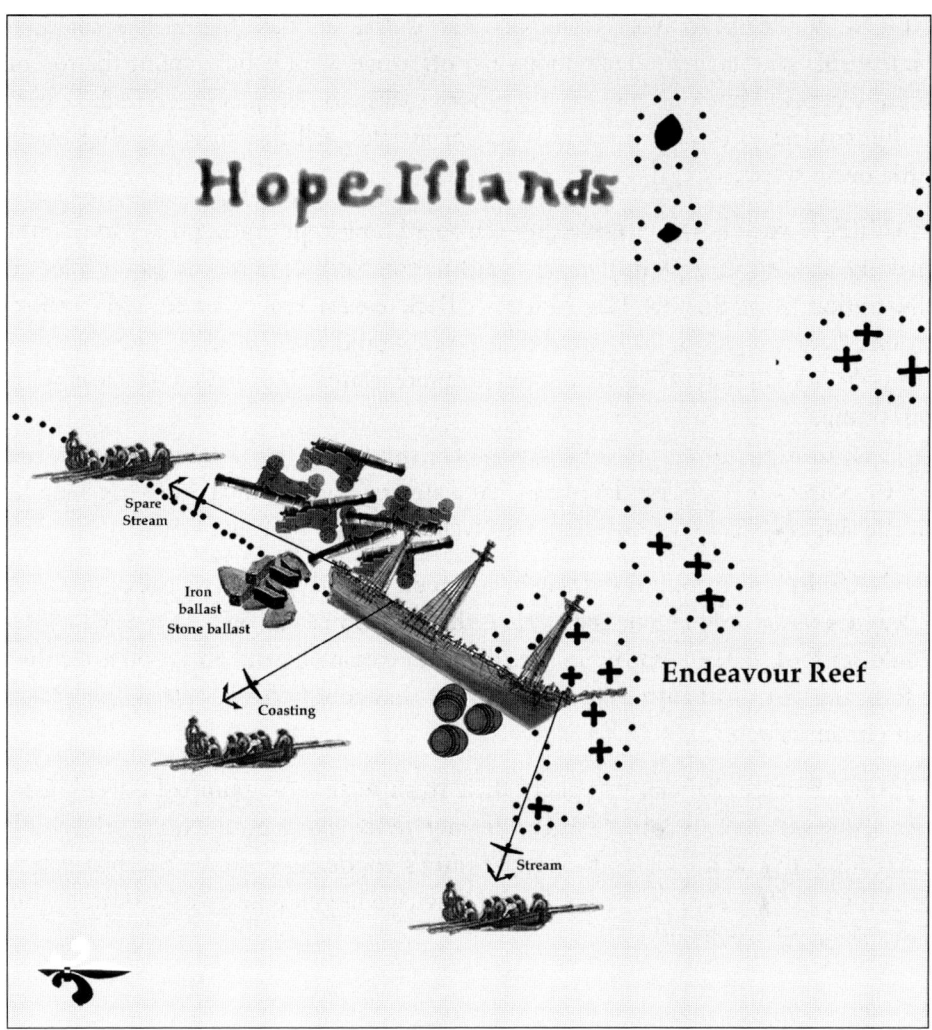

COOK: Meanwhile carried out the **spare stream anchor** to the south west and hove tort on't. At 8 found the ship had sprung a leak. Got three pumps to work, the fourth being choked.

At 11 o'clock in the a.m. being high water as we thought we tried to heave her off without success, she not being afloat by a foot or more notwithstanding by this time we had thrown overboard 40 or 50 ton weight; as this was found not sufficient we continued to lighten her by every method we could think of. Hove the six carriage guns that was upon the deck overboard.

BANKS: All this time the seamen worked with surprising cheerfulness and alacrity; no grumbling or growling was to be heard throughout the ship, no not even an oath – though the ship in general was as well furnished with them as most in his majesties service.

COOK: As the tide fell the ship began to make water as much as two pumps could free.

PARKINSON: About noon, the ship heaved much on one side.

12th June

COOK: Fortunately a smooth sea all these twenty four hours gave us an opportunity to carry out **two bower anchors** the one on the starboard quarter and the other right a stern.

PARKINSON: Five anchors were carried out and dropped at different parts.

COOK: Got blocks and tackles upon the cables brought the falls in abaft and hove taught. By this time it was 5 o'clock in the p.m., the tide we observed now began to rise.

BANKS: … as it rose the ship worked violently upon the rocks so that she began to make water and increased very fast.

COOK: … which obliged us to set the third pump to work, as we should have done the fourth also, but could not make it work.

BANKS: The fourth absolutely refused to deliver a drop of water.

PARKINSON: Every man on board assisted, the Captain, Mr Banks and all the officers, not excepted; relieving one another every quarter of an hour.

BANKS: Now in my opinion I entirely gave up the ship and packing up what I thought I might save prepared myself for the worst.

COOK: At 9 p.m. the ship righted and the leak gained upon the pumps considerably. This was an alarming and I may say terrible circumstance and threatened immediate destruction to us as soon as the ship was afloat.

BANKS: The most critical part of our distress now approached; the ship was almost afloat and everything ready to get her into deep water but she leaked so fast that with all our pumps we could just keep her free. If, as was probable, she should make more water when hauled off she must sink, and we well knew that our boats were not capable of carrying us all ashore, so that some, probably the most of us, must be drowned; a better fate maybe than those would have who should get ashore without arms to defend themselves from the Indians, or provide themselves with food, on a country where we had not the least reason to hope for subsistence, so barren had we always found it; debarred from a hope of ever again seeing their native country or conversing with any but the most uncivilized savages perhaps in the world.

COOK: However I resolved to risk all and heave her off.

BANKS: A dreadful time now approached and the anxiety in everybody's countenance was visible enough; the capstan and windlass were manned and they began to heave; fear of death now stared us in the face.

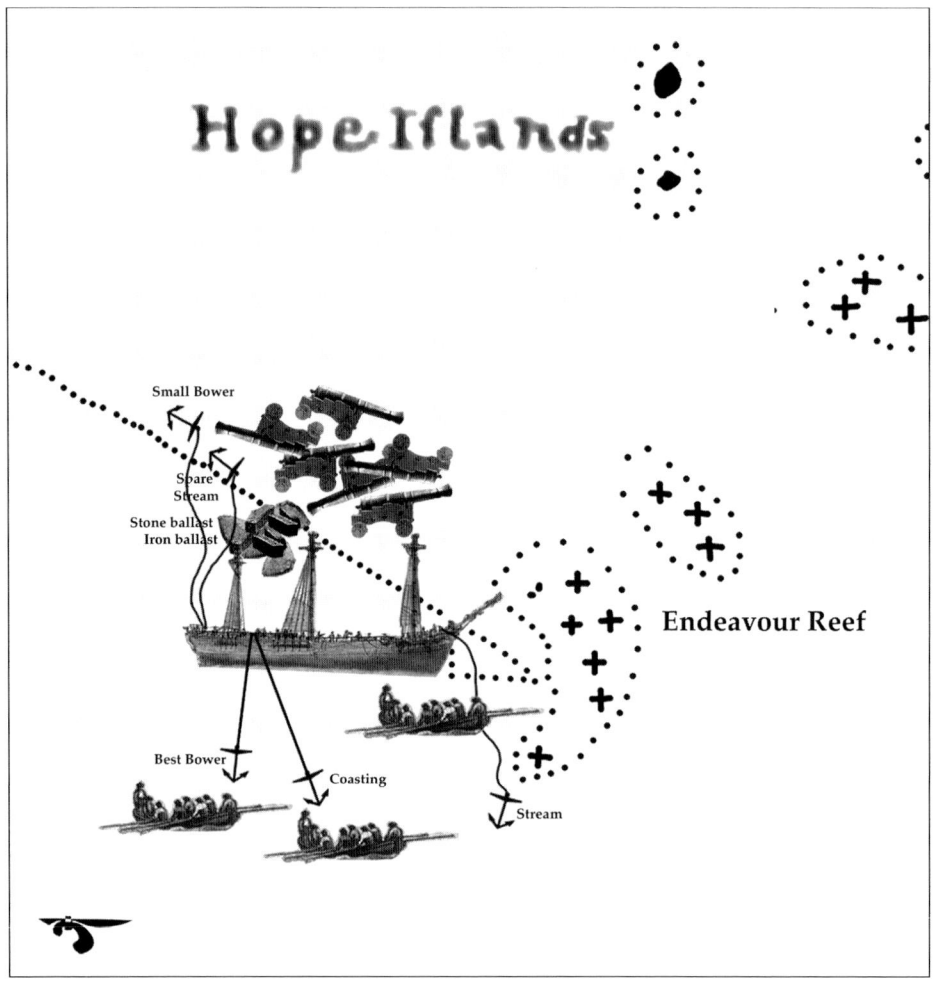

COOK: About twenty past 10 o'clock the ship floated and we hove her off into deep water …

BANKS: … where to our great satisfaction she made no more water than she had done.

PARKINSON: This desirable event gave us spirits; which, however, proved but the transient gleam of sun shine, in a tempestuous day.

COOK: A mistake soon after happened which caused fear to operate upon every man in the ship.

BANKS: The people who had been 24 hours at exceeding hard work now began to flag; myself unused to labour was much fatigued and had laid down to take a rest, was awaked about 12 with the alarming news of the ship having four feet water in her hold.

PARKINSON: ... the water increased faster than we could throw it out; and we expected, every minute, that the ship would sink, or that we should be obliged to run her again upon the rocks.

COOK: The man which attend the well [the place in the ship's hold for pumps] took ye depth of water above the ceiling, he being relieved by another who did not know in what manner the former had sounded, took the depth of water from the outside plank the difference being 16 or 18 Inches and made it appear that the leak had gained this upon the pumps in a short time, this mistake was no sooner cleared up than acted upon every man like a charm.

BANKS: Rest was no more thought of but the pumps went with an unwearied vigour till water was all out, which was done in a much shorter time than was expected. We now began again to have some hopes and to talk of getting the ship into some harbour.

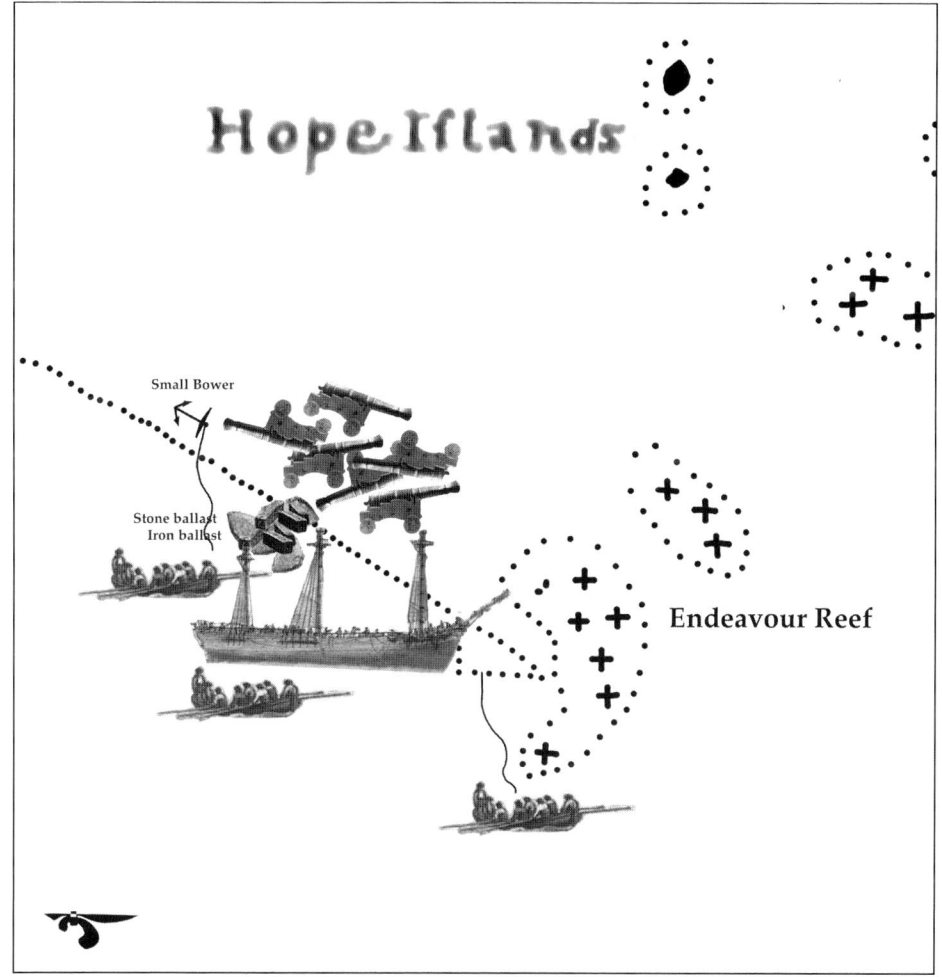

140   The Endeavour Journals

COOK: We now hove up the **best bower** but found it impossible to save the **small bower** so cut it away at a whole cable.

BANKS: But got **the other three small anchors** far more valuable to us than the bowers.

COOK: I sent the longboat to take up the **stream anchor**, got the anchor but lost the cable among the rocks.

HMS Bark Endeavour's salvaged and restored small bower anchor, and one of the six carriage guns hove overboard on 12th June 1770.
Courtesy of James Cook Museum Cooktown

Endeavour's cannon and anchor.[3]

PARKINSON: When we threw the guns overboard, we fixed buoys to them, intending, if we escaped, to have heaved them up again, but, on attempting it, we found it was impracticable.

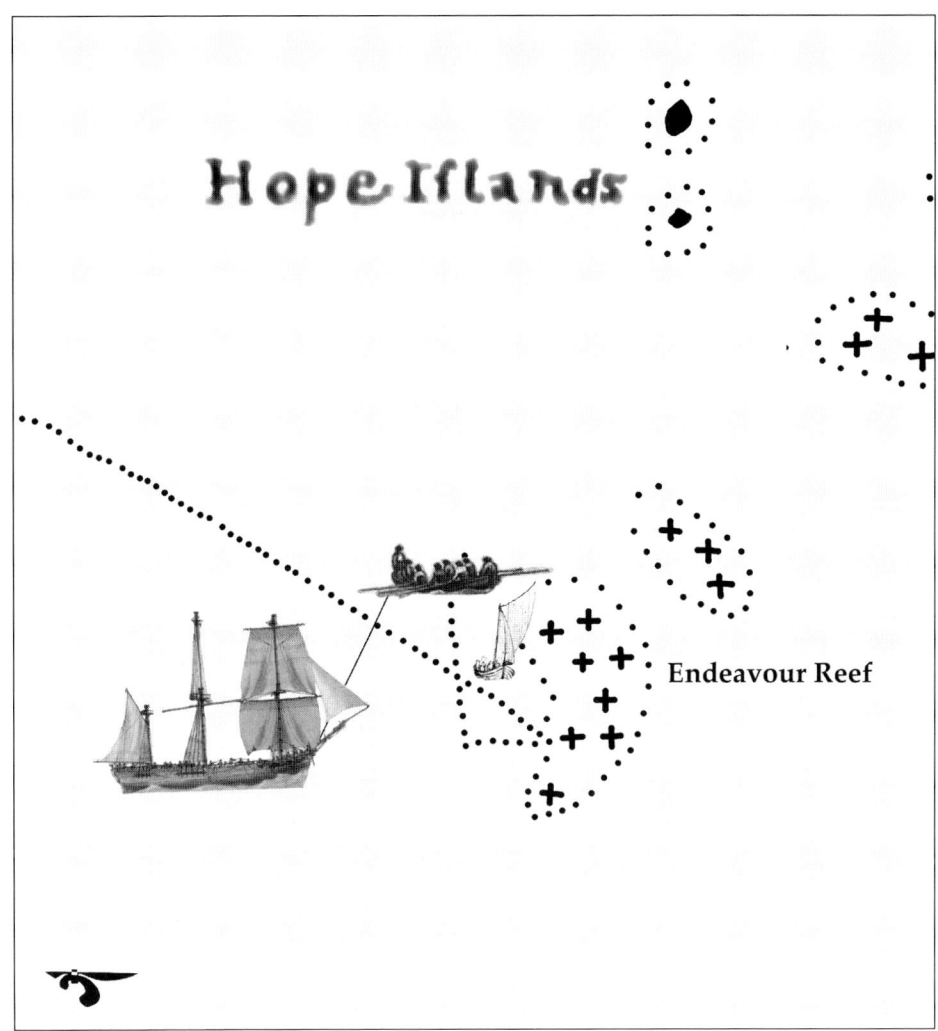

BANKS: One of our midshipmen now proposed an expedient which no one else in the ship had seen practised, though all had heard of it by the name of fothering the ship. He was immediately set to work with 4 or 5 assistants.

COOK: The manner this is done is thus, we mix ockam & wool together, / but ockam alone would do / and chop it up small and then stick it loosly by hand fulls all over the sail and throw over it sheep's dung or other filth. Horse dung for this purpose is the best; the sail thus prepared is hauled under the Ships bottom by ropes and if the place of the leak is uncertain it must be hauled from one part of her bottom to another until the place is found where it takes effect; while the sail is under the Ship the ockam &c. is washed off and part of it carried along with the water into the leak and in part stops up the hole. Mr Monkhouse, one of my midshipmen [Jonathan Munkhouse] was once in a merchant ship which sprung a leak and which made 48 inches water an hour by this means was brought home from Virginia to London; to him I gave the direction of this who executed it very much to my satisfaction.

Got up the fore topmast and foreyard, warped the Ship to the SE and at 11 got under sail and stood in for land with a light breeze at ESE. Sent the pinnace ahead to tow, and a small boat to lye on the edge of the bank.

13th June

COOK: In the p.m. had light airs at ESE with which we kept edging in for the land. Got up the Main topmast and Main yard. Having got the sail ready for fothering the Ship we put it over under the Starboard fore chains where we suspected the ship had suffered most.

BANKS: In about half an hour to our great surprise the ship was pumped dry and upon letting the pumps stand she was found to make very little water.

COOK: This fortunate circumstance gave new life to everyone on board. It is much easier to conceive than to describe the satisfaction felt by everybody on board on this occasion. Every man's hopes enlarged.

BANKS: … in an instant raised from almost despondency to the greatest hopes.

PARKINSON: Providentially too at this instant a breeze sprang up and we steered towards the land.

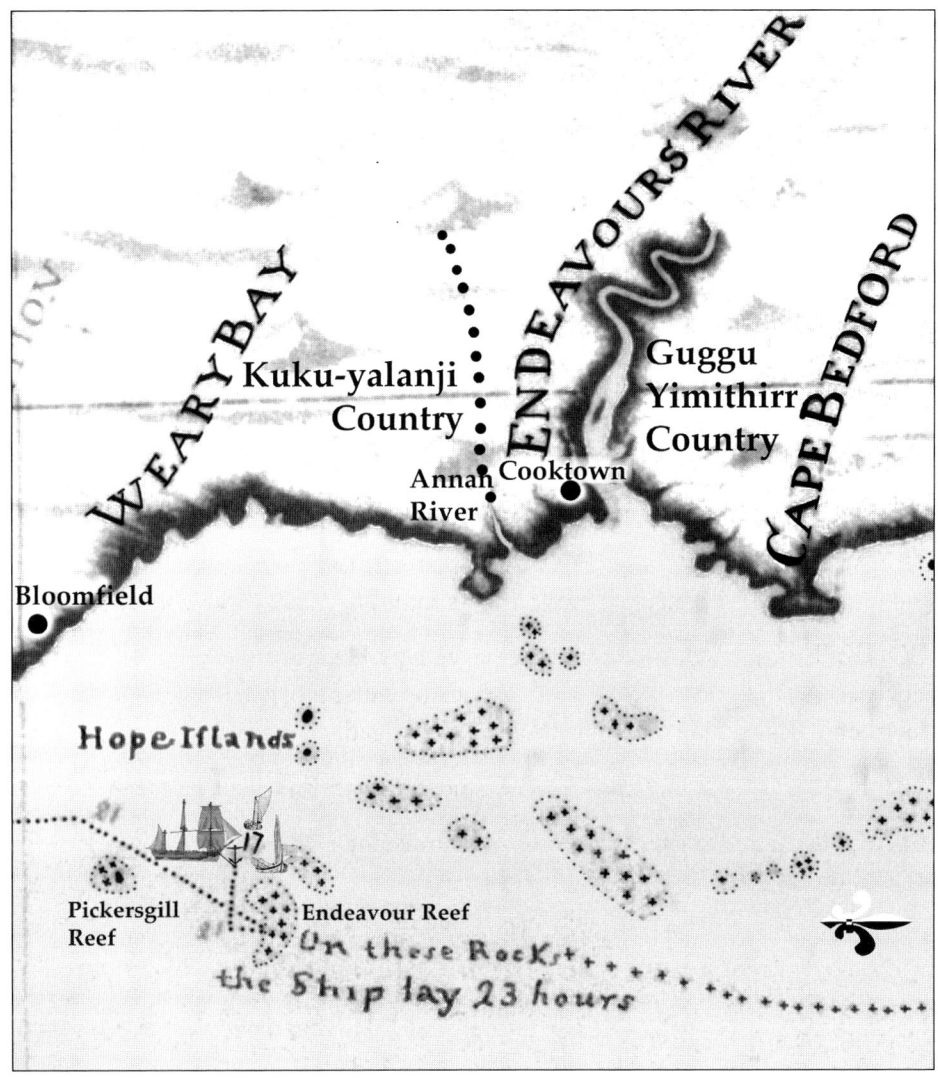

Portion of Cook's chart 'Cape Tribulation to Endeavours Streights'.

COOK: We now thought of nothing but ranging along shore in search of a harbour where we could repair the damages we had sustained. In justice to the Ships Company I must say that no men ever behaved better than they have done on this occasion animated by the behaviour of every gentleman on board every man seemed to have a just sense of the danger we were in and exerted himself to the very utmost.

BANKS: Every man exerted his utmost for the preservation of the ship, contrary to what I have universally heard to be the behaviour of sea men who have commonly as soon as a ship is in a desperate situation began to plunder and refuse all command. This was no doubt owing entirely to the cool and steady conduct of the officers, who during the whole time never gave an order which did not shew them to be perfectly composed and unmoved by the circumstances howsoever dreadful they might appear.

COOK: At 6 o'clock [p.m.] we **anchored** in **17** fathom water about 5 or 6 leagues from the land.

BANKS: One pump and that not half worked kept the ship clear all night.

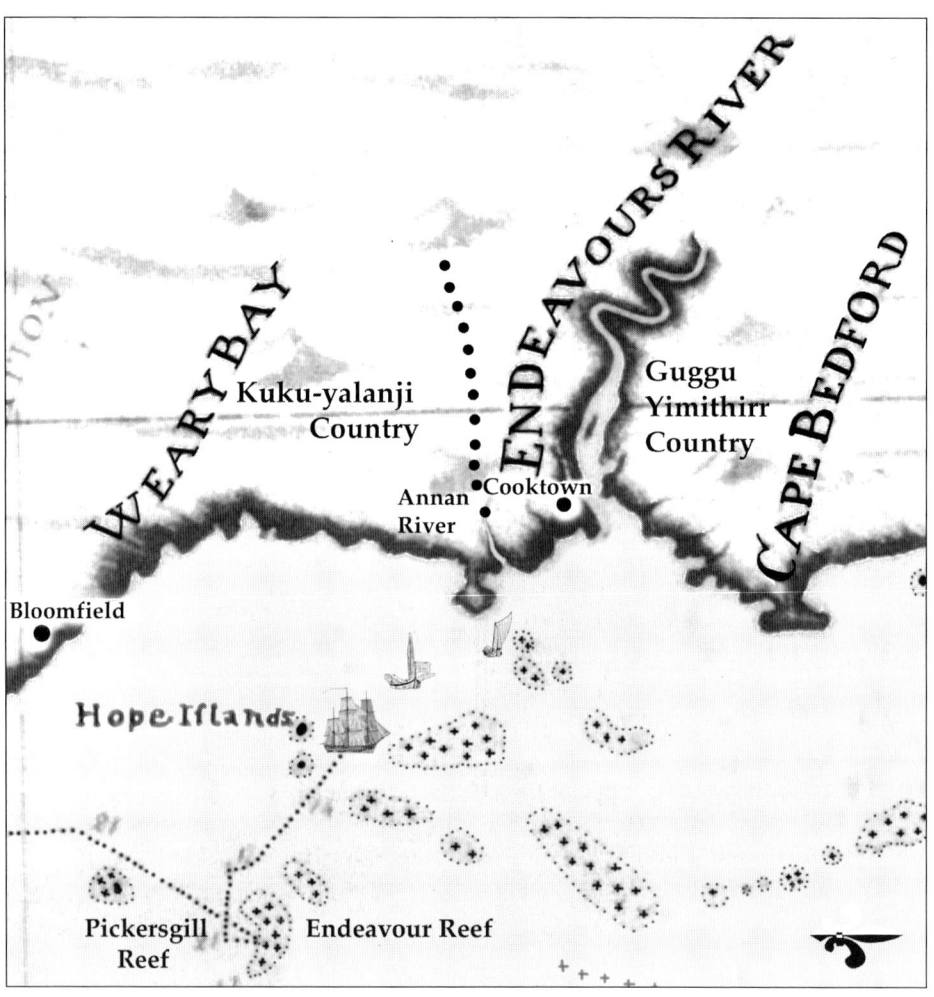

BANKS: In the morn we weighed with a fine breeze of wind and steered along shore among innumerable shoals.

COOK: At 9 o'clock we passed close without two small low islands. I have named them <u>Hope Islands</u> because we were always in hopes of being able to reach these Islands.

14th June

COOK: In the p.m. sent the Master [Robert Molineux] with two boats as well to sound ahead of the Ship as to look out for a harbour where we could repair our defects and put the Ship into a proper trim both of which she now very much wanted.

This day I restor'd Mr Magra to his Duty as I did not find him guilty of the crimes laid to his charge. [Cook had earlier suspected Magra of participating in the malicious wounding of Mr Orton, Cook's clerk, on 23rd May.]

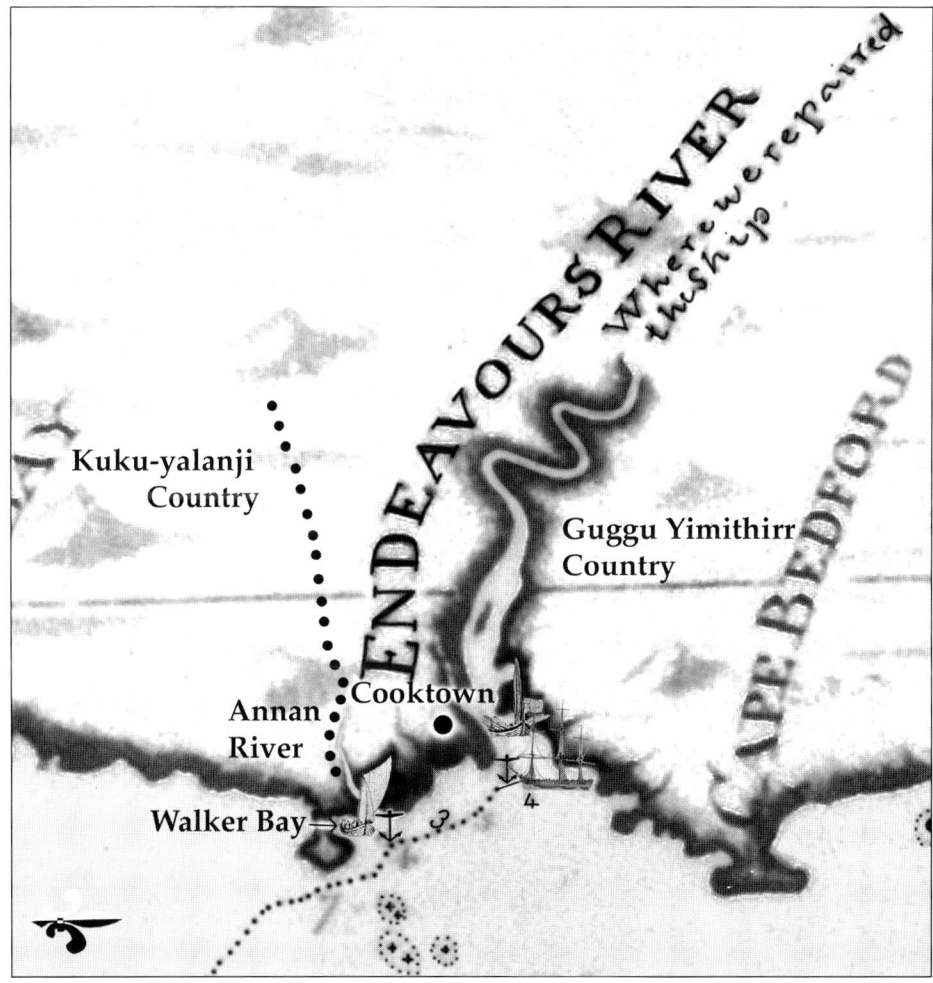

Detail of Cook's chart 'Cape Tribulation to Endeavours Streights'.

COOK: At 3 o'clock saw an opening that had the appearance of a harbor [now Walker Bay]. Stood off and on while the boats were examining it who found that there was not sufficient depth of water for the ship.

BANKS: Nothing was met with which could possibly suit our situation, bad as it was, having nothing but a lock of wool between us and destruction, so at night we came to an **anchor** [anchorage marked on chart above].

The pinnace however which had gone far ahead was not returned, nor did she till 9 o'clock, when she reported that she had found just the place we wanted [Endeavours River].

COOK: … about 2 leagues to leeward. In consequence of this information we at 6 in the a.m. weighed and run down to it, first sending two boats ahead to lay upon the shoals that lay in our way and notwithstanding this precaution we were once in **3** fathom water with the ship.

BANKS: A boat was sent ahead to shew us the way into the harbour, but by some mistake of signals we were obliged to come to an anchor again off the mouth of it without going in, where it soon blew too fresh for us to weigh.

COOK: The Ship would not work having missed stays twice, and being entangled among shoals I was afraid of being drove to leeward before the boats could place themselves, and therefore **Anchored** in **4** fathom about a Mile from the shore [anchorage marked on chart above].

PARKINSON: Now it began to blow hard which prevented us from getting into the bay until the 18th.

BANKS: We now began to consider our good fortune; had it blown as fresh the day before yesterday or before that we could never have got off but must inevitably have been dashed to pieces on the rocks.

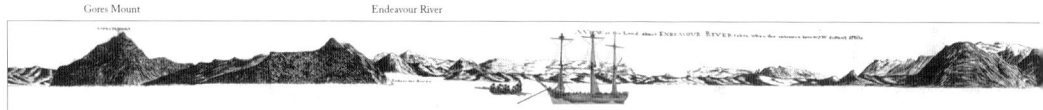

Portion of 'A VIEW of the Land about ENDEAVOUR RIVER taken when the enterance bore WSW distance 1 Mile'.[4] Charles Praval – copied from a lost original by Parkinson or Spöring.

BANKS: The Captain and myself went ashore to view the harbour [Endeavours River] and found it indeed beyond our most sanguine wishes: it was the mouth of a river the entrance of which was to be sure narrow enough and shallow, but when once in the ship might be moored afloat so near the shore that by a stage from her to it all her Cargo might be got out and in again in a very short time.

Fear of Death Now Stared Us in the Face 147

The meeting with so many natural advantages in a harbour so near us at the time of our misfortune appeared almost providential; we had not in the voyage before seen a place so well suited for our purpose as this was; we therefore returned onboard in high spirits and raised the spirits of our friends on board as much as our own by bringing them the welcome news of our approaching security.

15th June

COOK: As it blowed too fresh to break the Ship loose to run into the harbour we got down topgallant yards, unbent the Main sail and some of the small Sails, got down the Fore topgallant Mast and the Jibb boom and spritsail yard in, intending to lighten the Ship forward as much as possible in order to lay her ashore to come at the leak.

BANKS: It blew too fresh tonight for us to weigh the anchor, I even think as fresh as it has ever done since we have been upon the Coast. We thought much of our good fortune in having fair weather upon the rocks when upon the brink of such a gale. Our people were now however pretty well recovered from their fatigues having had plenty of rest, as the ship since she was Fothered has not made more water than one pump half worked will keep clear.

16th June

BANKS: At night we observed a fire ashore near where we were to lay, which made us hope that the necessary length of our stay would give us an opportunity of being acquainted with the Indians who made it [Guugu Yimithirr people].

BANKS: In the morn it was a little more moderate and we attempted to weigh but were soon obliged to vere away all that we had got, the wind freshening upon us so much.

148   The Endeavour Journals

Fires were made upon the hills and we saw 4 Indians through our glasses who went away along shore, in going along which they had made two more fires for what purpose we could not guess [Guugu Yimithirr people].

Tupia whose bad gums were very soon followed by livid spots on his legs and every symptom of inveterate scurvy, notwithstanding acid, bark and every medicine our Surgeon could give him, became now extremely ill; Mr Green the astronomer was also in a very poor way, which made everybody in the Cabin very desirous of getting ashore and impatient at our tedious delays.

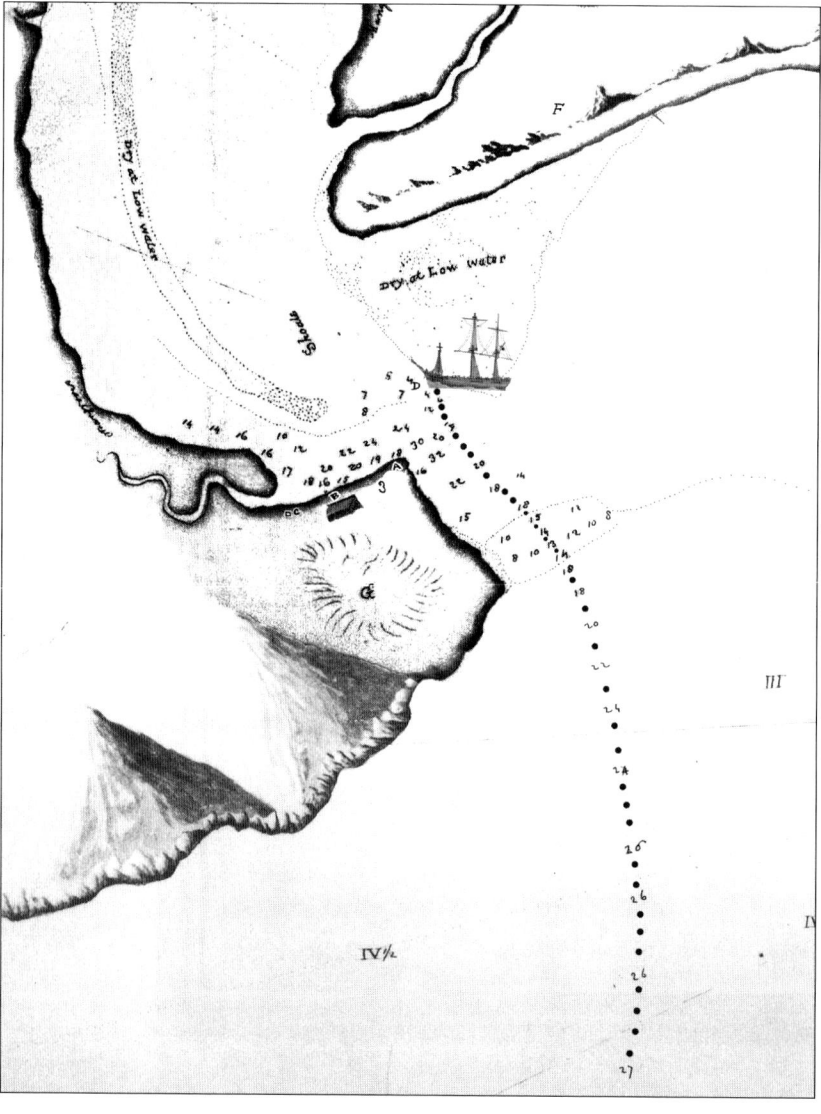

Portion of 'A Plan of the River on the East Coast of New Holland where his Maj: bark Endeavour repaired her bottom after Running on a reef of rocks where she lay 24 hours June the 10th 1770'.[5]

17th June

COOK: Most part strong gales at SE and Cloudy hazy weather with showers of rain.

At 6 in the a.m. being pretty moderate we weighed and run in to the harbour in doing of which we run the Ship ashore twice.

One of three pencil drawings of *Endeavour* by Sydney Parkinson.[6] (The location where Parkinson drew this sketch of *Endeavour* remains uncertain.)

COOK: While the Ship held fast we got down the foreyard, fore topmast, booms &c. overboard and made a raft of them alongside.

BANKS: In the meantime Dr Solander and myself began our Plant gathering.

[Banks and Dr Solander may have landed on the North Shore beach. The sandbar where the ship stranded is part of the North Shore beach where there are no mangroves to hinder a landing. In 1770, the South Shore of Endeavour River was barricaded with mangroves.]

Chapter 9

Nor Have We Seen One Since We Have Been in Port

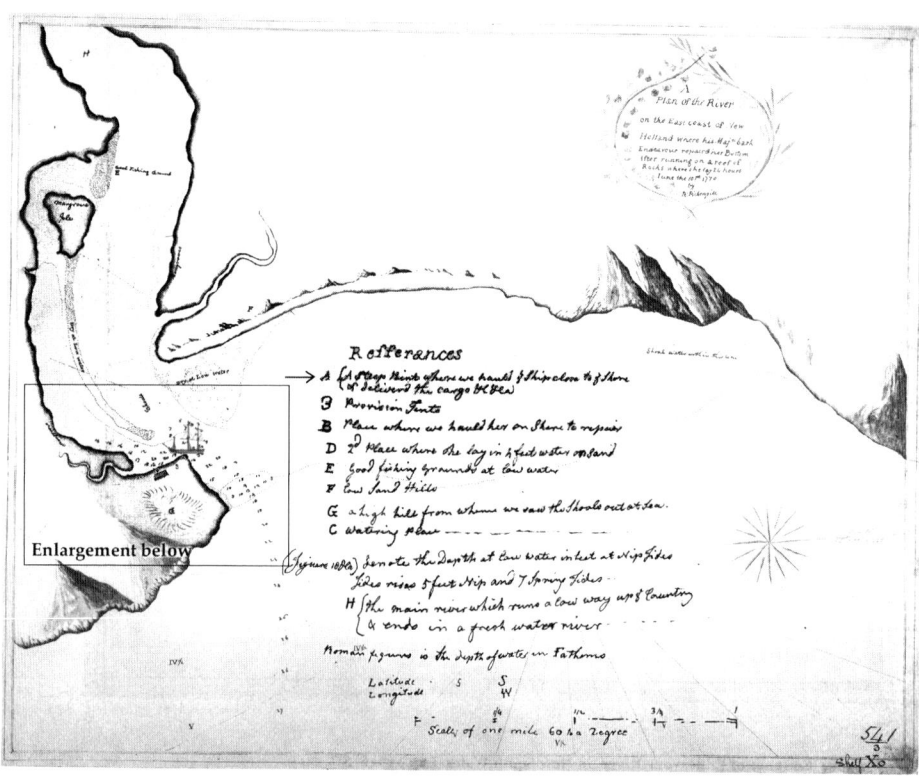

'A Plan of the River...' By Richard Pickersgill (Master's Mate).[1]

18th June

COOK: Fresh gales and cloudy with showers of rain. At 1 p.m. the ship floated and we warped her into the harbour and moored her alongside of a steep beach on the south side ['**A**' on Refferances above – **See enlargement below** – Guugu Yimithirr country].

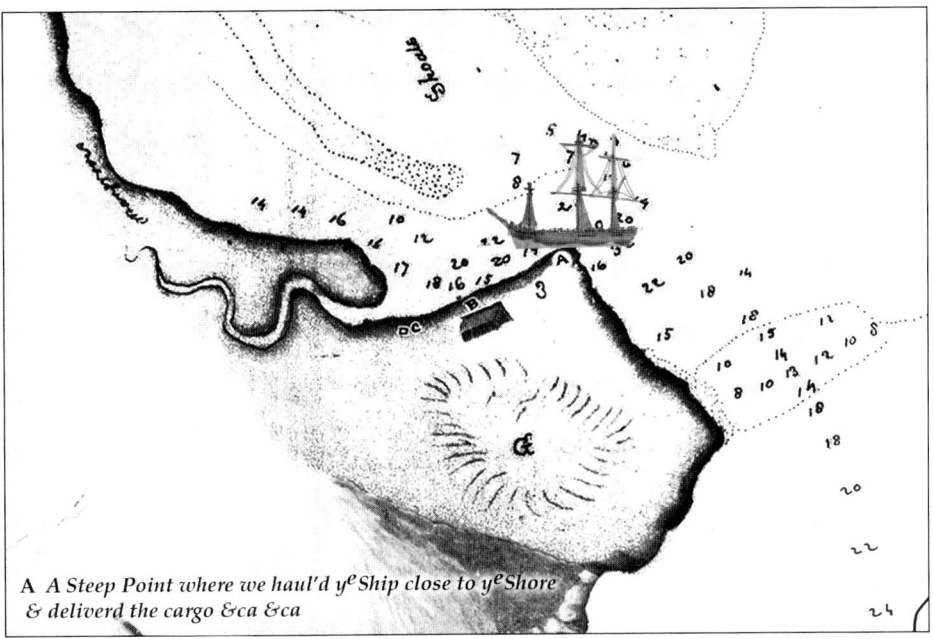

A *A Steep Point where we haul'd y^e Ship close to y^e Shore & deliverd the cargo &ca &ca*

Enlargement.

BANKS: The ship was moored within 20 feet of the shore afloat and before night much lumber was got out of her.

COOK: In the a.m. made a stage from the Ship to the shore.

BANKS: … which much facilitated our undertaking.

Sydney Parkinson. Portion of 'A view of the Endeavour River, on the coast of New Holland, where the ship was laid on shore, in order to repair the damage which she received on the rock'. Engraving by Will Byrne from 'A journal of a voyage to the South Seas: in His Majesty's ship, the Endeavour: faithfully transcribed from the Papers of the late Sydney Parkinson, draughtsman to Joseph Banks, esq. on his late expedition with Dr. Solander around the world / embellished with views and designs delineated by the author.' Hawkesworth (1773).[3]

PARKINSON: We set up tents ashore, unloaded her, carried all the cargo and provisions into them, and there lodged and accommodated our sick.

COOK: The sick amounted at this time to 8 or 9 afflicted with different disorders but none very dangerously ill.

BANKS: Tupia who had employed himself since we were here in angling and had lived entirely on what he caught was surprisingly recovered. Poor Mr Green still very ill. Myself walking in the country saw old frames of Indian houses and places where they had dressed shellfish in the same manner as the Islanders. Weather blowing hard with showers.

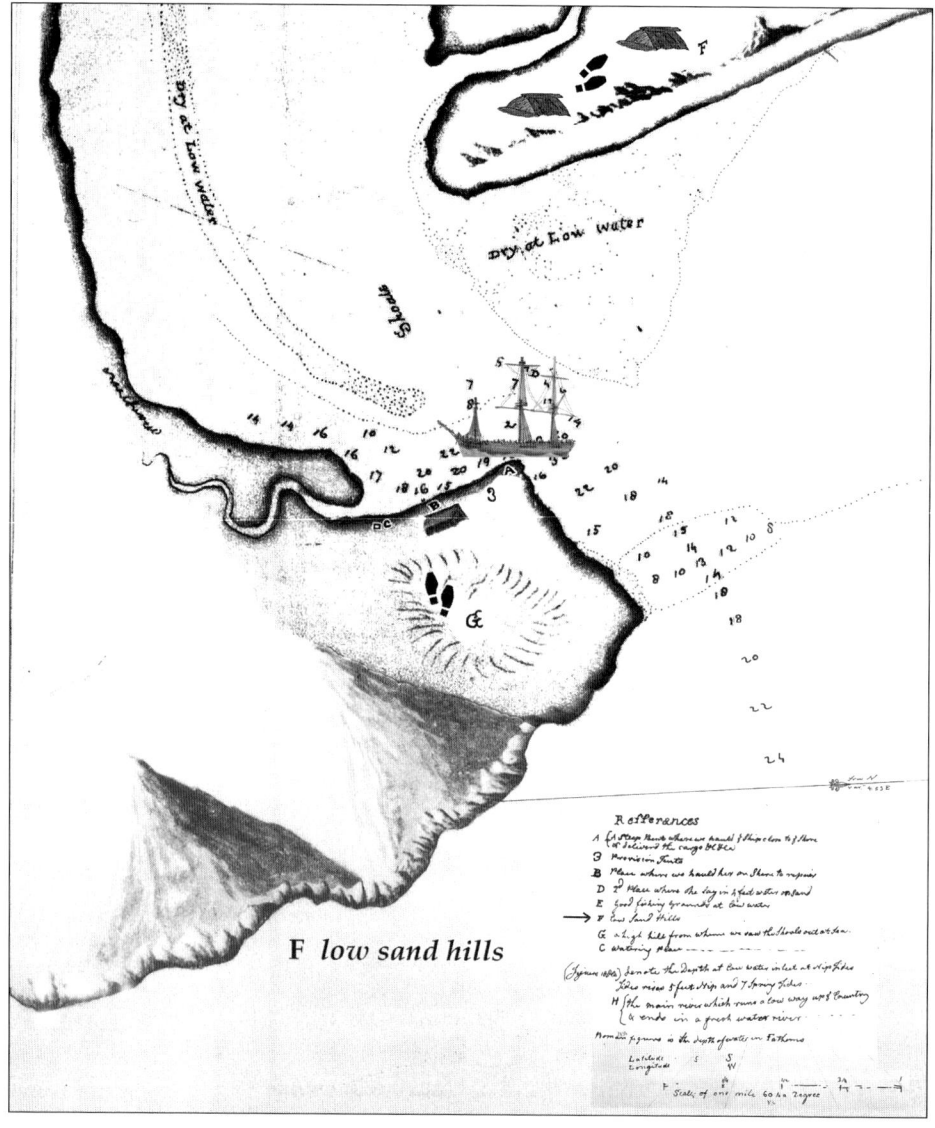

Portion of Richard Pickersgill's 'A Plan of the River ...'.

19th June

COOK: This afternoon I went upon one of the highest hills over the harbour ['G' on plan above – Now 'Grassy Hill'] I had a perfect view of the inlet or river and adjacent country which afforded a very indifferent prospect [Guugu Yimithirr country]. (Colour Plate No. 32) The low lands near the river is all overrun with mangroves among which the salt water flows every tide and the high land appears to be barren and stony.

BANKS: I went over the water today to spy the land which there was **sand hills** ['F' on plan above – Guugu Yimithirr country – Gamay clan]. On them I saw some Indian houses which seemed to have been inhabited since those on this side, tho not very lately. There were vast flocks of pigeons and crows; of the former which were very beautiful we shot several [Topknot Pigeon – *Lopholaimus antarcticus*]. (Colour Plate No. 33)

The latter [crows] exactly like those in England were so shy that we could not come near them by any means. A Crow in England tho in general sufficiently wary is I must say a fool to a New Holland crow and the same may be said of almost if not all the Birds in the countrey.

COOK: In the a.m. got the 4 remaining guns out of the hold and mounted them on the quarter deck. Got a spare anchor and anchor stock a shore and the remaining part of the stores and ballast that were in the hold.

20th June

COOK: This day got out all the officers stores and the ground tier of water. Many of the butts were quite rotten and would not bear the rolling. Now having nothing in the fore and main hold but the coals and a little ballast.

BANKS: In the evening hard rain. In the a.m. the weather cleared up so we began to gather and dry plants of which we had hopes of as many as we could muster during our stay.

21st June

COOK: In the p.m. landed the powder. After the coals was trimmed away from over the leak we could hear the water gushing in a little abaft the foremast and about 3 feet from her keel; this determined me to clear the hold completely. Accordingly very early in the morning we went to work to get out the coals which was employment for all hands.

BANKS: Fine clear weather: began today to lay plants in sand.

'A View of Endeavour River, where the ship was laid ashore, in order to repair the damage, which she received on the rock.' Engraving by Will Byrne, after a lost drawing by Parkinson. Hawksworth 1773. (*Courtesy of Wells' Cathedral Library*)

Present-day Cooktown from similar perspective. (© *John MacDonald*)

Endeavour River – 1770 shore settlement, now present-day Cooktown.[4]

[Parkinson made the original drawing, upon which the above engraving is based, from the same North Shore sandbar where the ship grounded entering the harbour on 17th June 1770. The sandbar dries at low tide. The photograph was taken from the same position.]

22nd June

COOK: At 4 in the p.m. having got out most of the coals, cast loose the moorings and warped the Ship a little higher up the harbour to a place I had pitched upon to lay ashore for stopping the leak. At 2 o'clock in the a.m. the tide left her which gave us an opportunity to examine the leak which we found to be at her floor heads a little before the starboard fore chains.

BANKS: In the morn I saw her leak which was very large: in the middle was a hole large enough to have sunk a ship with twice our pumps but here providence had visibly worked in our favour, for it was in great measure plugged up by a stone which was as big as a man's fist.

COOK: A large piece of Coral rock was sticking in one hole and several pieces of the fothering, small stones, sand &c. had made its way in and lodged between the timbers which had stopped the water from forcing its way in in great quantities.

PARKINSON: … that same rock therefore, that endangered us, yielded us the principal means of our redemption. We lost no time, but immediately set about repairing the ship's bottom.

COOK: Part of the sheathing was gone from under the larboard bow [port side] – part of the false keel was gone and the remainder in such a shattered condition that we should be much better off was it gone also – her fore foot and some part of her Main keel was also damaged but not materially. What damage she may have received abaft we could not see but believe not much as the Ship makes but little water while the Tide keeps below the leak forward – At 9 o' clock the Carpenters went to work upon the ship while the Smiths were busy making bolts nails &c. [See Glossary of Terms for definitions.]

BANKS: Myself employed all day in laying in Plants. The people who were sent to the other side of the water to shoot pigeons saw an animal as large as a greyhound, of a mouse colour and very swift; they also saw many Indian houses and a brook of fresh water.

23rd June

COOK: In the a.m. I sent a boat to haul the seine who returned at noon having made three hauls and caught only three fish, and yet we see plenty jumping in the harbour but can find no method of catching them.

BANKS: The people who went over the River saw the animal again and described him much in the same manner as yesterday.

24th June

BANKS: Gathering plants and hearing descriptions of the animal which is now seen by everybody.

COOK: I saw myself this morning a little way from the Ship one of the animals. I could have taken it for a wild dog, but for its walking or running in which it jumped like a hare or deer; another of them was seen today by some of our people who saw the first, they describe him as having very small legs and the print of a foot like that of a goat, but this I could not see myself because the ground the one I saw was upon was too hard and the length of the grass hindered my seeing its legs.

BANKS: A seaman who had been out in the woods brought home the description of an animal he had seen composed in so Seamanlike a style that I cannot help mentioning it; it was says he 'About as large and much like a one gallon cagg [keg], as black as the Devil and had 2 horns on its head, it went but slowly but I dare not touch it'.

After taking some pains I found out that the animal he had seen was no other than the Large Bat [Flying fox – *Pteropus sp.*]. (Colour Plate No. 34)

The *Endeavour* careened in Endeavour River in 1770.[5]

25th June

COOK: At low-water in the p.m. while the Carpenters were busy in repairing the sheathing and plank under the larboard bow I got people to go under the ships bottom to examine all her larboard side she only being dry forward but abaft were 9 feet water, they found part of the sheathing off abreast of the Main mast about her floor heads and a part of one plank a little damaged. There were three people who went down who all agreed in the same story, the master was one who was positive that she had received no material damage besides the loss of the sheathing, this alone will be sufficient to let the worm into her bottom which may prove of bad consequence – however, we must run all risk for I know of no method to remedy this but by heaving her down which would be a work of emince labour & time, if not impractical in our present situation.

BANKS: In gathering plants today I myself had the good fortune to see the beast so much talked of, tho but imperfectly; he was not only like a grey hound in size and running but had a long tail, as long as any grey hounds; what to liken him to I could not tell, nothing certainly that I have seen at all resembles him.

26th June

COOK: At low-water in the p.m. the Carpenters finished under the larboard bow and every other place the tide would permit them to come at. Lashed some casks under the Ships bows [visible in drawing above] in order to float her at high-water in the night. Attempted to heave her off but could not, she not being afloat partly owing to some of the casks not holding that were lashed under her. I am much afraid that we shall not be able to float her now the tides are taking off.

BANKS: Since the ship has been hauled ashore the water that has come into her has of course all gone backwards and my plants which were for safety stowed in the bread room were this day found under water. Nobody had warned me of this danger which had never once entered into my head; the mischief was however now done so I set to work to remedy it to the best of my power. The day was scarce long enough to get them all shifted &c: many were saved but some entirely lost and spoiled.

27th June

BANKS: Some of the Gentlemen who had been out in the woods Yesterday brought home the leaves of a plant which I took to be *Arum esculentum*, the same I believe as is called Coccos in the West Indies. In consequence of this I went to the place and found plenty [Kale/Taro – *Colocasia esculenta*]. (Colour Plate No. 35)

COOK: The tops we found made good greens and eat exceeding well when boiled but the roots were so Acrid few besides myself could eat them.

BANKS: Tupia by roasting his Coccos very much in his Oven [earth oven] made them lose entirely their acridity; the Roots were so small that we did not think them at all an object for the ship so resolved to content ourselves with the greens which are called in the West Indies Indian Kale. I went with the seamen to shew them the Place and they gathered a large quantity.

In the same place grew plenty of Cabbage trees, and a kind of Wild Plantain whose fruit was so full of stones that it was scarce eatable [Wild plantain – *Musa acuminata subsp. Banksia*]. (Colour Plate No. 36)

The Wild Plantain trees, tho their fruit does not serve for food, are to us a most material benefit; we made Baskets of their stalks (a thing we learned of the Islanders) in which our plants which would not otherwise keep home remain fresh for 2 or 3 days; indeed in a hot climate it is hardly Practicable to go on without such baskets which we call by the Island name of *Papa Mya*.

In the same place grew another fruit about as large as a small golden pippin but flatter, of a deep purple colour; these when gathered off from the tree were very hard and disagreeable but after being kept a few days became soft and tasted much like indifferent Damsons [Burdekin plum – *Pleiogymium cerasiferum*]. (Colour Plate No. 37)

The Pinnace by Sydney Parkinson.[6]

COOK: I went in the pinnace up the harbour and made several hauls with the seine but caught only 20 and 30lb of fish which were given to the sick and such as were weak and ailing.

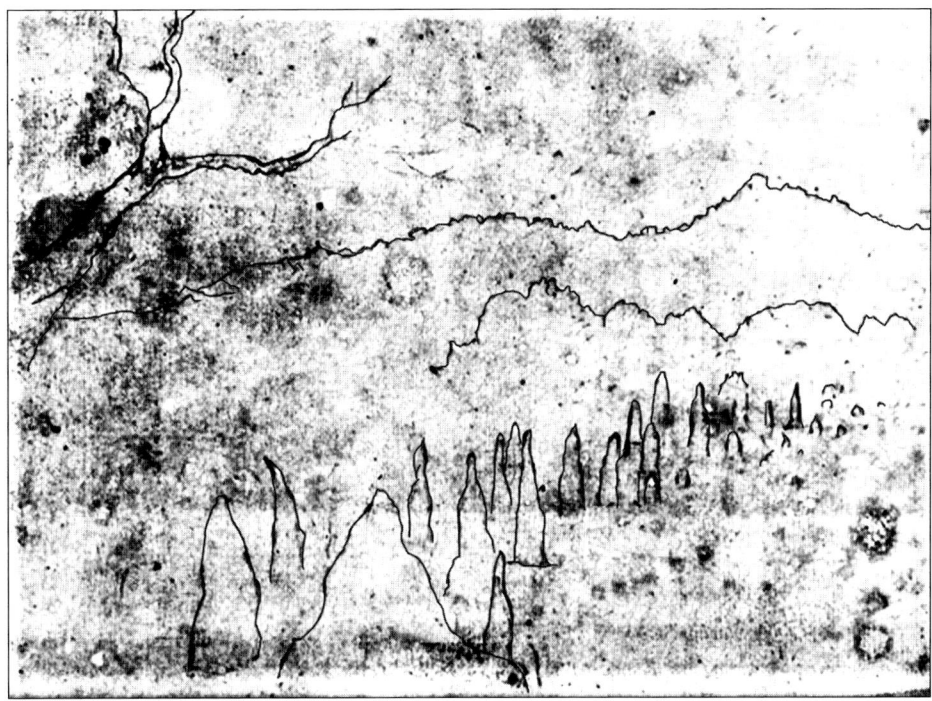

Nests of the White Ant by Sydney Parkinson – Endeavour River.[7]

28th June

BANKS: We have ever since we have been here observed the nests of a kind of Ants much like the White ants in the East indies but to us perfectly harmless; they were always pyramidical, from a few inches to 6 feet in height and very much resembled stones which I have seen in English Druidical monuments. Today we met with a large number of them of all sizes ranged in a small open place which had a very pretty effect.

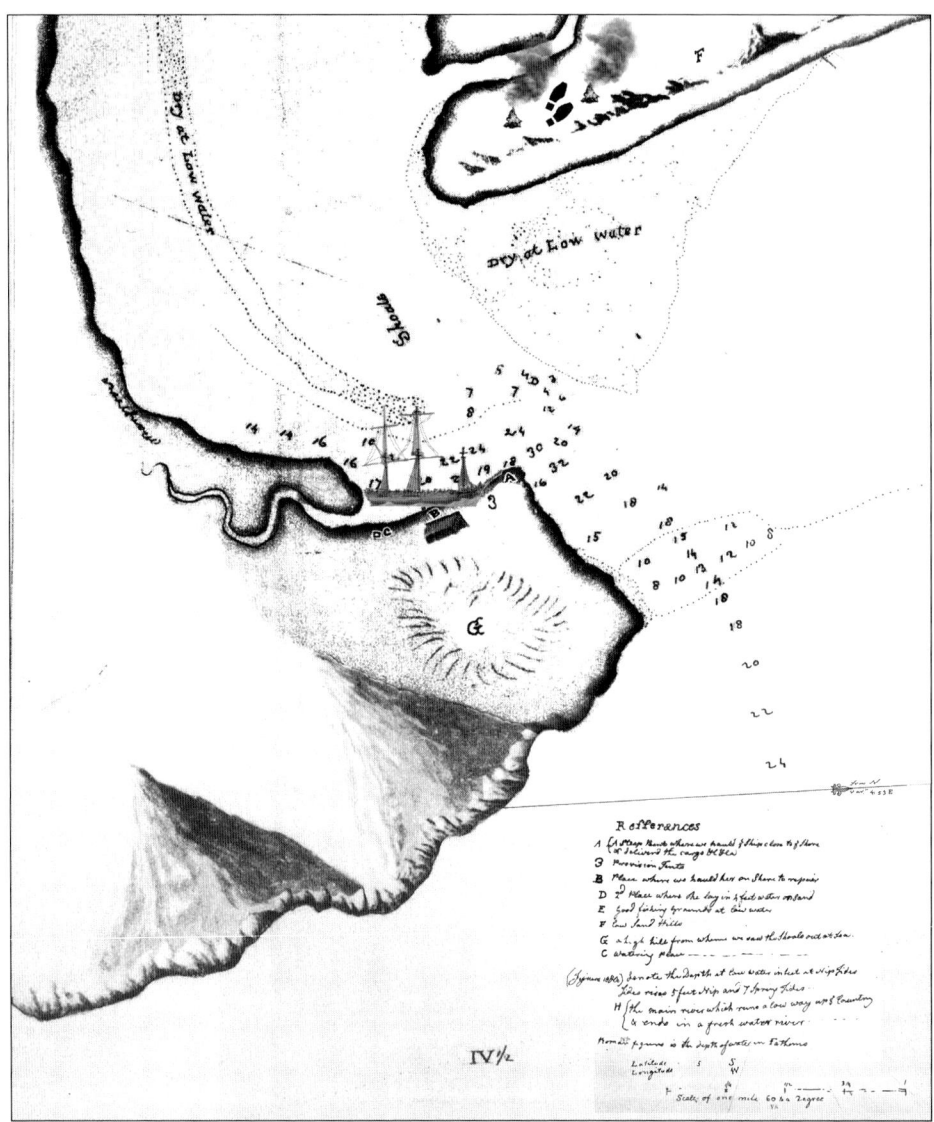

Detail of Pickersgill's Plan of Endeavour River.[8]

29th June

COOK: Some of our people who were out yesterday on the north side of the river met with a place where the natives had just been, as there were fires still burning but they saw nobody nor have we seen one since we have been in port [Guugu Yimithirr people – Gamay clan].

30th June

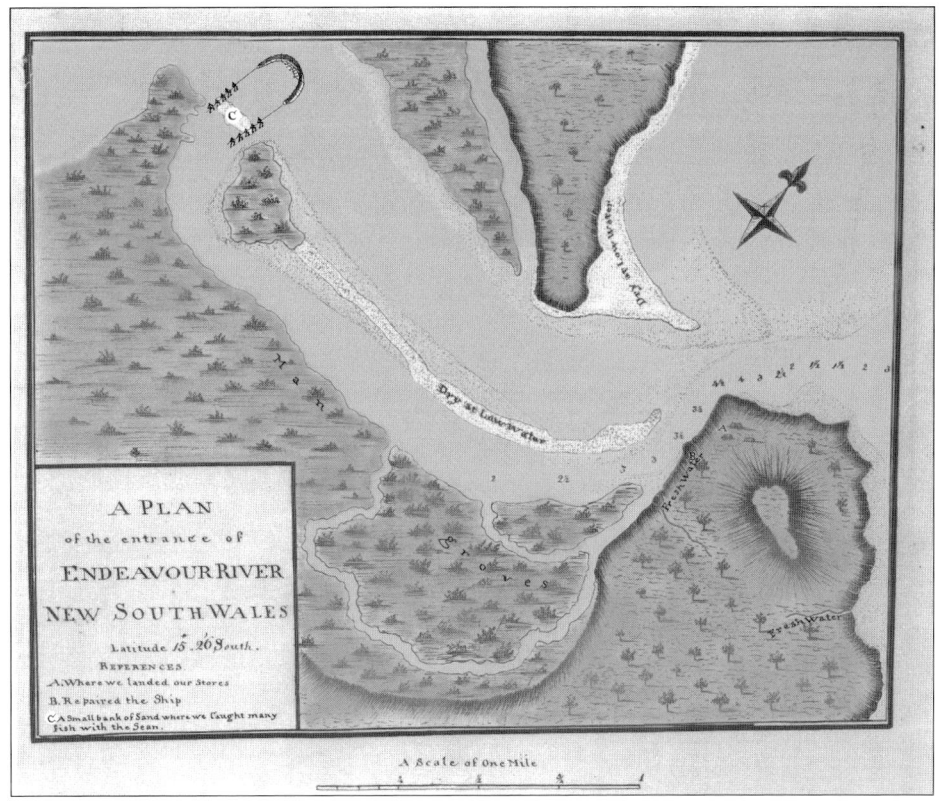

'C' – 'A small bank of Sand where we Caught many fish with the Seine'.⁹

COOK: In the p.m. the Boat returned from hauling the seine ['C' on plan above] having caught as many fish as came to a pound and a half a man [Threadfin salmon – *Eleutheronema tetradactylum* was a good food fish caught in Endeavour River]. (Colour Plate No. 38)

In the a.m. I sent her out again. I likewise sent some of the young gentlemen to take a Plan of the harbour [plan above].

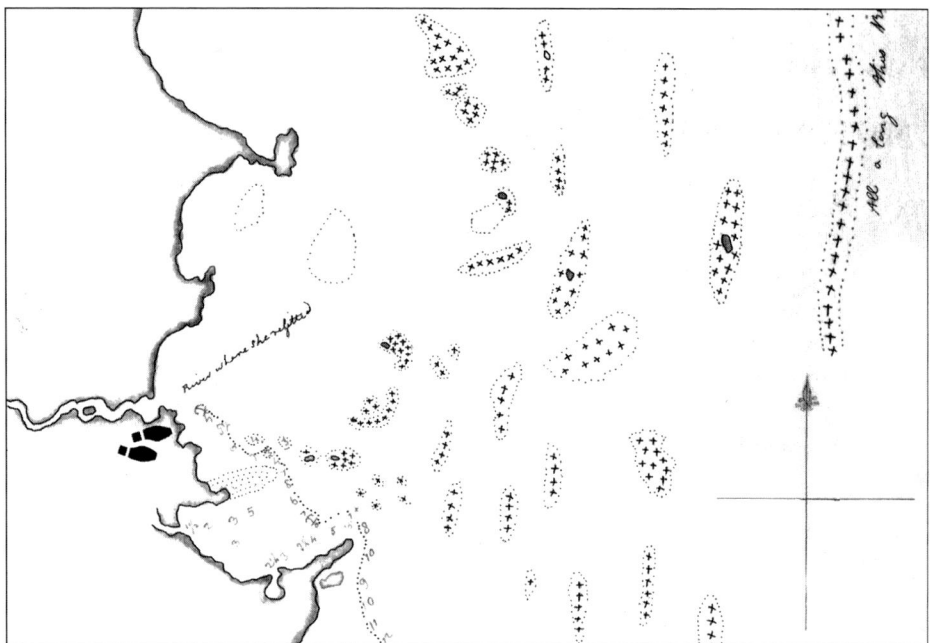

Portion of 'A draft of part of the shoals seen and sailed through by His Maj Bark the *Endeavour* in 1770...' by Richard Pickersgill.[10]

COOK: I went myself upon the hill which is over the south point to take a view of the sea [now Grassy Hill].

BANKS: When we came there the prospect was indeed melancholy: the sea everywhere full of innumerable shoals, some above and some under water, and no prospect of any straight passage out.

COOK: The only hope I have of getting clear of them is to the northward where there seems to be a passage for as the winds blow constantly from the SE we shall find it difficult if not impractical to return back to the southward.

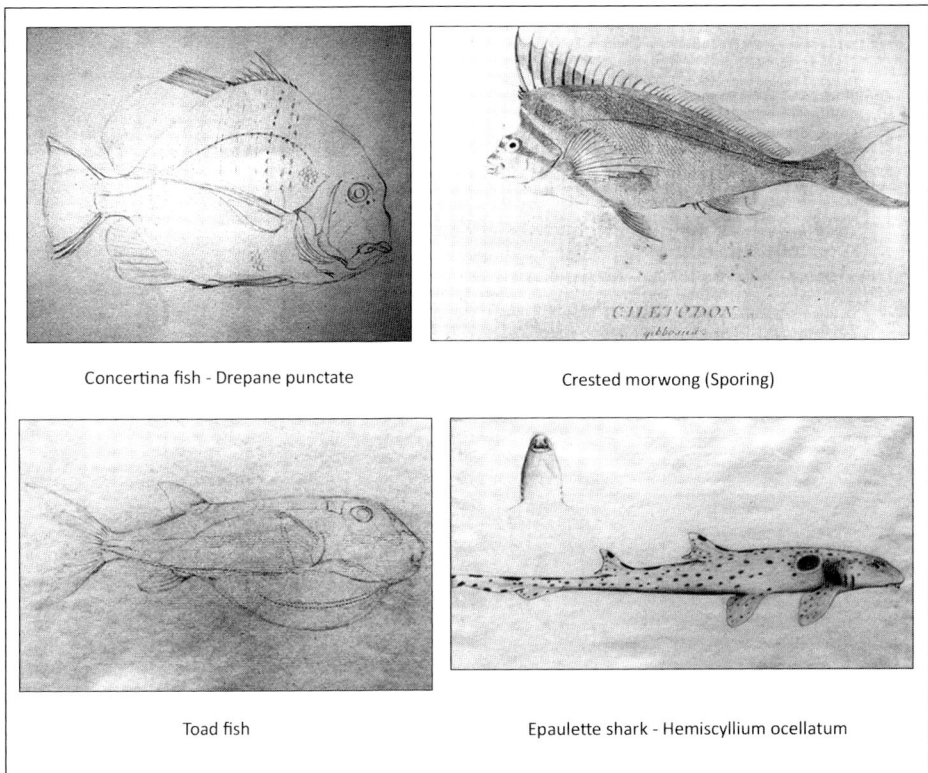

Other fish species caught at Endeavour River and sketched by Sydney Parkinson and Herman Spöring.[11]

1st July

COOK: In the p.m. the People returned from hauling the seine having caught as much fish as came to two and a half pound a man, no one on board having more than another, the few greens we got I caused to be boiled among the Pease and makes a very good mess, which together with the fish is a great refreshment to the People.

Chapter 10

Their Unaccountable Timidity

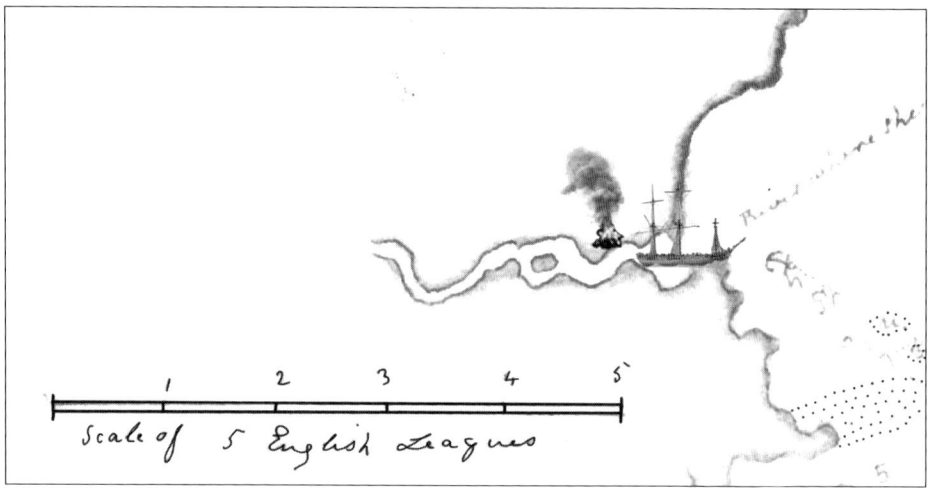

Detail of Master's Mate, Richard Pickersgill's chart.[1]

1st July (cont'd)

BANKS: The Indians had a fire about a league off up the river.

2nd July

COOK: Early in the a.m. I sent the Master [Robert Molineux] in the Pinnace out of the harbour to Sound about the shoals in the offing and look for a channel to the Northward.

BANKS: Our Plants dry better in Paper Books than in Sand, with this precaution, that one person is entirely employed in attending them who shifts them all once a day, exposes the Quires in which they are to the greatest heat of the sun and at night covers them most carefully up from any damps, always careful not to bring them out too soon in a morning or leave them out too late in the evening.

The *Endeavour* careened in Endeavour River in 1770.[2] (Engraving by Will Byrne, after a lost drawing by Parkinson. Hawkesworth – 1773)

3rd July

COOK: At high water attempted to heave the Ship off but did not succeed.

SHIP'S LOG: A party of men having been out all night in quest of some beasts that we have seen frequently, returned without success.

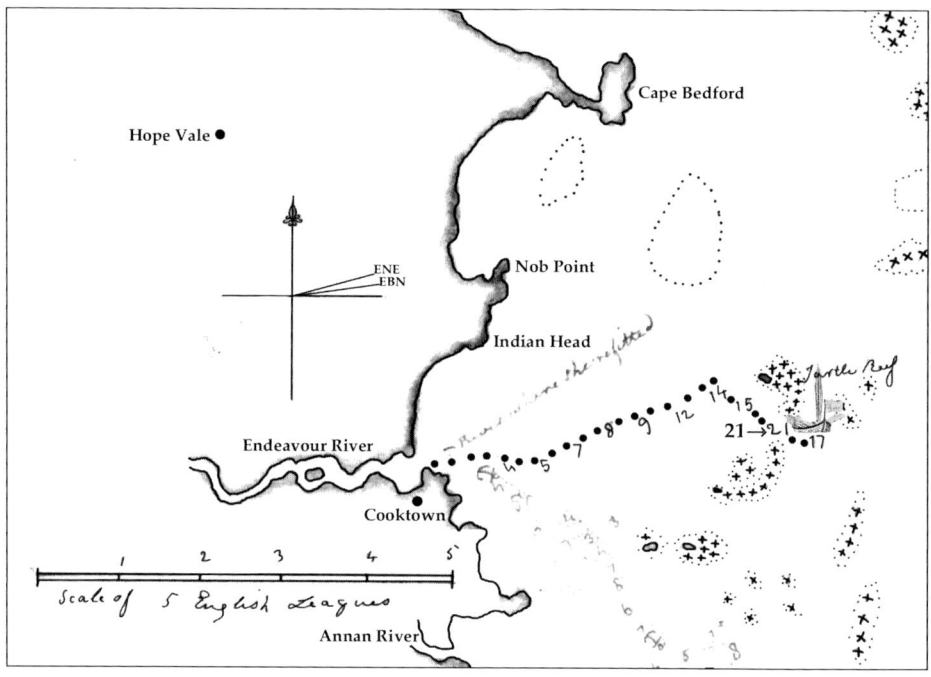

Molineux's passage. (Portion of Master's Mate, Richard Pickersgill's chart.)[3]

COOK: At Noon the Master returned and reported that he had found a passage out to sea between the shoals which passage lies out ENE or EBN from the Rivers mouth. He told me that he was 5 Leagues out at Sea having at that distance off **21** fathom water [marked on chart above] and judged himself to be without all the shoals which I very much doubted.

BANKS: He having found a way by which he past most of the shoals that we could see but not all. This Passage was also to windward of us so that we could only hope to get there by the assistance of a land breeze, of which we have had but one since we lay in the Place, so this discovery added little comfort to our situation.

He had in his return landed on a dry reef [Turtle Reef on chart above – now Boulder Reef] where he found vast plenty of shell fish so that the boat was completely loaded, chiefly with a large kind of cockles [the giant clam – *Tridacna gigas*] (Colour Plate No. 39) one of which was more than 2 men could eat. Many indeed were larger; the Cockswain of the Boat a little man declared that he saw on the reef a dead shell of one so large that he got into it and it fairly held him. The large ones of this kind had 10 or 15lb [4.5 or 6.8 kgs] of meat in them; it was indeed rather strong but I believe a very wholesome food and well relished by the people in general.

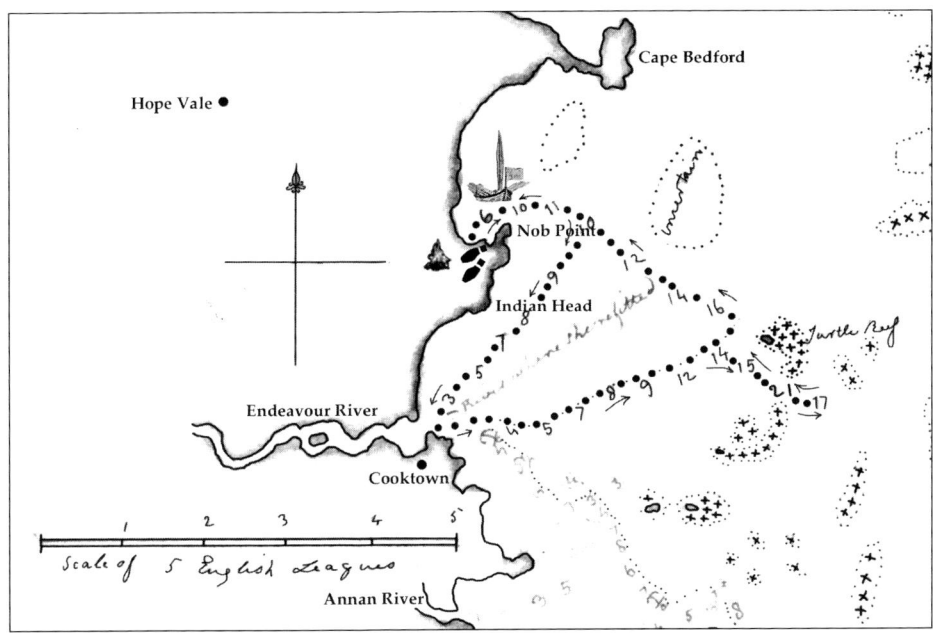

Molineux's passage. (Portion of Master's Mate, Richard Pickersgill's chart.)[4]

COOK: After this he came in shore and stood to the northward where he met with a number of shoals laying a little distant from the shore – about 9 o'clock in the evening he landed in a bay about 3 Leagues to the northward of this place [bay behind Nob Point, marked on chart above] where he disturbed some of the natives whome he supposed to be at supper; they all fled upon his approach and left him some fresh Sea Eggs [Blue-black urchin – *Echinothrix diadema*] (Colour Plate No. 40) and a fire ready lighted behind them but there was neither house nor hut near.

4th July

COOK: At high water hove the ship afloat.

BANKS: An Allegator was seen swimming alongside of her for some time.

PARKINSON: There were many alligators on the coast, some of them very large, and we frequently saw them swimming round the ship.

BANKS: As I was crossing the harbour in my small boat we saw many sholes of Gar fish leaping high out of the water, some of which leaped into the boat and were taken [Garfish – Barred Longtom – *Ablennes hains*]. (Colour Plate No. 41)

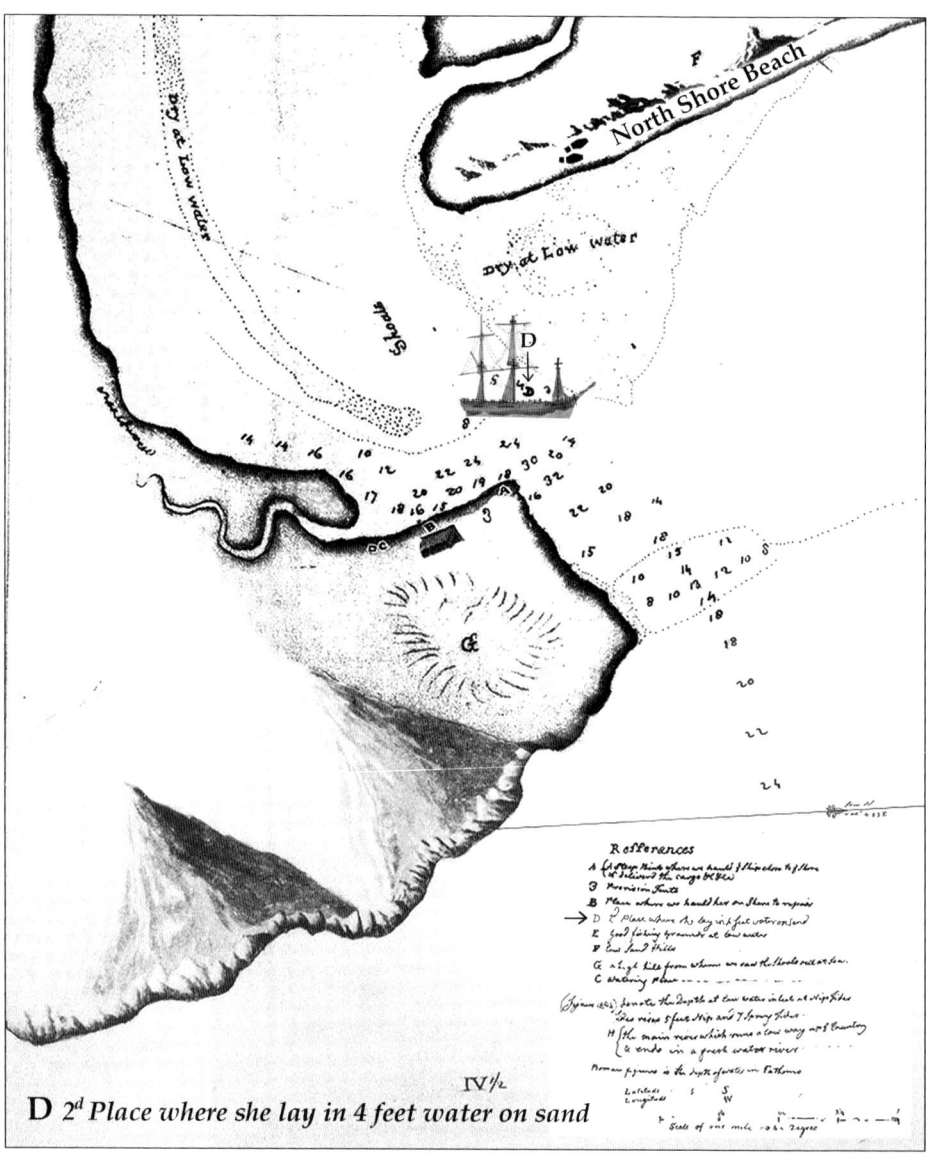

Richard Pickersgill's 'Plan of the River...'.[5]

5th July

COOK: In the p.m. warped the Ship over and at high-water laid her ashore on the sandbank on the south side of the river ['**D**' on plan above].

BANKS: The ship has been a good deal strained by laying so long as she has done with her head aground and her stern afloat; so much so that she has sprung a plank between decks abreast the main chains. She was laid ashore in order if possible to examine if she had got any damage near that place.

COOK: I was afraid to lay her broad side to the shore where she lay before because the ground lies with too great a descent.

BANKS: [a.m.] Went to the other side of the harbour and walked along a sandy beach open to the trade wind [North shore beach]. Here I found innumerable fruits, many of Plants I had not seen in this countrey; among them were some Cocoa nuts that had been opened (as Tupia told us) by a kind of Crab, called by the Dutch *Beurs Krabbe* (*Cancer latro*)that feeds upon them [Coconut-opening crab – *Birgus latro*]. (Colour Plate No. 42)

All these fruits were incrusted with sea productions and many of them Covered with Barnacles, (Colour Plate No. 43) a sure sign that they have come far by sea, and as the trade wind blows almost right on shore they must have come from some other countrey – probably that discovered by Quiros and called Terra del Esprito Santo [later called New Hebrides, now Vanuatu] as the Latitudes according to his own account agree pretty well.

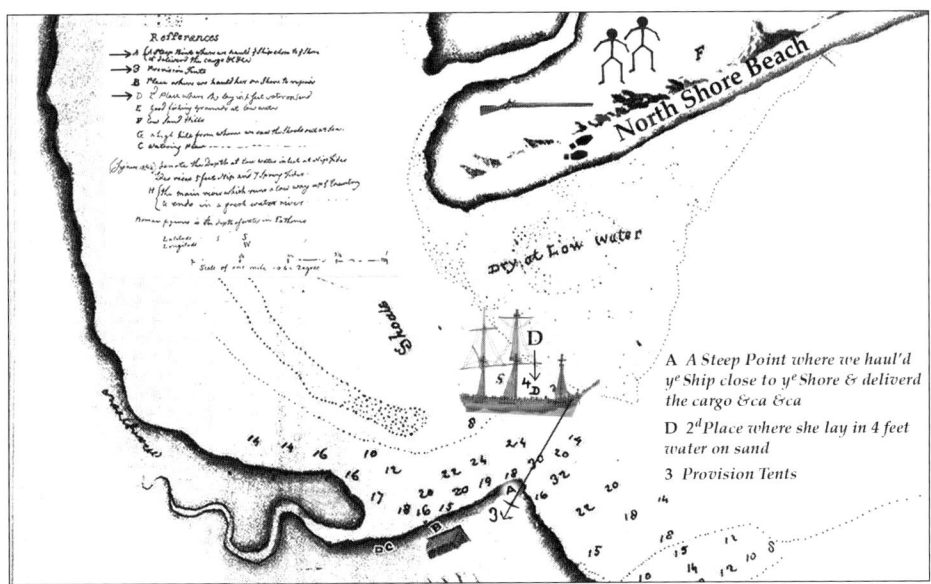

Portion of Richard Pickersgill's 'Plan of the River...'.⁶

BANKS: Tupia who parted from us and walked away a shooting, on his return told us that he had seen 2 people who were digging in the ground for some kind of roots; on seeing him they ran away with great precipitation [Guugu Yimithirr people].

6th July

COOK: At low water in the p.m. had hardly 4 feet water under the ship ['D' marked on plan above] yet could not repair the sheathing that was beat off the place being all under water, one of the Carpenters crew, a Man I could trust, went down and examined it and found three streaks of the sheathing gone about 7 or 8 feet long and the Main plank a little rubbed, this account agrees with the report of the Master and others that were under her bottom before. The Carpenter who I look upon to be well skilled in his profession and a good judge of these matters was of opinion that this was of little consequence and as I found that it would be difficult if not impractical for us to get under her bottom to repair it, I resolved to spend no more time about it.

Accordingly bent the coasting cable to a bower anchor at the tents ['3' on plan] to assist in heaving her off at High-water. Hove off the bank and hove her alongside the bank where we cleared the ship ['A' on plan]. In the a.m. got a stage made from the ship to the shore. Got on board 8 tons of water for the ground tier of the after hold. Some hands employed about the rigging.

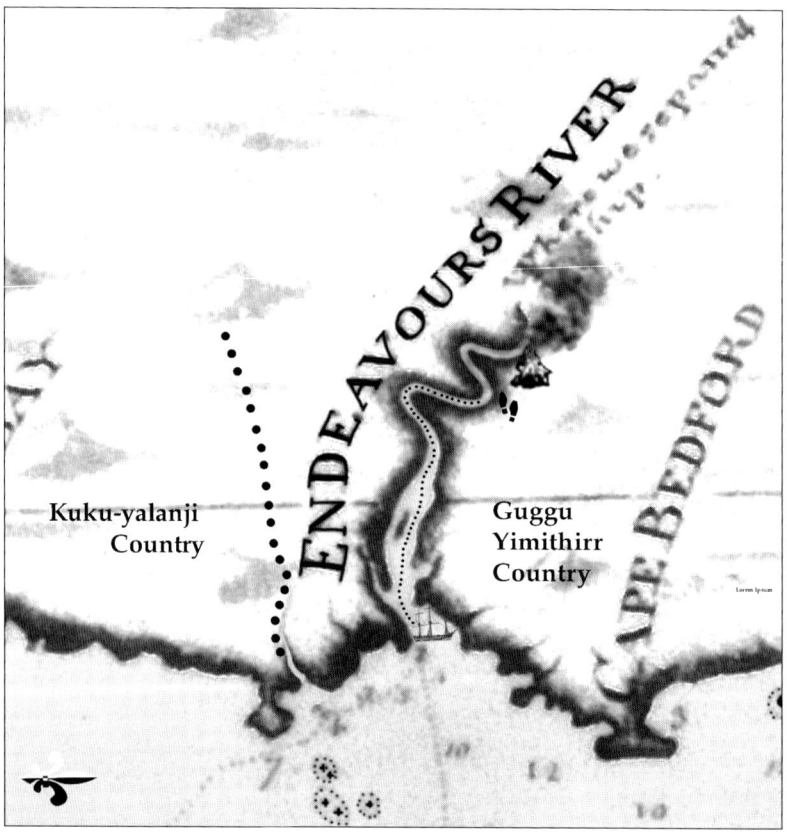

Portion of Cook's chart 'Cape Tribulation to *Endeavours* Streights'.[7]

Their Unaccountable Timidity 171

COOK: In the morning Mr Banks and Lieutenant Gore with three men in a small boat went up the River with a view to stay 2 or 3 days to try to kill some of the animals we have so often seen about this place.

BANKS: We went for about 3 leagues among Mangroves. (Colour Plate No. 44)

From hence we proceeded up the river which contracted itself much and lost most of its mangroves. The banks were steep and covered with trees of a beautiful verdure particularly what is called in the West Indies Mohoe or Bark tree [Sea hibiscus – *Hibiscus tiliaceus*]. (Colour Plate No. 45)

The land within was generally low, covered thick with long grass, and seemed to promise great fertility were these people to plant and improve it.

In the course of the day Tupia saw a wolf, so at least I guess by his description (Colour Plate No. 46) and we saw 3 of the animals of the country but could not get one also a kind of Batts as large as a Partridge but these also we were not lucky enough to get.

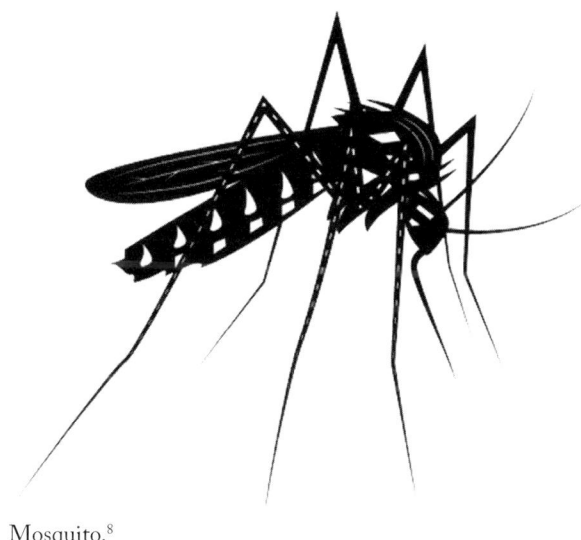

Mosquito.[8]

[7th July Ship's time]

BANKS: At night we took up our lodging close to the banks of the river and made a fire, but the Musquetos, whose peaceful dominions it seems we had invaded, spared no pains to molest us as much as was in their Power: they followed us into the very smoke, nay almost into the fire, which hot as the Climate was we could better bear the heat of than their intolerable stings.

Between the hardness of our beds, the heat of the fire and the stings of these indefatigable insects the night was not spent so agreeably but that day was earnestly wished for by all of us; at last it came and with its first dawn we set out in search of game.

We walked many miles over the flats and saw 4 of the animals, 2 of which my greyhound fairly chased, but they beat him owing to the length and thickness of the grass which prevented him from running while they at every bound leaped over the tops of it.

We observed much to our surprize that instead of going upon all fours this animal went only upon two legs, making vast bounds just as the Jerbua (*Mus jaculus*) does.

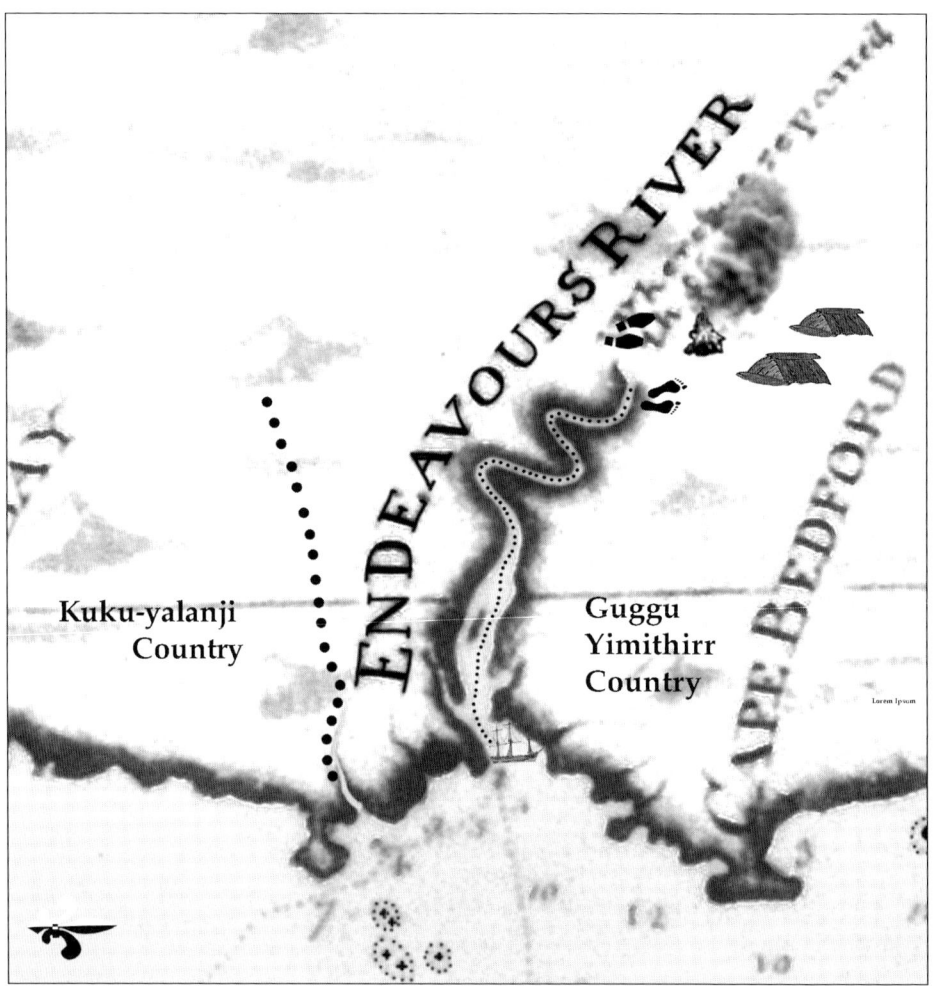

Portion of Cook's chart 'Cape Tribulation to *Endeavour*s Streights'.⁹

BANKS: We returned about noon and pursued our course up the river, which soon contracted itself into a fresh water brook where however the tide rose pretty considerably.

[8th July Ship's time]

BANKS: Towards evening it was so shallow being almost low water that we were obliged to get out of the boat and drag her, so finding a convenient place for sleeping in we resolved to go no farther. Before our things were got up out of the boat we observed a smoke about a furlong from us: we did not doubt at all that the natives, who we had so long had a curiosity to see well, were there so three of us went immediately towards it hoping that the smallness of our numbers would induce them not to be afraid of us; when we came to the place however they were gone, probably upon having discovered us before we saw them. The fire was in an old tree of touchwood; their houses were there, and branches of trees broken down with which the children had been playing not yet withered; their footsteps also upon the sand below the high tide mark proved that they had very lately been there.

Near their oven, in which victuals had been dressed since morn, were shells of a kind of clam [Fresh water mussel – *Velesunio wilsoni*] (Colour Plate No. 47) and roots of a wild yam which had been cooked in it.

Thus were we disappointed of the only good chance we have had of seeing the people since we came here by their unaccountable timidity.

The land about this place was not so fertile as lower down, the hills rose almost immediately from the river and were barren, stony and sandy much like those near the ship. The river near us abounded much in fish who at sun set leaped about in the water much as trouts do in Europe but we had no kind of tackle to take them with.

Night soon coming on we repaired to our quarters, which was upon a broad sand bank under the shade of a bush where we hoped the Musquetos would not trouble us. Our beds of plantain leaves spread on the sand as soft as a mattress, our cloaks for bedclothes and grass pillows, but above all the entire absence of Musquetos made me and I believe all of us sleep almost without intermission; had the Indians came they would certainly have caught us all napping but that was the least in our thoughts.

At day light in the Morn the tide serving we set out for the ship.

In our passage down met several flocks of Whistling Ducks of which we shot some [Whistling tree duck – *Dendrocygna arcuate*]. (Colour Plate No. 48)

We saw also an Allegator of about 7 feet long come out of the Mangroves and crawl into the Water [Estuarine crocodile – *Crocodilus porosus*]. (Colour Plate No. 49)

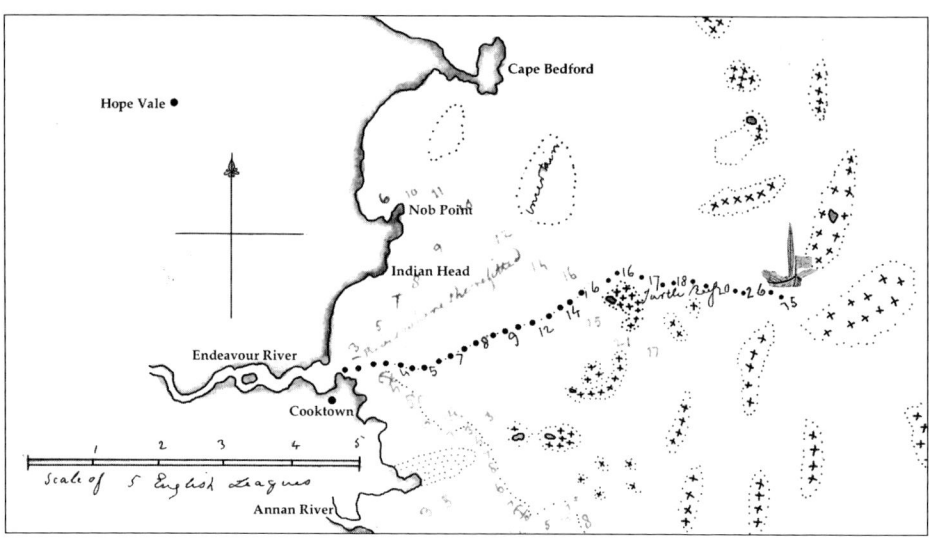

Molineux's passage and Turtle Reef.[10]

8th July

COOK: [a.m.] Early I sent the Master in a boat out to sea to sound again about the shoals because the account he had given of the Channell before mentioned was to me by no means satisfactory.

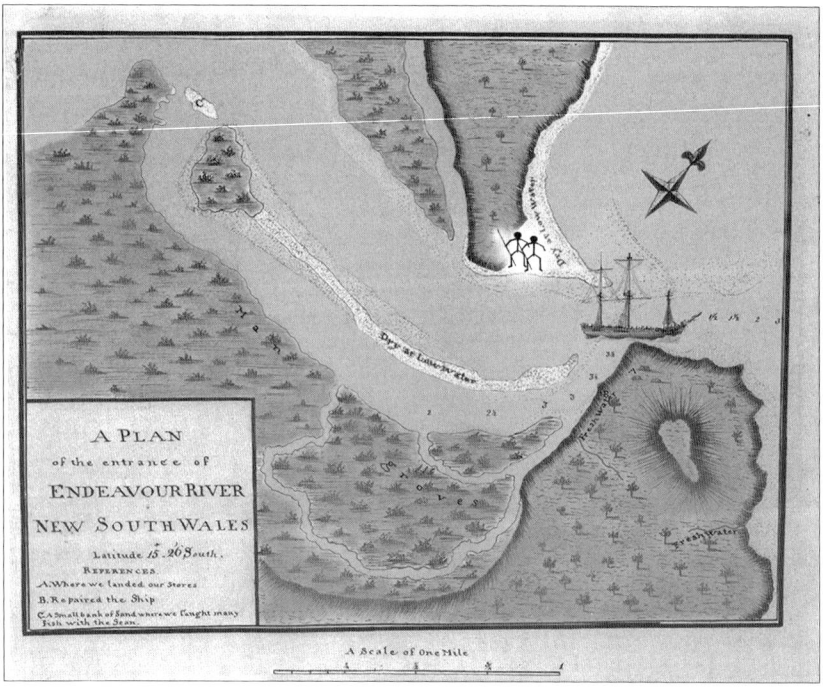

'A PLAN of the entrance of ENDEAVOUR RIVER NEW SOUTH WALES.'[11]

9th July

BANKS: By 4 o'clock [p.m.] we arrived at the ship where we heard that the Indians had been near them but not come to them; Yesterday they had made a fire about a mile and a half off, and this morn 2 had appeared on the beach opposite to the ship [Guugu Yimithirr people – Gamay clan].

COOK: In the evening the Master returned having been seven Leagues out at sea and at that distance off saw shoals without him and was of opinion that there was no getting out to sea that way. In his return he touched upon one of the shoals the same as he was upon the first time he was out [Turtle Reef on chart – now Boulder Reef]. Here he saw a great number of turtle three of which he caught weighing 791 pounds. This occasioned my sending him out again this morning provided with proper gear for striking them he having before nothing but a boat hook.

Endeavour River turtle by Sydney Parkinson.[12] (Green turtle – *Chelonia sp.*)

COOK: This day all hands feasted upon turtle for the first time.

BANKS: The promise of such plenty of good provisions made our situation appear much less dreadful; were we obliged to wait here for another season of the year when the winds might alter we could do it without fear of wanting provisions: this thought alone put everybody in vast spirits.

Chapter 11

Two of Them Embarked and Came Towards the Ship

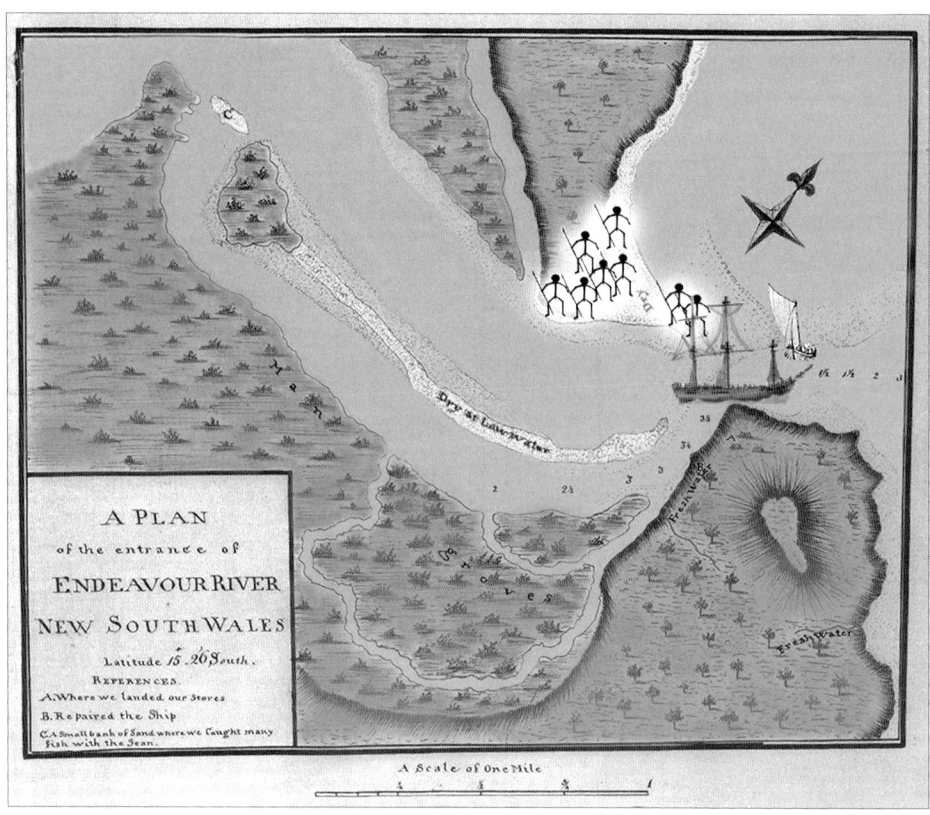

Cook's Plan of Endeavour River.[1]

10th July

COOK: In the p.m. saw seven or eight of the natives [Guugu Yimithirr people – Gamay clan] on the opposite side of the River and two of them came down upon the sandy point opposite the ship but as soon as I put off in a boat in order to speak with them they run away as fast as they could.

BANKS: Myself went turtling in hopes to have loaded our long boat, but by a most unaccountable conduct of the officer not one turtle was taken. At night returned with my small boat leaving the large one upon the reef who I was sure would catch no turtle.

COOK: The Master [Robert Molineux] was so obstinate that he would not return; which obliged me to send Mr Gore out in the yawl this morning to order the boat and the people in.

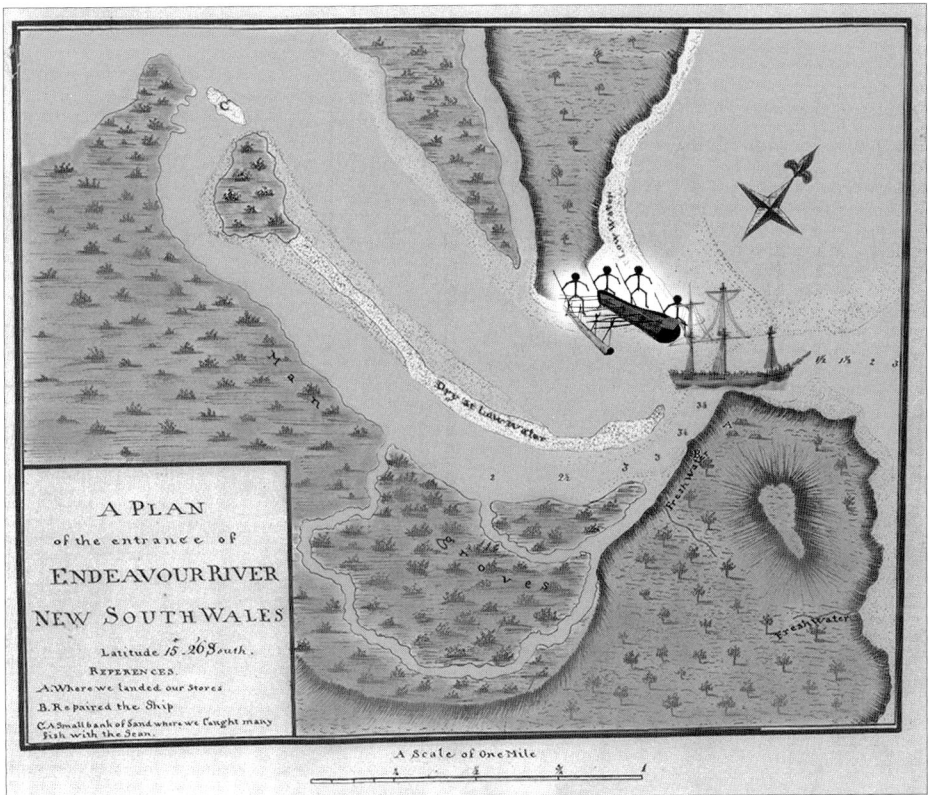

COOK: In the a.m. 4 of the Natives came down to the sandy point on the north side of the harbour, having along with them a small wooden Canoe with outriggers in which they seemed to be employed striking fish. Some were for going over in a boat to them but this I would not suffer but let them alone without seeming to take any notice of them.

BANKS: Two of them embarked and came towards the ship but stopped at the distance of a long Musket shot [approximately 300 yards or 274 metres], talking much and very loud to us. We hollowd to them and waving made them all the signs we could to come nearer; by degrees they ventured almost insensibly nearer and nearer till they were quite alongside, often holding up their Lances as if to shew us that if we used them ill they had weapons and would return our attack. Cloth, Nails, Paper, &c. &c. was given to them all which they took and put into the canoe without shewing the least signs of satisfaction: at last a small fish was by accident thrown to them on which they expressed the greatest joy imaginable, and instantly putting off from the ship made signs that they would bring over their comrades.

Canoe of the type used at Endeavour River in 1770. (W. E. Roth)[2]

COOK: They went away and brought over the other two and came again alongside nearer then they had done before and took such trifles as we gave them. After this they landed close to the Ship and all 4 went ashore carrying their arms with them.

BANKS: Each carrying in his hand 2 Lances and his stick to throw them with. Tupia went towards them, they stood all in a row in the attitude of throwing their Lances; he made signs that they should lay them down and come forward without them; this they immediately did and sat down with him upon the ground. We then came up to them and made them presents of Beads, Cloth &c. which they took and soon became very easy, only Jealous if any one attempted to go between them and their arms.

COOK: One of these men was something above the Middle age, the other three were young, none of them were above 5 & ½ feet high and all their limbs proportionately small, they were wholly naked their skins the colour of wood soot or a dark chocolate and this seemed to be their natural colour, their hair was black, lank and cropped short and neither woolly nor frizzled, nor did they want any of their fore teeth as Dampier has mentioned those did he saw on the western side of this Country. Some part of their bodies had been painted with red and one of them had his upper lip and breast paint with streaks of white which he called Carbanda: their features were far from being disagreeable, the voices were soft and tuneable, and they could easily repeat many words after us, but neither us nor Tupia could understand one word they said.

BANKS: At dinner time we made signs to them to come with us and eat but they refused; we left them and they going into their canoe paddled back to where they came from.

END OF FIRST MEETING

11th July

COOK: In the night Mr Gore and the Master returned with the longboat and brought with them one turtle and a few shell fish, the Yawl Mr Gore left upon the shoal with six men to endeavour to strike more turtle.

Altho these Shoals lay within sight of the coast and abound very much with shell fish and other small fish which are to be caught at low-water in holes in the rocks – yet the natives never visit them for if they did we must have seen of these large shells on shore about their fire places, the reason I do suppose is that they have no boats that they dare venture so far out to Sea in.

The Longboat by Sydney Parkinson.³

BANKS: Indians came over again today, 2 that were with us yesterday and two new ones who our old acquaintance introduced to us by their names, one of which was Yaparico.

Tho we did not yesterday observe it they all had the septum or inner part of the nose bored through with a very

Drawing by Charles Praval.⁴ [Praval who joined the voyage later in Batavia, had never seen a New Hollander. He concocted this image from written descriptions to be found in Cook and Banks's journals.]

large hole, in which one of them had stuck the bone of a bird as thick as a man's finger and 5 or 6 inches long, an ornament no doubt tho to us it appeared rather an uncouth one. It sticks across their face making in the eyes of Europeans a most ludicrous appearance, tho no doubt they esteem even this as an addition to their beauty which they purchased with hourly inconvenience; for when this bone was in its place, or as our seamen termed it 'their spritsail yard was rigged across', it completely stopped up both nostrils so that they spoke in the nose in a manner one should think scarce intelligible. Besides these extraordinary bones they had necklaces made of shells neatly enough cut and strung together.

COOK: They had likewise holes in their ears but no ornaments hanging to them, they had bracelets upon their arms made of hair and like hoops of small cord; they sometimes must wear a kind of fillet about their heads for one of them had applied some part of an Old shirt which I had given them to this use.

BANKS: They brought with them a fish which they gave to us in return I suppose for the fish we had given them yesterday. Their stay was but short for some of our gentlemen being rather too curious in examining their canoe they went directly to it and pushing it off went away without saying a word.

END OF SECOND MEETING

12th July

COOK: At 2 o'clock in the a.m. the yawl came on board and brought 3 turtle and a large skeat and as there was a probability of succeeding in this kind of fishery I sent her out again after breakfast.

BANKS: Indians came again today and ventured down to Tupia's tent, where they were so well pleased with their reception that three staid while the fourth went with the canoe to fetch two new ones; they introduced their strangers (which they always made a point of doing) by name and had some fish given them. They received it with indifference, signed to our people to cook it for them, which was done, and they eat part and gave the rest to my Bitch [springer spaniel retriever 'Lady']. They staid the most part of the morning but never ventured to go above 20 yards [18 metres] from their canoe. The ribbands by which we had tied medals round their necks the first day we saw them were covered with smoke; I suppose they lay much in the smoke to keep off the Musquetos.

They are a very small people or at least this tribe consisted of very small people, in general about 5 feet 6 [167cm] in height and very slender; one we measured 5 feet 2 [157cm] and another 5 feet 9 [175cm], but he was far taller than any of his fellows.

PARKINSON: Their bones were so small, that I could more than span their ankles; and their arms too, above the elbow joint.

BANKS: I do not know by what deception we were to a man of opinion, when we saw them run on the sand about a quarter of a mile from us, that they were taller and larger than we were.

PARKINSON: Though of a diminutive size, they ran very swiftly, and were very merry and facetious.

BANKS: What their absolute colour is difficult to say, they were so completely covered with dirt, which seemed to have stuck to their hides from the day of their birth without their once having attempted to remove it; I tried indeed by spitting upon my finger and rubbing but altered the colour very little, which as nearly as might be resembled that of chocolate. The smoke and dirt with which they were all cased over, I suppose served them instead of cloths.

Their hair was strait in some and curled in others; they always wore it cropped close round their heads; it was of the same consistence with our hair, by no means woolly or curled like that of Negroes. Their eyes were in many lively and their teeth even and good; They were all of them clean limbed, active and nimble. Clothes they had none, not the least rag, those parts which nature willingly conceals being exposed to view completely uncovered; yet when they stood still they would often or almost always with their hand or something they held in it hide them in some measure at least, seemingly doing that as if by instinct.

PARKINSON: They became, at length, more free when only three of us were present, and made signs for us to take off some of our garments, which we did accordingly. They viewed them with surprize; but they seemed to have had no idea of clothes; nor did they express a desire for any; and a shirt, which we gave them, was found afterwards torn into rags.

BANKS: They painted themselves with white and red, the first in lines and bars on different parts of their bodies, the other in large patches.

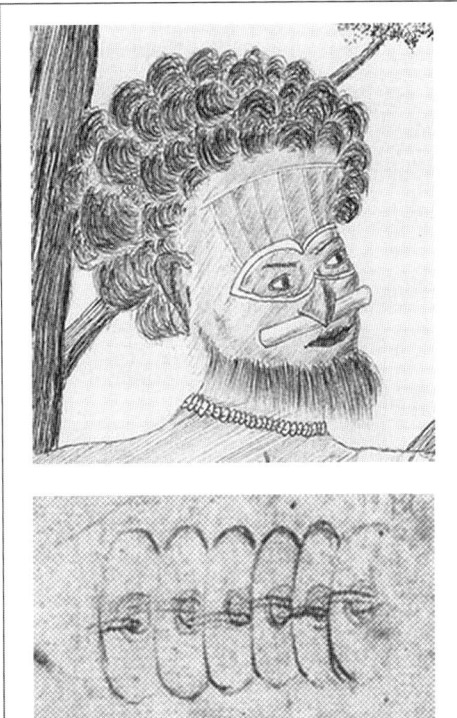

Sydney Parkinson's study of shell necklace.[5]

PARKINSON: On their breasts and hips were corresponding marks like ridges, or seams, raised above the rest of the flesh, which looked like the cicatrices of ill-healed wounds.

Some of them had necklaces made of oval pieces of bright shells, which lay imbricated over one another, and linked together by two strings [as in portrait and detail].

BANKS: Their ornaments were few: bracelets wore round the upper part of their arms, consisting of strings lapped round with other strings as what we call gymp in England, a string no thicker than a packthread tied round their bodies which was sometimes made of human hair, or that of the Beast called by them Kangooroo, a piece of Bark tied over their forehead, and the preposterous bone in their noses which I have before mentioned were all that we observed.

PARKINSON: They had also a bag that hung by their necks, which they carried shell-fish in.

BANKS: One had indeed one of his Ears bored, the hole being big enough to put a thumb through, but this was peculiar to that one man and him I never saw wear in it any ornament.

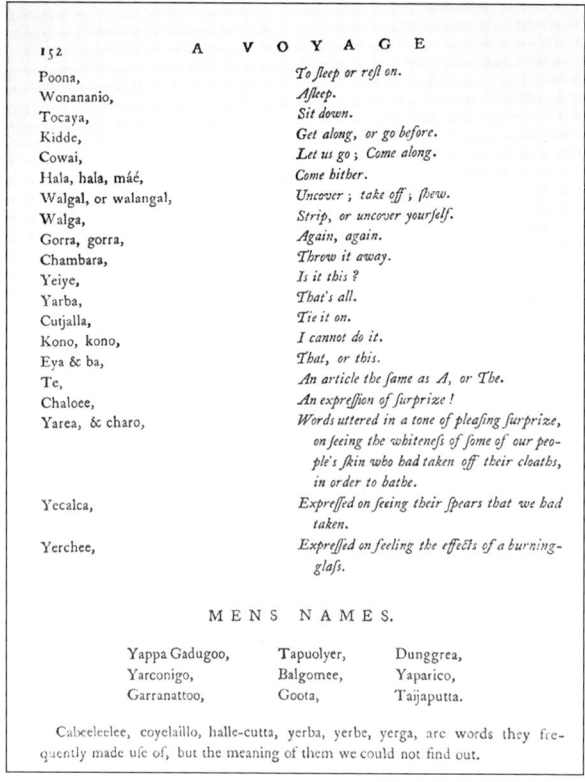

Part of a list of 250 Guguu Yimithirr words, and men's names collected by Sydney Parkinson at Endeavour River.[6]

Two of Them Embarked and Came Towards the Ship

BANKS: Their language was totally different from that of the Islanders; it sounded more like English in its degree of harshness tho it could not be called harsh neither.

PARKINSON: They articulated their words very distinctly, though, in speaking, they made a great motion with their lips, and uttered their words vociferously, especially when they meant to shew their dissent or disapprobation. When they were pleased, and would manifest approbation, they said Hee, with a long flexion of the voice, in a high and shrill tone.

BANKS: They almost continually made use of the word *Chircau*, which we conceived to be a term of admiration as they still used it whenever they saw anything new; also Cherr, tut tut tut tut tut, which probably have the same signification.

PARKINSON: They often said Tut, tut, many times together, but we knew not what they meant by it, unless it was intended to express astonishment. At the end of this Tut, they sometimes added Urr, and often whistled when they were surprised. As a mark of dissent, they said Aipa, several times, and this was the only word, that we could distinguish, to accord with the Otaheitean language [Tahitian].

BANKS: Their Canoe was not above 10 feet long and very narrow built, with an outrigger fitted much like those at the Islands only far inferior; they in shallow waters set her on with poles, in deep paddled her with paddles about 4 feet long; she just carried 4 people so that the 6 who visited us today were obliged to make 2 embarkations. [Among those plants gathered at Endeavour River was *Cochlospermum gillivraei*, a specimen of the flowering Kapok tree, the timber used by the Guugu Yimithirr to build their canoes.] (Colour Plate No. 50)

PARKINSON: To throw the water out of their canoes, they used a large shell called the Persian-crown [Common baler shell – *Melo amphora*]. (Colour Plate No. 51)

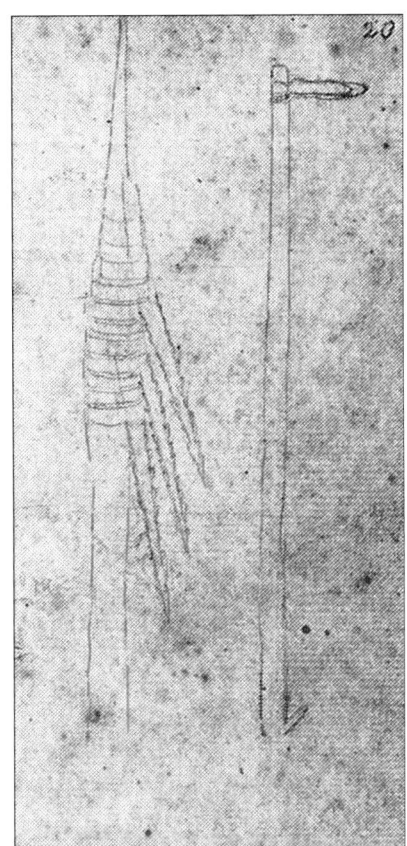

Sydney Parkinson's study of Endeavour River spear tip and spear thrower.[7]

BANKS: Their Lances were much like those we had seen in Botany Bay, only they were all of them single pointed, and some pointed with the stings of stingrays and bearded with two or three beards of the same, which made them indeed a terrible weapon; the board or stick with which they flung them was also made in a neater manner. [Among those plants collected at Endeavour River was *Xanthorrhoea resinosa*; the flowering stem was used by the Guugu Yimithirr to make spear shafts.] (Colour Plate No. 52)

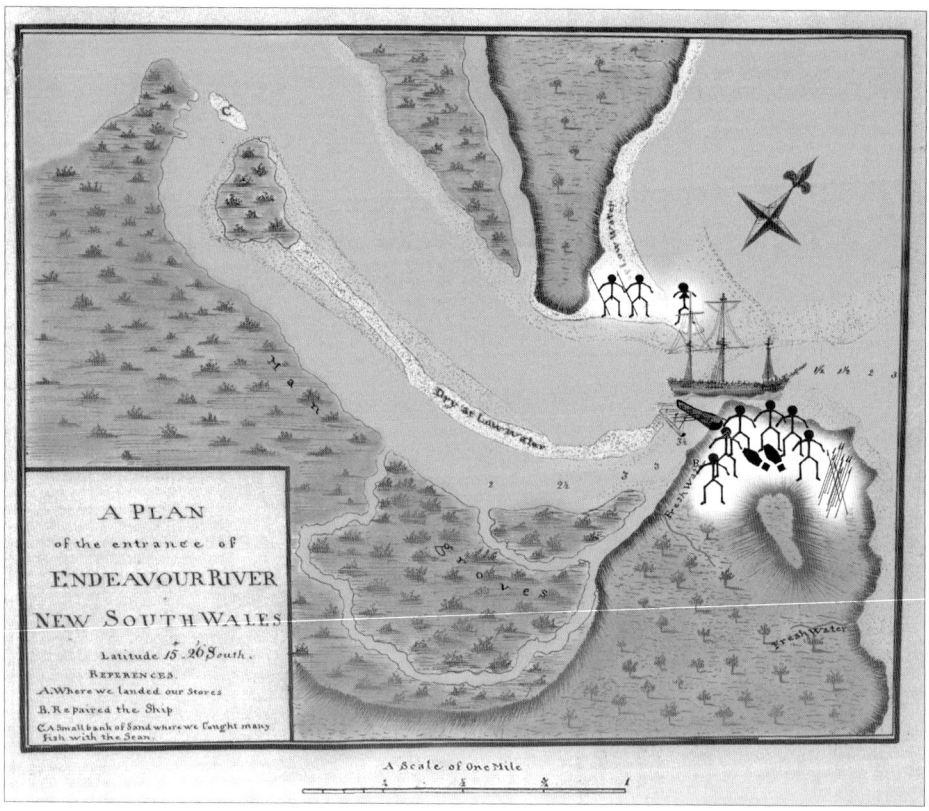

BANKS: After having staid with us the greatest part of the morning they went away as they came. While they staid 2 more and a young woman made their appearance upon the Beach.

COOK: On the point of sand on the other side of the River about 200 Yards from us, we could very clearly see with our glasses that the woman was as naked as ever she was born even those parts which I always before now thought Nature would have taught a woman to conceal were uncovered.

END OF THIRD MEETING

The Yawl by Sydney Parkinson.[8]

COOK: At noon the Yawl returned with one turtle and a large sting-ray.

13th July

COOK: In the p.m. got the bread aired, five hundred and sixty pounds of which was so bad that it could not be eat, and about the same quantity very much damaged, which I believe was owing to the casks being wet when the ship was on the rocks.

BANKS: Two Indians came in their Canoe to the ship, staid by her a very short time and then went along shore striking fish.

END OF FOURTH MEETING

John Gore.[9]

Eastern wallaroo – *Macropus robustus*.[10] (Representation only)

14th July

BANKS: Our second lieutenant [John Gore] who was a shooting today had the good fortune to kill the animal [a male Eastern wallaroo – *Macropus robustus*][11] that had so long been the subject of our speculations. To compare it to any European animal would be impossible as it has not the least resemblance of any one I have seen. Its fore legs are extremely short and of no use to it in walking, its hind again as disproportionally long; with these it hops 7 or 8 feet at each hop in the same manner as the Gerbua, to which animal indeed it bears much resemblance except in Size, this being in weight 38lb and the Gerbua no larger than a common rat.

COOK: The head neck and shoulders of this animal was very small in proportion to the other parts. The head and ears I thought was something like a Hare's. The tail was nearly as long as the body, thick next the rump and tapering towards the end. The fore legs were 8 inch long and the hind 22. The Skin is covered with a short hairy fur of a dark Mouse or Grey Colour.

Detail of Charles Praval's 'A view of the land about Endeavour River...'.[12] Copied by Charles Praval from a lost original by Parkinson or Spöring.

[It is probable John Gore shot their first animal, a male eastern wallaroo – *Macropus robustus*, on the rocky foothills of 'Gores Mount', which was named in Gore's honour (marked on coastal profile above). Phillip Parker King RN, later in 1819, renamed it 'Mount Cook' in honour of Cook.

Nathaniel Dance (1735–1811) made a painting (wash drawing) of the skull and lower jaw of the 38lb Eastern wallaroo, *Macropus robustus*, obtained by Cook's party from Endeavour River on 14th July 1770].[13] (Colour Plate No. 53)

PARKINSON: An animal of a kind nearly approaching the *mus* genus, about the size of a grey-hound, that had a head like a fawns; lips and ears, which it throws back, like a hares; on the upper jaw six large teeth; on the under one two only; with a short and small neck, near to which are the fore-feet, which have five toes each, and five hooked claws; the hinder legs are long, especially from the last joint, which, from the callosity below it, seems as if it lies flat on the ground when the animal descends any declivity; and each foot had four long toes, two of them behind, placed a great way back, the inner one of which has two claws; the two other toes were in the middle, and resembled a hoof, but one of them was much larger than the other. The tail, which is carried like a grey-hounds, was almost as long as the body, and tapered gradually to the end. The chief bulk of this animal is behind; the belly being largest, and the back rising toward the posteriors. The whole body is covered with short ash-coloured hair.

[The Eastern wallaroo's habits are predominantly nocturnal, which probably accounts for the ship's log entry for 3rd July 1770: 'A party of men having been out all night in quest of some beasts that we have seen frequently, returned without success.'][14]

15th July

BANKS: The Beast which was killed yesterday was today Dressed for our dinners and proved excellent meat.

PARKINSON: The flesh of it tasted like a hare's but has a more agreeable flavour.

COOK: Kangooroo, or Kanguru so called by the Natives.

Chapter 12

Our Very Good Friends

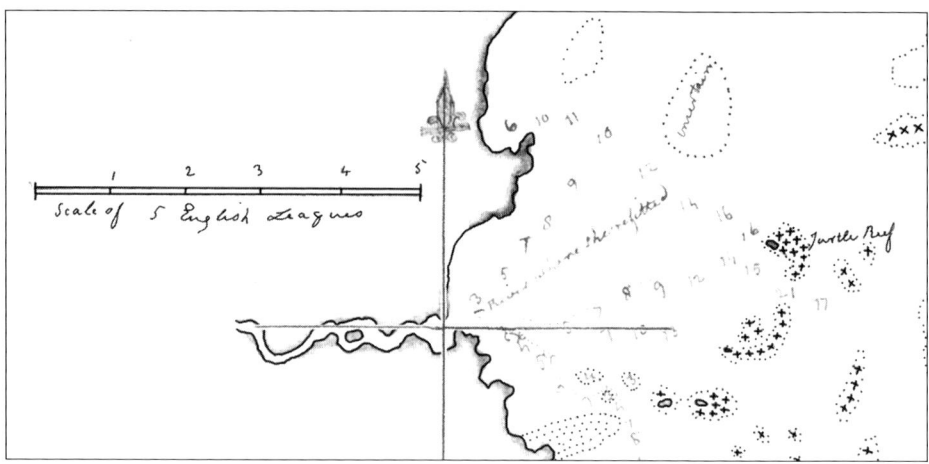

Turtle Reef (now Bolder Reef) where all their turtles were harvested.[1]

16th July

BANKS: In the evening the Boat returned from the reef bringing 4 Turtles, so we may now be said to swim in Plenty.

Our Turtles are certainly far preferable to any I have eat in England, which must proceed from their being eat fresh from the sea before they have either wasted away their fat, or by unnatural food which is given them in the tubs where they are kept given themselves a fat of not so delicious a flavour as it is in their wild state. Most of those we have caught have been green turtle from 2 to 300lb weight [Green turtle – *Chelonia mydas*]. (Colour Plate No. 54)

These when killed were always found to be full of Turtle Grass (a kind of Conferva I believe) [Turtle grass – *Thalassia hemprichii*]. (Colour Plate No. 55)

Two only were Loggerheads which were but indifferent meat; in their stomachs were nothing but shells [Loggerhead turtle – *Caretta caretta*]. (Colour Plate No. 56)

Our Very Good Friends 189

A VIEW of the Land about ENDEAVOUR RIVER taken when the entrance bore WSW distance 1 Mile

1770 and present-day North Shore Beach and Mount Saunders.[2]

COOK: In the a.m. I went upon one of the high hill on the north side of the River [Mount Saunders] from which I had an extensive view of the inland country which consisted of hills valleys and large plains, agreeably diversified with woods and lawns.

BANKS: As the ship was now nearly ready for her departure Dr Solander and myself employed ourselves in winding up our Botanical Bottoms [finish our botanical work], examining what we wanted, and making up our complement of specimens of as many species as possible.

17th July

COOK: In the evening the Pinnace returned with three turtle, two of which the Yawl caught and sent in.

BANKS: Tupia who was over the water by himself saw 3 Indians, who gave him a kind of longish roots about as thick as a man's finger and of a very good taste [Native yam – *Dioscorea transversa*]. (Colour Plate No. 57)

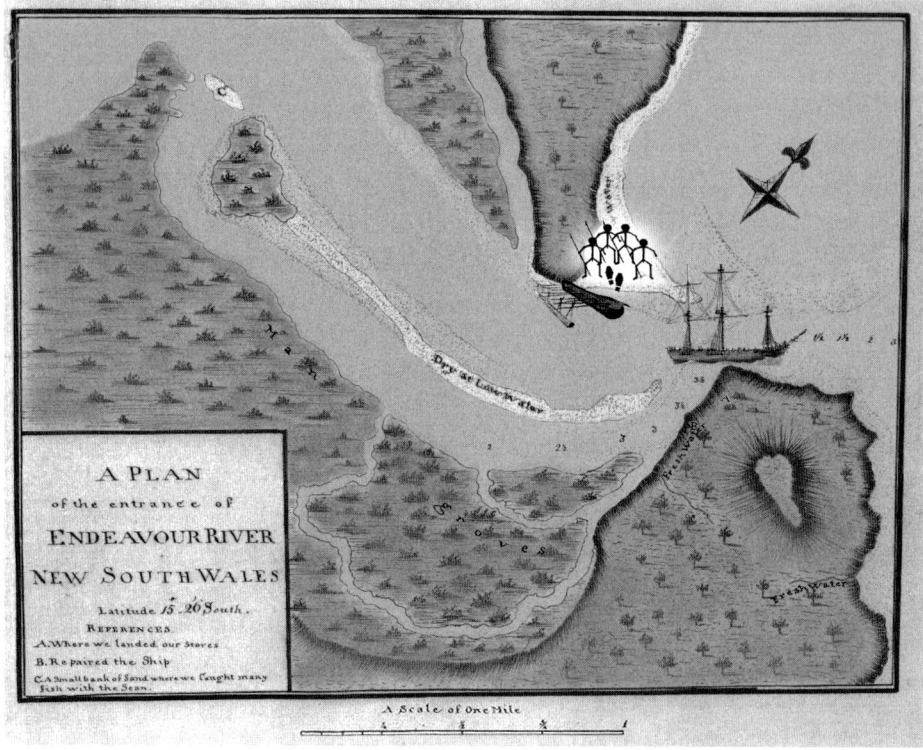

Cook's 'A Plan of the entrance of Endeavour River – New South Wales'.[3]

BANKS: On his return the Captn Dr Solander and myself went over in hopes to see them and renew our connections. We met with four in a canoe who soon after came ashore and came to us without any sign of fear.

COOK: Although we had not seen any of them before. Two of these wore necklaces made of shells which they seemed to Value as they would not part with them.

BANKS: After receiving the beads &c. that we had given them they went away; we attempted to follow them hoping that they would lead us to their fellows where we might have an opportunity of seeing their Women; they however by signs made us understand that they did not desire our company.

END OF FIFTH MEETING

18th July

COOK: In the p.m. I sent the Master and one of the mates in the Pinnace to the northward to look for a Channel that way clear of the shoals. Hove the ship into the stream a little more to prevent her taking the ground. In the evening the Yawl came in with three turtle and early in the a.m. she went out again. Employed getting everything on board. About 8 o'clock [a.m.] we were visited by several of the natives who now became more familiar than ever.

BANKS: They seemed to have lost all fear of us and became quite familiar; one of them at our desire threw his Lance which was about 8 feet in Length – it flew with a degree of swiftness and steadiness that really surprised me, never being above 4 feet from the ground and stuck deep in at the distance of 50 paces.

COOK: They will hit a Mark at the distance of 40 or 50 yards, with almost, if not as much certainty as we can do with a Musquet, and much more so than with a ball.

BANKS: After this they ventured on board the ship and soon became our very good friends, so the Captain and me left them to the care of those who stayed on board and went to a high hill about Six miles from the ship [Rocky Mount].

END OF SIXTH MEETING

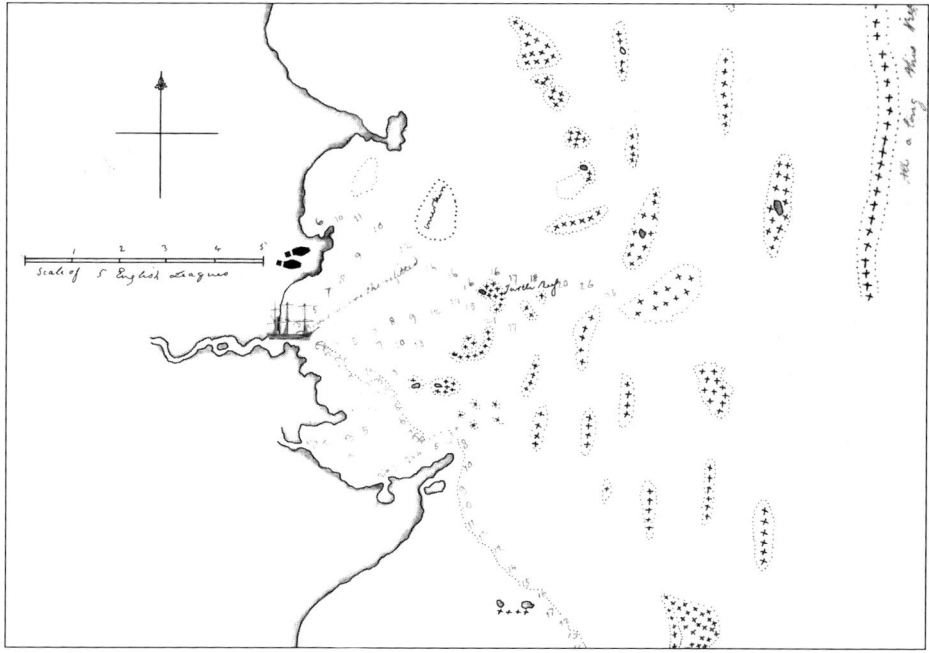

Portion of Master's Mate, Richard Pickersgill's chart – 'A draft of part of the shoals...'.[4]

BANKS: Here [Rocky Mount] we overlooked a great deal of sea to Leeward, which afforded a melancholy prospect of the difficulties we were to encounter when we came out of our present harbour: in which every direction we turned our eyes shoals innumerable were to be seen and no such thing as any passage to sea but through the winding channels between them, dangerous to the last degree.

COOK: After this we returned to the Ship and found several of the natives on board, at this time we had 12 turtle upon our decks which they took more notice of then anything else in the ship as I was told by the officers for their curiosity was satisfied before I got on board and they went away soon after.

Chapter 13

A Countenance Full of Disdain

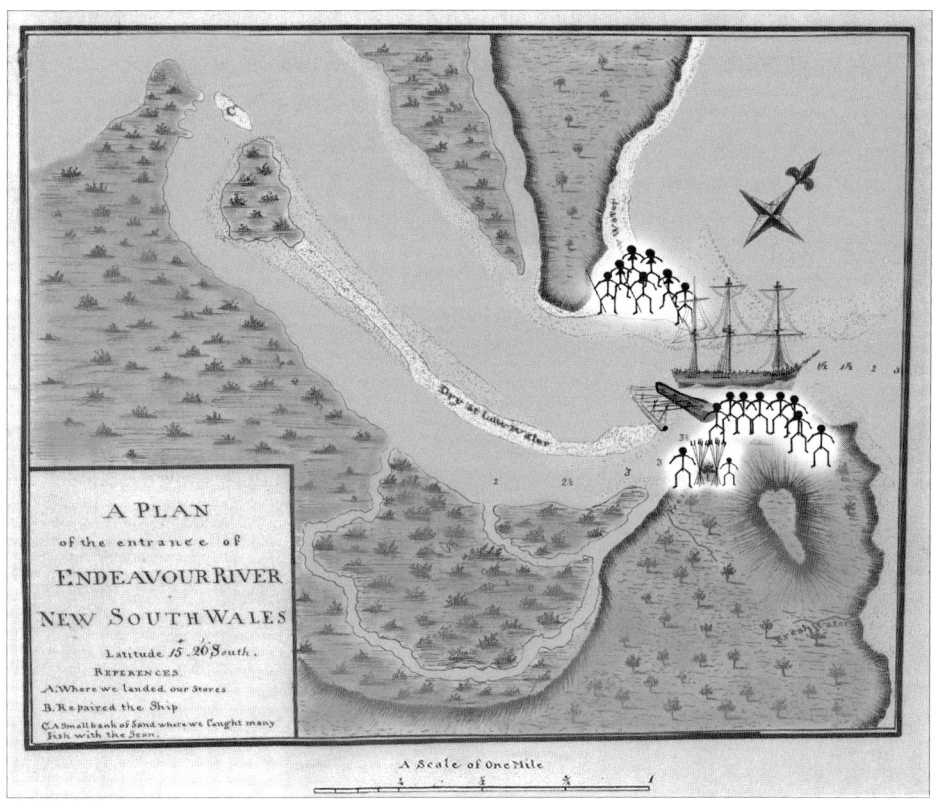

Cook's 'A Plan of the entrance of Endeavour River – New South Wales'.[1]

19th July

BANKS: Ten Indians visited us today and brought with them a larger quantity of Lances than they had ever done before, these they laid up in a tree leaving a man and a boy to take care of them and came on board the ship.

COOK: The most of them came from the other side of the River where we saw six or seven more, the most of them women and like the men quite naked.

194　The Endeavour Journals

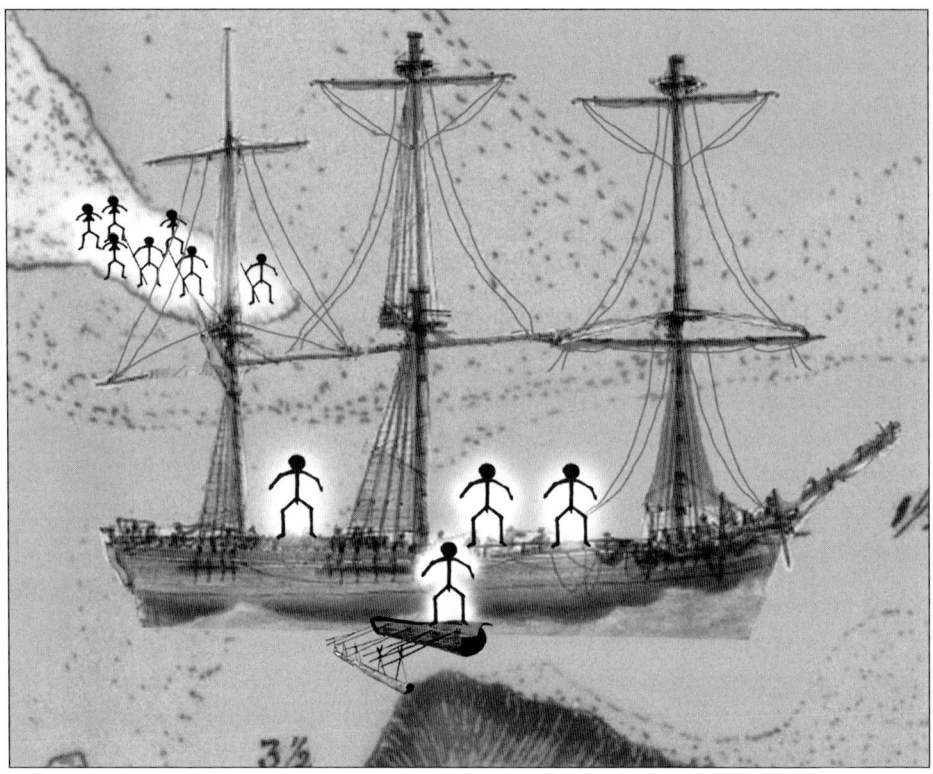

Detail of image above.

BANKS: They soon let us know their errand which was by some means or other to get one of our Turtle of which we had 8 or 9 laying upon the decks. They first by signs asked for One and on being refused shewd great marks of Resentment; one who had asked me on my refusal stamping with his foot pushed me from him with a countenance full of disdain and applied to someone else; as however they met with no encouragement in this they laid hold of a turtle and hauled him forwards towards the side of the ship where their canoe lay. It however was soon taken from them and replaced. They nevertheless repeated the experiment 2 or 3 times.

COOK: Being disappointed in this they grew a little troublesome, and were for throwing everything overboard they could lay their hands upon.

PARKINSON: They shewed a great antipathy to our tame birds, and attempted to throw one of them over-board.

COOK: As we had no victuals dressed at this time I offered them some bread to eat, which they rejected with scorn as I believe they would have done anything else excepting turtle.

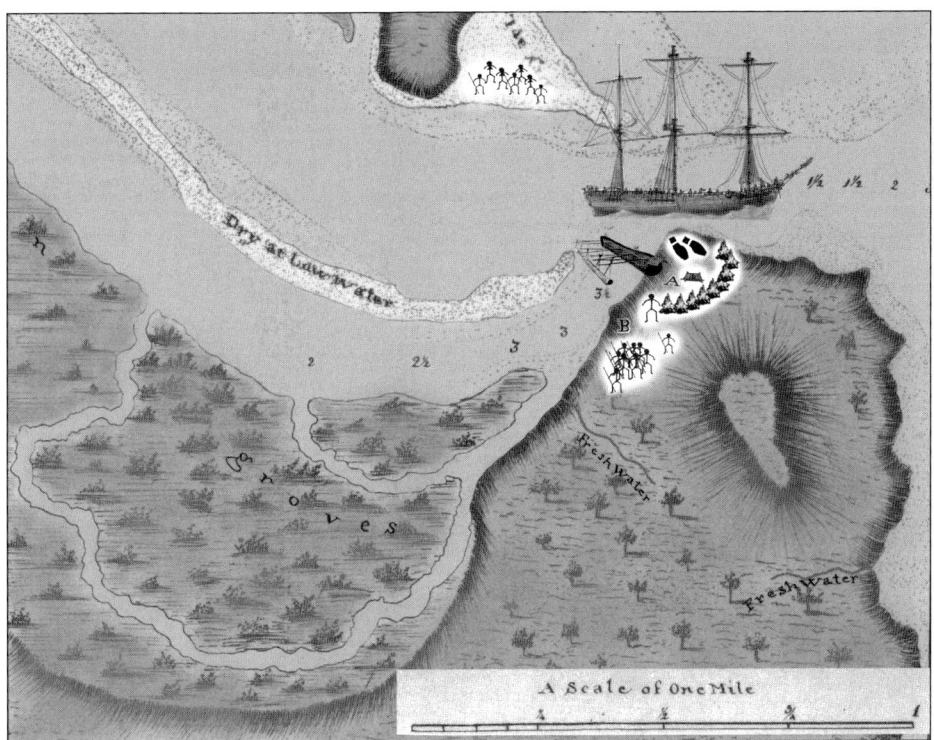

BANKS: After meeting with so many repulses all in an instant leaped into their Canoe and went ashore where I had got before them just ready to set out plant gathering.

COOK: Myself and five or six of our people being a shore at the same time, immediately upon their landing one of them took a handful of dry grass and lighted it at a fire we had a shore and before we well know'd what he was going about he made a large circuit round about us and set fire to the grass in his way and in an Instant the whole place was in flames.

BANKS: The grass which was 4 or 5 feet high and as dry as stubble burnt with vast fury.

COOK: Luckily at this time we had hardly anything ashore besides the forge and a sow with a Litter of young pigs one of which was scorched to death in the fire.

BANKS: A Tent of mine which had been put up for Tupia when he was sick was in the way of it so I leaped into a boat to fetch some people from the ship in order to save it, and quickly returning hauled it down to the beach Just time enough.

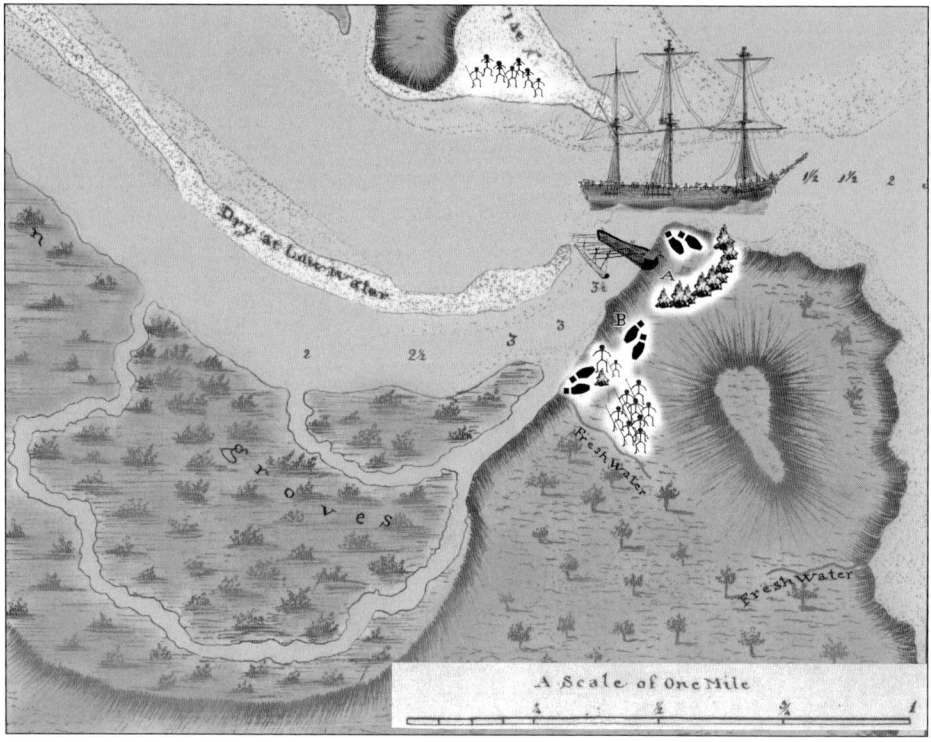

BANKS: The Captain in the meantime followed the Indians to prevent their burning our Linen and the seine which lay on the grass just where they were gone.

COOK: They all went to a place where some of our people were washing and where all our nets and a good deal of linen were laid out to dry, here with the greatest obstinacy they again set fire to the grass which I and some others who were present could not prevent.

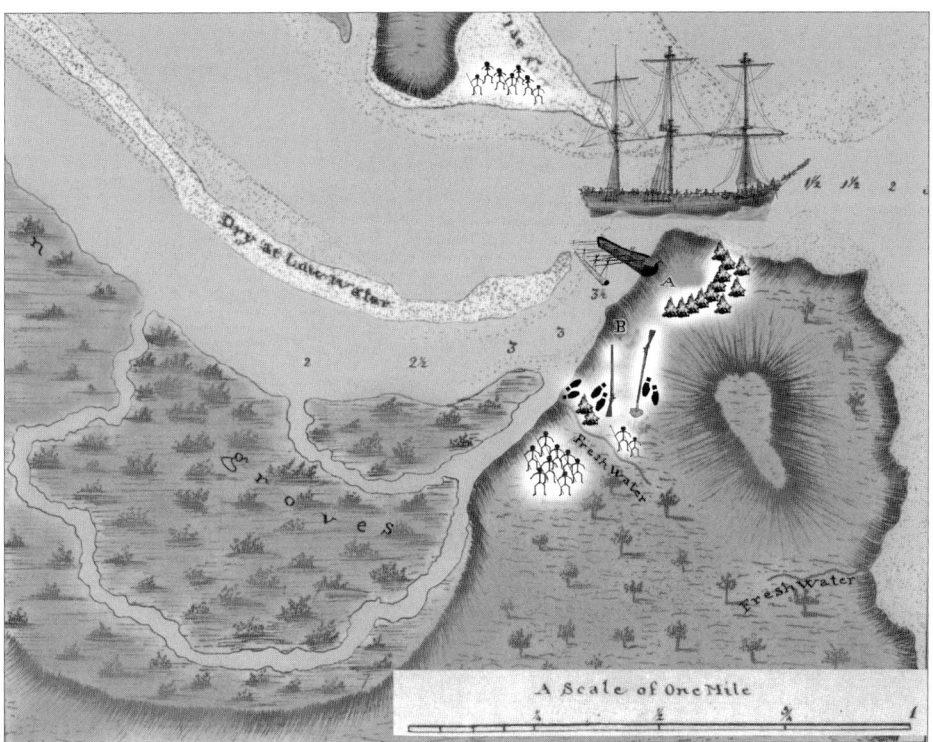

BANKS: He had no musket with him so soon returned to fetch one for no threats or signs would make them desist. Mine was ashore and another loaded with shot, so we ran as fast as possible towards them and came just time enough to save the seine by firing at an Indian who had already fired the grass in two places just to windward of it.

COOK: I was obliged to fire a musket load with small shot at one of the ring leaders.

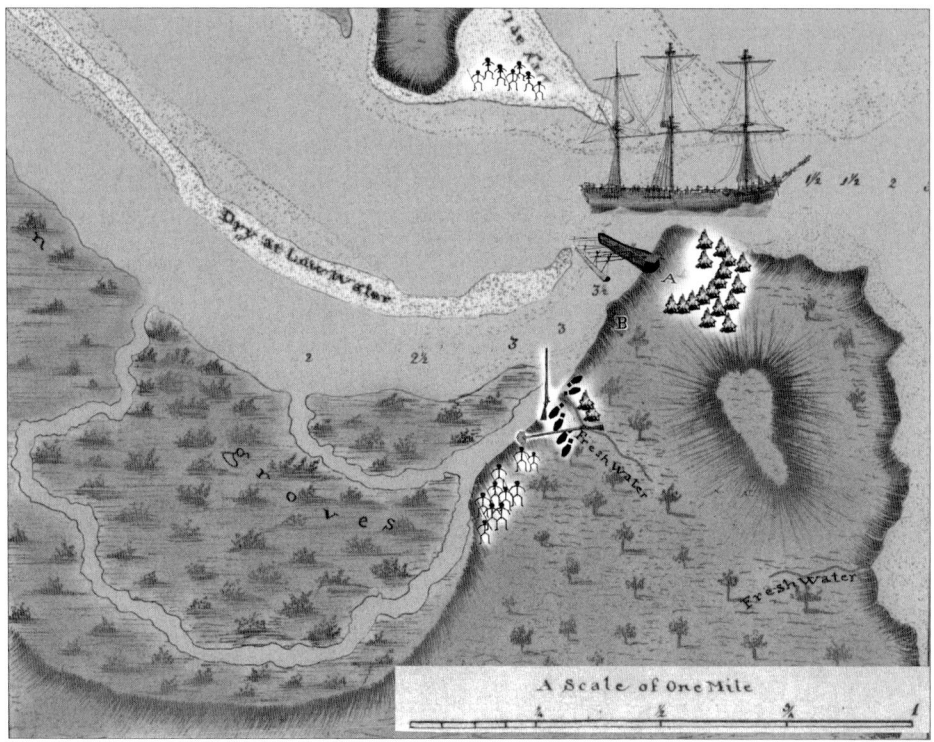

BANKS: On the shot striking him, tho he was full 40 yards from the Captain who fired, he dropped his fire and ran nimbly to his comrades who all ran off pretty fast.

The Captain then loaded his musquet with a ball and fired it into the Mangroves abreast of where they ran to shew them that they were not yet out of our reach.

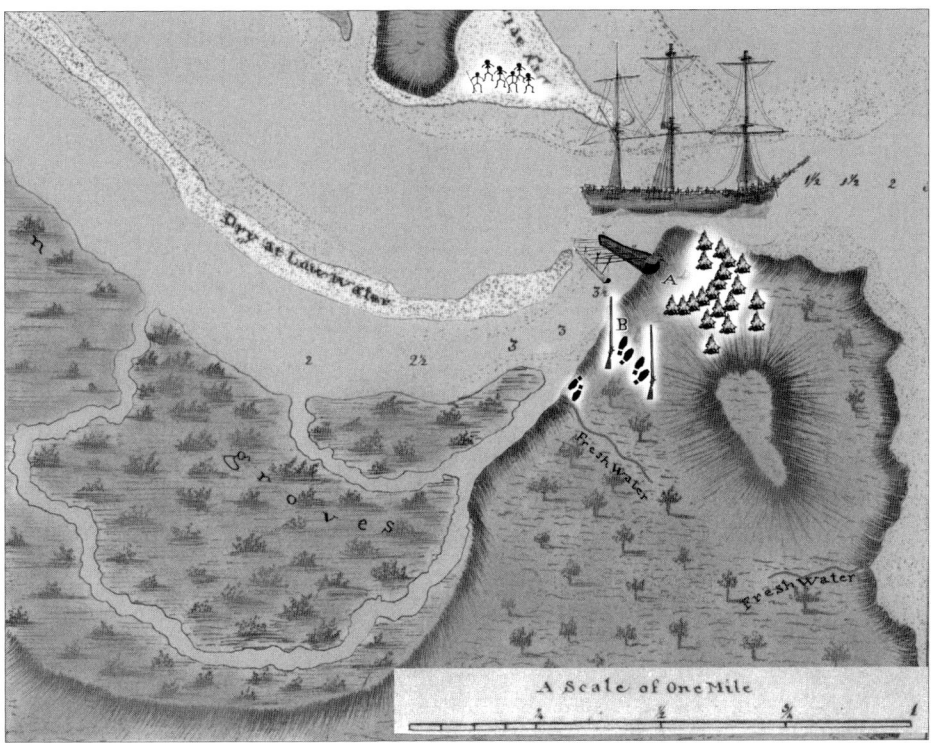

BANKS: They ran on quickening their pace on hearing the Ball and we soon lost sight of them. We then returned to the seine where the people who were ashore had got the fire under.

COOK: As we were apprised of this last attempt of theirs we got the fire out before it got head, but the first spread like wild fire in the woods and grass.

Notwithstanding my firing in which one must have been a little hurt because we saw a few drops of blood on some of the linen he had gone over, they did not go far from us for we soon after heard their voices in the woods.

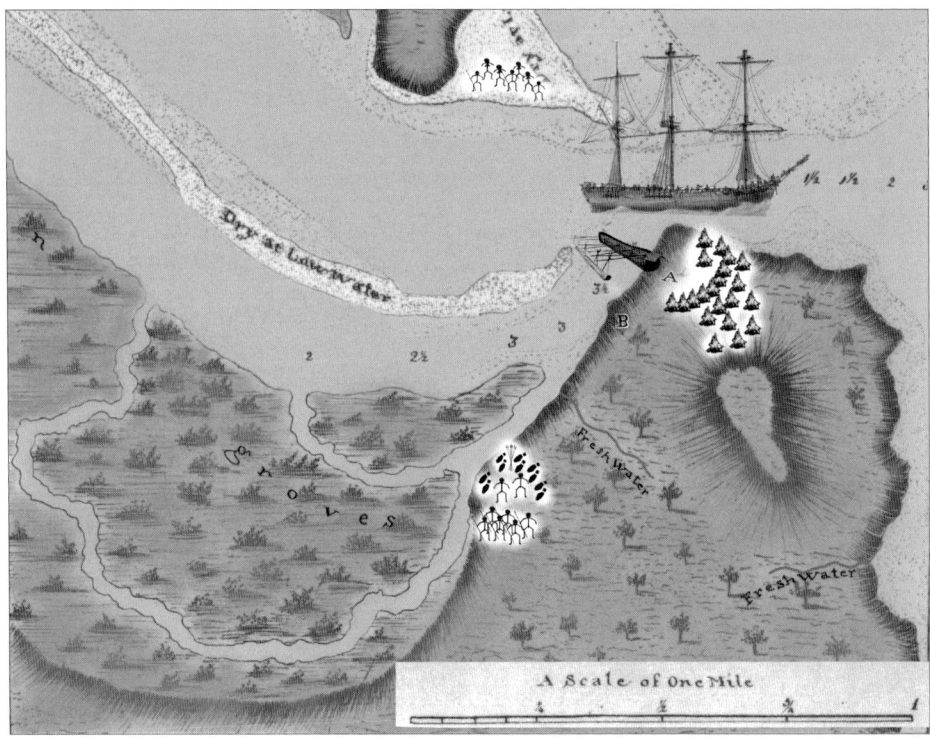

BANKS: We thought we were free'd from these troublesome people but we soon heard their voices returning on which, anxious for some people who were washing that way, we ran towards them.

COOK: Mr Banks and I and 3 or 4 More went to look for them and very soon met them coming toward us.

BANKS: On seeing us come with our musquets they again retired leisurely after an old man had ventured quite to us and said something which we could not understand.

COOK: As they had each 4 or 5 darts a piece and not knowing their intention we seized upon six or seven of the first darts we met with.

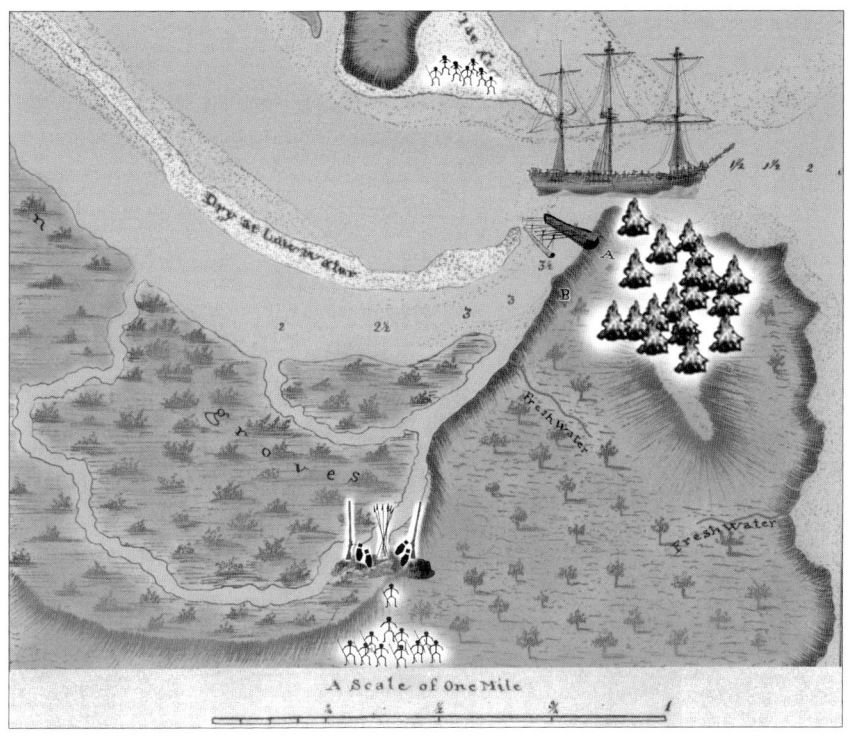

COOK: This alarmed them so much that they all made off and we followed them for near half a Mile.

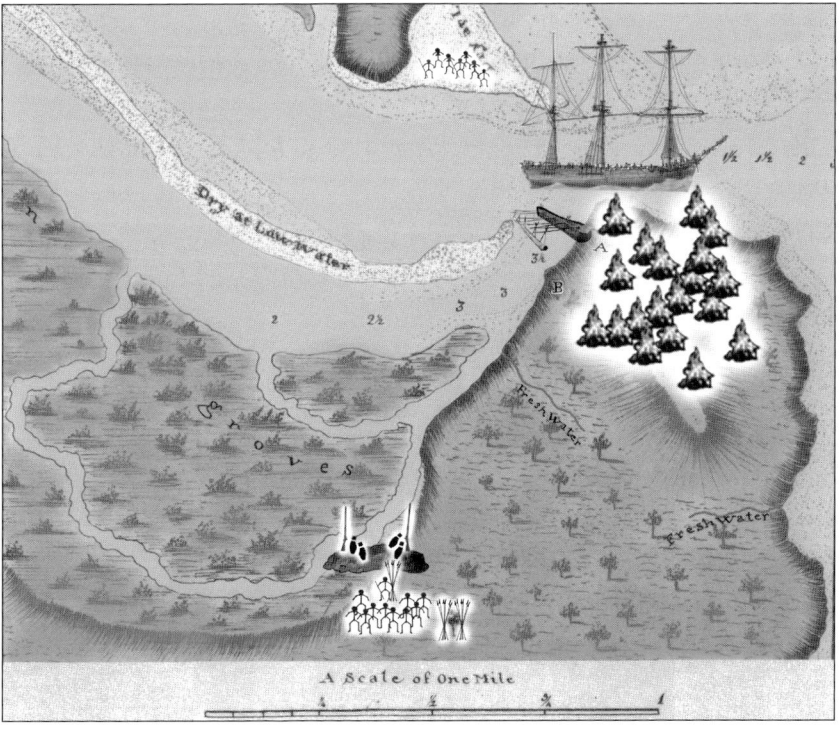

BANKS: Then meeting with some rocks from whence we might observe their motions we sat down and they did so too about 100 yards from us. The little old man now came forward to us carrying in his hand a lance without a point. He halted several times and as he stood employed himself in collecting the moisture from under his arm pit with his finger which he every time drew through his mouth.

BANKS: We beckoned to him to come: he then spoke to the others who all laid their lances against a tree and leaving them came forwards likewise and soon came quite to us.

COOK: They came to us in a very friendly manner. We now returned the darts we had taken from them which reconciled everything.*

BANKS: They had with them it seems 3 strangers who wanted to see the ship but the man who was shot at and the boy were gone, so our troop now consisted of 11.

COOK: The man which we supposed to have been struck with small shott could not be much hurt as he was at a great distance when I fired.

BANKS: The Strangers were presented to us by name and we gave them such trinkets as we had about us; then we all proceeded towards the ship, they making signs as they came along that they would not set fire to the grass again and we distributing musquet balls among them and by our signs explaining their effect.

When they came abreast of the ship they sat down but could not be prevailed upon to come on board, so after a little time we left them to their contemplations; they stayed about two hours and then departed.

COOK: And soon after set the woods on fire about a Mile and a half and two miles from us.

BANKS: We had great reason to thank our good Fortune that this accident happened so late in our stay, not a week before this our powder which was put ashore when first we came in had been taken on board, and that very morning only the store tent and that in which the sick had lived were got on board. I had little Idea of the fury with which the grass burnt in this hot climate, nor of the difficulty of extinguishing it when once lighted: this accident will however be a sufficient warning for us if ever we should again pitch tents in such a climate to burn Everything round us before we begin.

END OF SEVENTH MEETING

---

* Reconciliation Rocks were placed on the Queensland State Heritage Register in 2021. Acknowledged to be the first act of reconciliation in modern Australian history.

Chapter 14

Lumber Not Worth Carriage

20th July

COOK: In the p.m. got everything on board the Ship, new berth'd her and let her swing with the tide.

BANKS: No Indians came near us but all the hills about us for many miles were on fire and at night made the most beautiful appearance imaginable.

COOK: In the night the Master returned with the Pinnace and reported that there was no safe passage for the Ship to the northward.

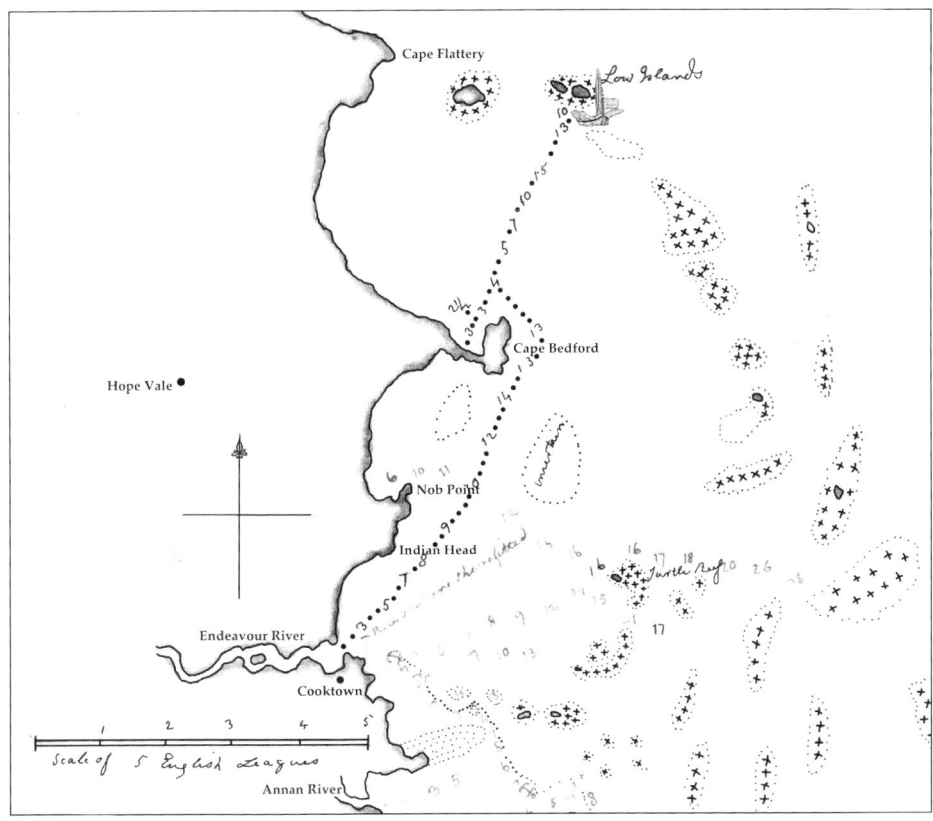

Molineux's passage Portion of Master's Mate Richard Pickersgill's chart 'A draft of part of the shoals...'.[1]

BANKS: The officer had met with nothing but shoals and not the least likelihood of a Passage that way, no very comfortable situation.

Our ship it is true was now repaired: Leaky she was from the strains she had got but the water she made was trifling. We were ready to sail with the first fair wind but where to go? – to windward was impossible, to leeward was a Labyrinth of Shoals, so that how soon we might have the ship to repair again or lose her quite no one could tell. Encounter the difficulty however we must and since our Bargain was a bad one make the Best of it.

COOK: At low water in the a.m. I went and sounded and buoy'd the bar, being now ready to put to sea the first opportunity.

21st July

COOK: Strong breezes at SE.

BANKS: No signs of the Indians today.

22nd July

COOK: Fresh breezes at SE and ESE. As the wind would not permit us to sail I sent the yawl out to catch turtle. In opening of one today we found sticking thro' both shoulder bones a wooden harpoon or turtle peg 15 Inches long bearded at the end such as we have seen among the natives, this proves to a demonstration that they strike turtle I suppose at the time they come ashore to lay their Eggs for they certainly have no boat fit to do this at sea or that will carry a turtle and this harpoon must have been a good while in as the wound was quite healed up.

23rd July

COOK: Fresh breezes in the SE quarter which so long as it continues will confine us in port.

Yesterday in the a.m. I sent some people into the country to gather greens (Colour Plate No. 58) one of which straggled from the rest and met with four natives by a fire on which they were broiling a fowl and the hind leg of one of the animals before spoke of.

BANKS: He came upon them as they sat down among some long grass on a sudden before he was aware of it.

COOK: He had the presence of mind not to run from them, being unarmed, least they should pursue him, but went and sit down by them.

BANKS: At first he was much afraid and offered them his knife, the only thing he had which he thought might be acceptable to them; they took it and after handing it from one to another returned it to him. They kept him about half an hour behaving most civilly to him, only satisfying their curiosity in examining his body.

COOK: They felt his hands and other parts of his body.

BANKS: … which done they made signs that he might go away.

COOK: … without offering the least insult.

BANKS: … which he did very well pleased.

COOK: … and perceiving that he did not go right for the ship, they directed him which way he should go.

Banksian cockatoo – *Calyptorhynchus magnificus magnificus* Parkinson's study – Endeavour River.[2]

BANKS: They had hanging on a tree by them, he said, a quarter of the wild animal and a cockatoo; but how they had been clever enough to take these animals is almost beyond my conception, as both of them are most shy especially the Cockatoos.

END OF EIGHTH MEETING

BANKS: In Botanizing today on the other side of the river we accidentally found the greatest part of the clothes which had been given to the Indians left all in a heap together, doubtless as lumber not worth carriage. Maybe had we looked farther we should have found our other trinkets, for they seemed to set no value upon any thing we had except our turtle, which of all things we were the least able to spare them.

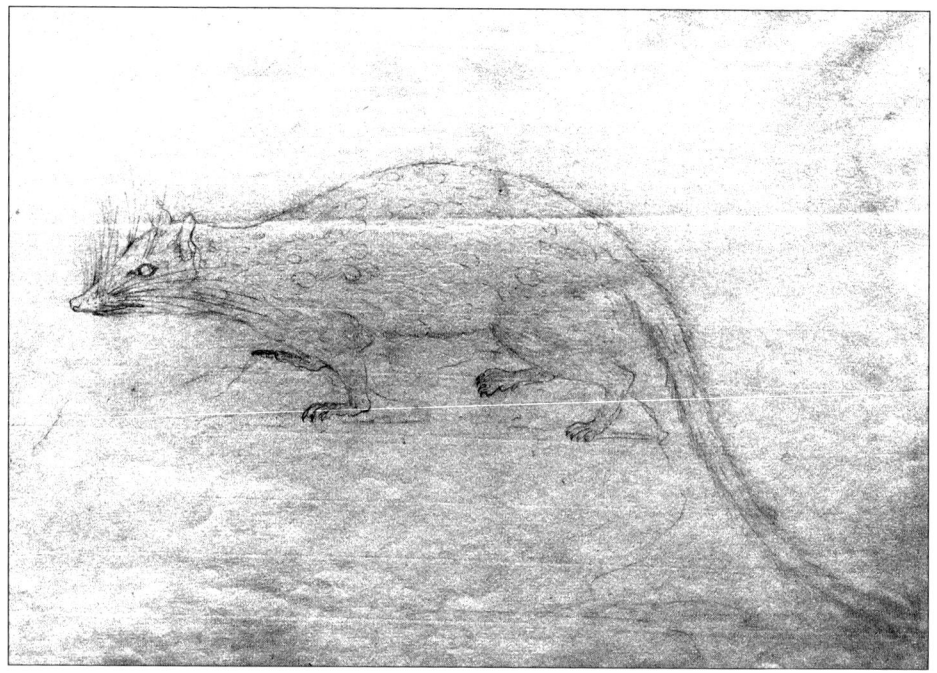

Quoll – *Dasyurus hallucatus* – unfinished drawing by Parkinson.[3]

[Specimen collected at Endeavour River in 1770. No exact date given.]

24th July

COOK: Wind and weather continues the same.

BANKS: The blowing weather which had hindered us from getting out several days still lasted, not at all to our satisfaction who had no one wish to remain longer in the place, which we had pretty well exhausted even of its natural history.

The Doctor and me were obliged to go very far for any thing new; to day we went several miles to a high hill where after sweating and broiling among the woods till night we were obligd to return almost empty. But the most vexatious accident imaginable befel us likewise: traveling in a deep vally, the sides of which were steep almost as a wall but coverd with trees and plenty of Brush wood, we found marking nuts (*Anacardium orientale*) laying on the ground [native cashew – now *Semecarpus australiensis*], (Colour Plate No. 59) and desirous as we were to find the tree on which they had grown, a thing that I believe no European Botanist has seen, we were not with all our pains able to find it; so after cutting down 4 or 5 trees and spending much time were obligd to give over our hopes.

[The juice of Marking Nuts makes an indelible black mark or stain on linen or other cloth – J.C. Beaglehole.]

25th July

COOK: Fresh gales at SE.

BANKS: The Captain who was up the river today found the Canoe belonging to our friends the Indians, which it seems they had left tied to some mangroves within a mile of the ship: themselves we could see by their fires were 5 or 6 miles off from us directly inland.

26th July

COOK: Wind and weather as yesterday.

Grey Queensland Ringtail Possum – *Pseudocheirus peregrinu*.[4] (Representation only)

BANKS: In botanizing to day I had the good fortune to take an animal of the Opossum (*Didelphis*) tribe [now *Pseudocheirus peregrinus*]: it was a female and with it I took two young ones.

27th July

COOK: Very fresh gales at SEbS.

Eastern Grey Kangaroo – *Macropus giganteus.*[5] (Representation only)

BANKS: This day was dedicated to hunting the wild animal. We saw several and had the good fortune to kill a very large one which weighed 84lb [a male Eastern grey kangaroo – *Macropus giganteus*].[6]

COOK: Mr Gore shott one which weighed 80lb [Banks says 84lb] and 54 exclusive of the entrails, skin and head, this was as large as the most we have seen.

Lumber Not Worth Carriage   209

'Kanguru' – unfinished pencil outline drawing of
a male animal springing by Sydney Parkinson.
Endeavour River.[7]

['Parkinson's drawing of the leaping kangaroo suggests he drew the adult male of 84lbs.' – Catalogue of the Natural History drawings commissioned by Joseph Banks on the *Endeavour* Voyage 1768 – 1771 Part 3: Zoology. Page 34. Natural History Museum, London.]

'Kanguru' – the second unfinished pencil outline drawing of a male animal crouched in a rocky landscape by Sydney Parkinson. Endeavour River.[8]

[The two different postures depicted (springing and crouched) suggest Parkinson intended to represent the movement of the animal].

28th July

BANKS: Botanizing with no kind of success. The Plants were now entirely completed and nothing new to be found, so that sailing is all we wish for if the wind would but allow us. Dined today upon the animal, who eat but ill, he was I suppose too old. His fault however was an uncommon one, the total want of flavour, for he was certainly the most insipid meat I eat.

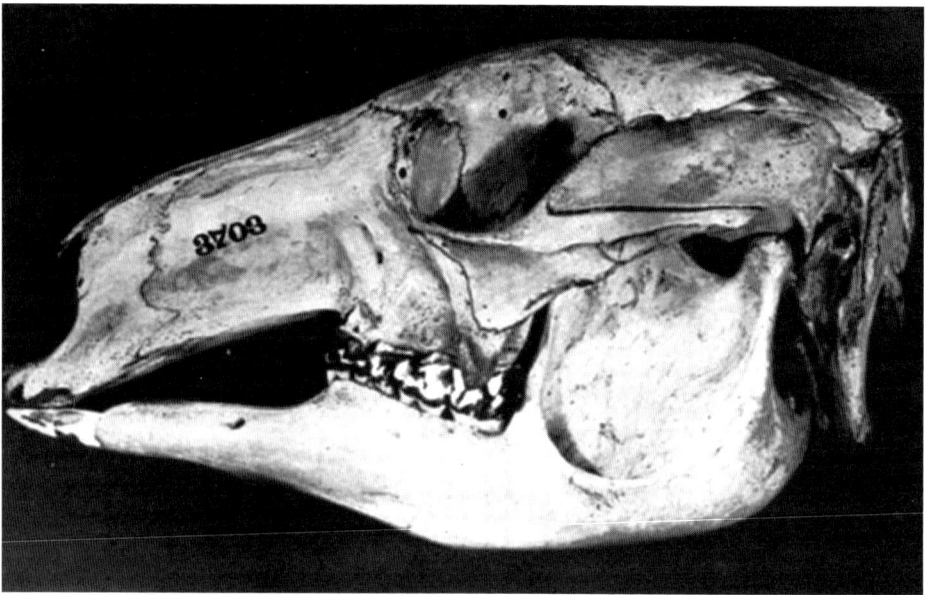

A photograph of the skull of the 84lb Eastern grey kangaroo, *Macropus giganteus*. Specimen obtained by Cook's party at Endeavour River on 27th July 1770. The skull specimen itself was destroyed by bombing in the Second World War.[9]

29th July

COOK: About 6 o'clock [a.m.] the sea breeze set in again which put an end to our sailing this day.

Eastern Grey kangaroo – *Macropus giganteus.*[10] (Representation only)

BANKS: Went out again in search of the animals: our success today was not however quite so good as the last time, we saw few and killed one very small one which weighed no more than 8 and a ½lb [a female juvenile Eastern grey kangaroo – *Macropus giganteus*].[11] My greyhound took him with ease tho the old ones where much too nimble for him.

'Possible the skins and skulls of all three specimens were retained. One of the smaller specimens was skinned and stuffed …' (Catalogue of the Natural History drawings commissioned by Joseph Banks on the *Endeavour* Voyage 1768 – 1771 Part 3: Zoology. Pages 28 and 34. Natural History Museum, London.)

Oliver Goldsmith (1791) in his An History of the Earth and Animated Nature, Volume 4, pp. 351–353 wrote 'the skin of that (kangaroo) which was stuffed and brought home by Mr Banks was not much above the size of a hare…'

H.B. Carter in his Sir Joseph Banks 1743–1820. British Museum (Natural History). Chapter 4, pp. 90–91 writes 'This circumstantial evidence strongly points to the young (8 & a half lb) female caught by Banks's greyhound on the 29th July as the material on which the Stubbs's painting was based and supplemented, of course, by Parkinson's pencil sketches in consultation with Banks who was the original collector of the stuffed specimen.' (Colour Plate No. 60)

30th July

BANKS: Ever since the ship was hauled off for sailing we have had Blowing weather till today, when it changed to little wind and rain which gave us some hopes; in the evening however the wind returned to its old Bias.

31st July

COOK: At 2 o'clock in the a.m. I had thoughts of trying to warp the ship out of the harbour but upon going first in a boat I found it blowed too fresh for such an attempt.

1st August

BANKS: The day was rainy with less wind but still not moderate enough for our undertakings.

2nd August

COOK: Wind and weather as yesterday or rather more stormy.

3rd August

BANKS: In the morn our people were dubious about trying to get out and I believe delayed it rather too long. At last however they began and warped ahead but desisted from their attempts after having run the ship twice ashore.

4th August

COOK: In the p.m. having pretty moderate weather I ordered the Coasting anchor and cable to be laid without the bar to be ready to warp out, that we might not lose the least opportunity that might offer, for laying in Port spends time to no purpose, consumes our provisions of which we are very Short in many articles, and we have yet a long Passage to make to the East Indies through an unknown and perhaps dangerous Sea; these circumstances considered, makes me very anxious of getting to sea.

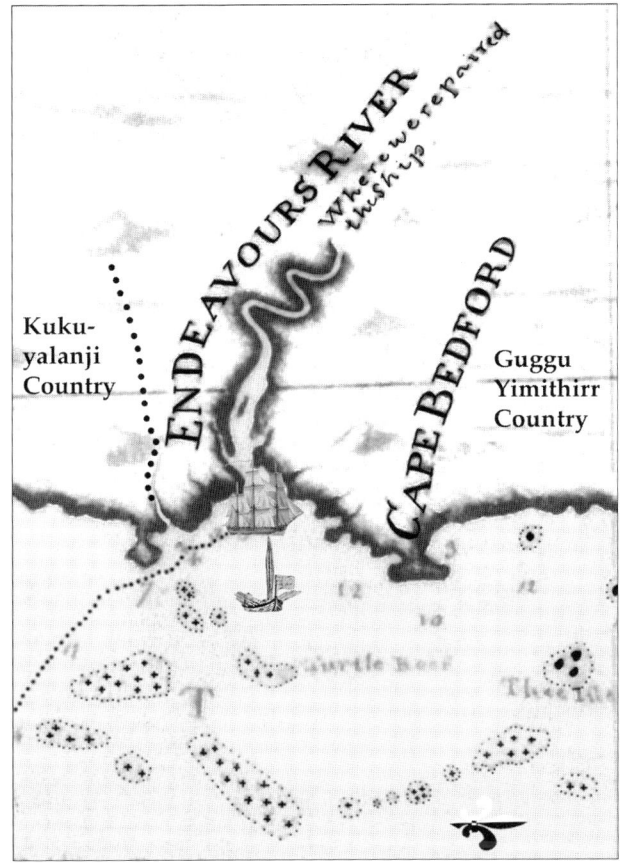

Portion of Cook's chart 'Cape Tribulation to *Endeavours* Streights'.[12]

COOK: The wind continued moderate all night and at 5 o'clock in the morning when it fell calm, gave us an opportunity to warp out. About 7 we got under sail having a light air from the land which soon died away and was succeeded by the Sea breeze from SEBS with which we stood off to Sea EBN having the Pinnace ahead sounding.

Chapter 15

Cape Flattery

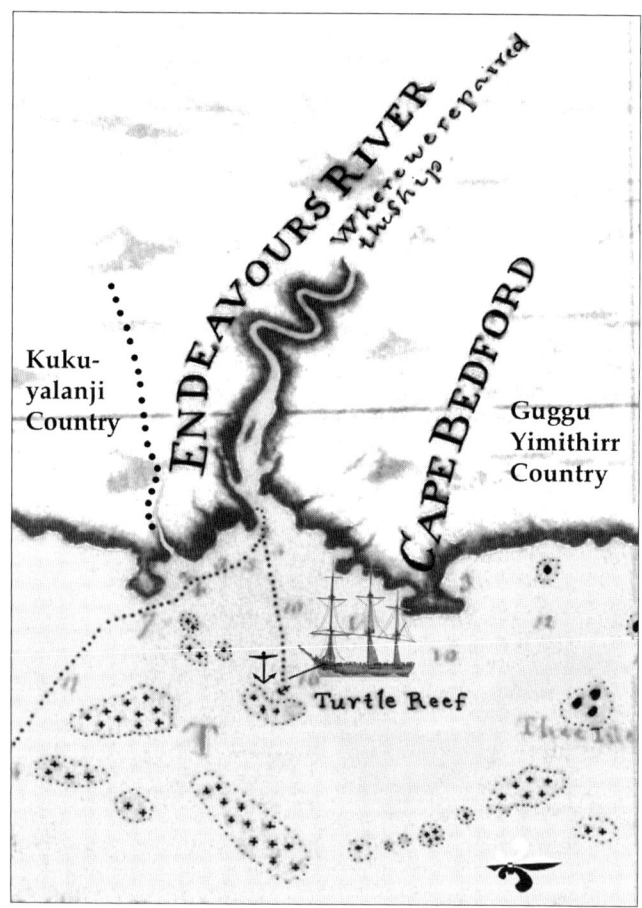

Portion of Cook's chart 'Cape Tribulation to *Endeavours* Streights'.[1]

4th August (cont'd)

BANKS: We sailed right out till we came to the turtle reef [now Boulder Reef].

COOK: We **anchored** for I did not think it safe to run in among the Shoals until I had well viewed them at low-water from the Mast head, that I might be better able to judge which way to steer for as yet I had not resolved whether I should beat back to the Southward round all the shoals or seek a passage to the Eastward or to the north all of which appeared to be equally difficult and dangerous.

Cape Flattery

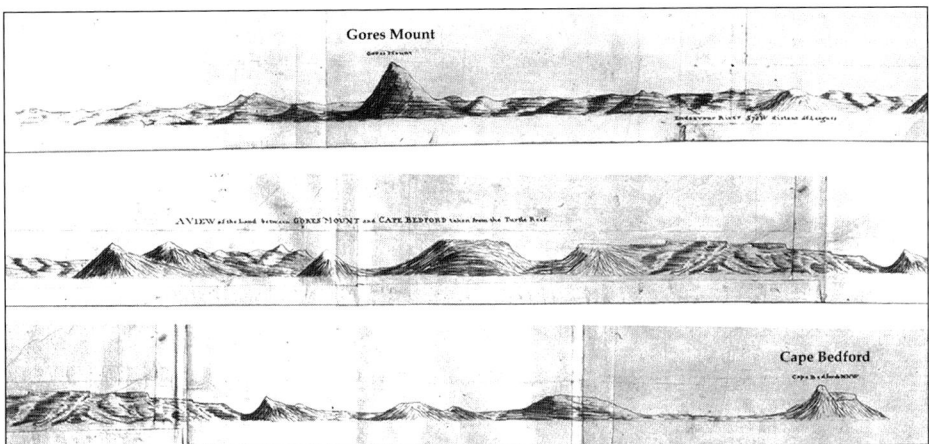

'A VIEW of the land between GORES MOUNT and CAPE BEDFORD taken from Turtle Reef' Charles Praval – copied from a lost original by Parkinson or Spöring.[2]

COOK: The Main land we had in sight I named <u>Cape Bedford</u>. We could see land to the NE of this Cape which made like two high Islands, the Turtle banks bore East distant 1 Mile.

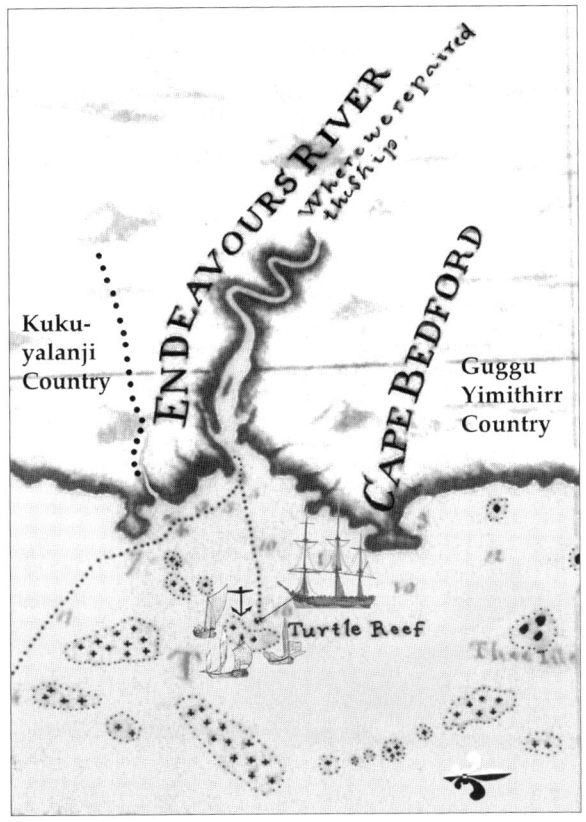

5th August

COOK: At Low-water from the Mast head I took a view of the shoals and could see several laying a long way without this one, a part of several of them appearing above water, but as it appeared pretty clear of shoal to the NE of the turtle reef, I came to a resolution to stretch out that way close upon a wind, because if we found no passage we could always return back the way we went.

As I did not intend to weigh until the Morning I sent all the boats to the reef to get what Turtle and shell fish they could. In the evening the boats returned with one turtle a sting-ray and as many large Clams as came to one and a half pound a man, in each of these Clams were about two pounds of meat.

BANKS: At night our people who fished caught an abundance of sharks.

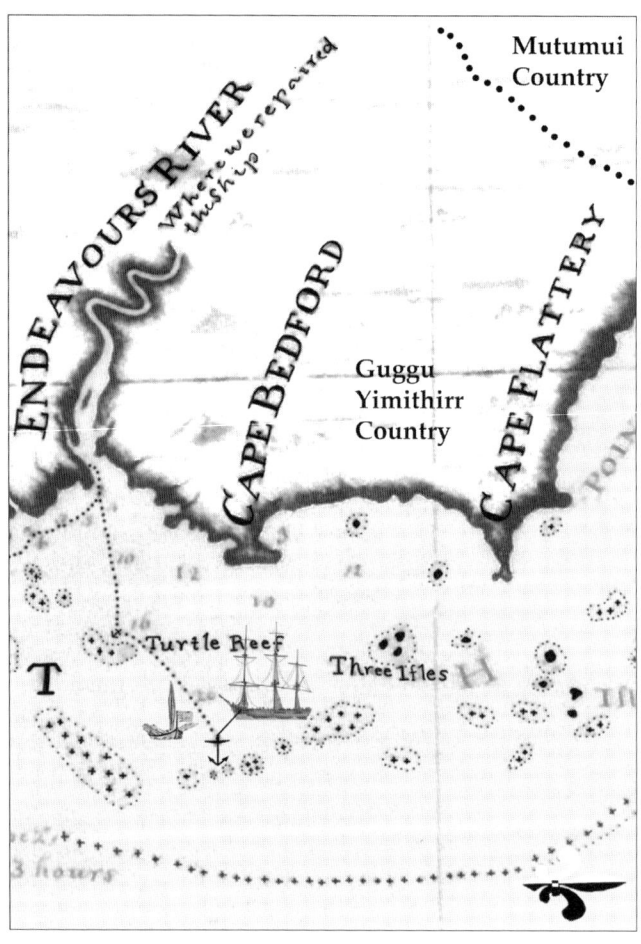

6th August

COOK: At 2 o'clock in the p.m. we got under sail, having the pinnace ahead sounding. We had not stood out long before we discovered shoals ahead and on both bows. As night was approaching I thought it safest to **anchor** [anchorage marked on chart].

In the morning we had a strong gale from the SE that instead of weighing as we intended we were obliged to bear away more Cable and to strike topgallant yards.

BANKS: It Blew so fresh that we could not move but lay still all day, not without anxiety least the anchor should not hold.

7th August

COOK: At Low-water in the p.m. I and several of the officers kept a look out at the Mast head to see for a passage between the Shoals but we could see nothing but breakers extending out to sea as far as we could see. We were surrounded on every side with shoals and no such thing as a passage to sea but through the winding channels between them dangerous to the highest degree. I was quite at a loss which way to steer for to beat back to the SE the way we came as the Master would have had me done would be an endless piece of work as the winds blow now constantly strong from that quarter without hardly any intermission. On the other hand if we do not find a passage to the northward we shall have to come back at last.

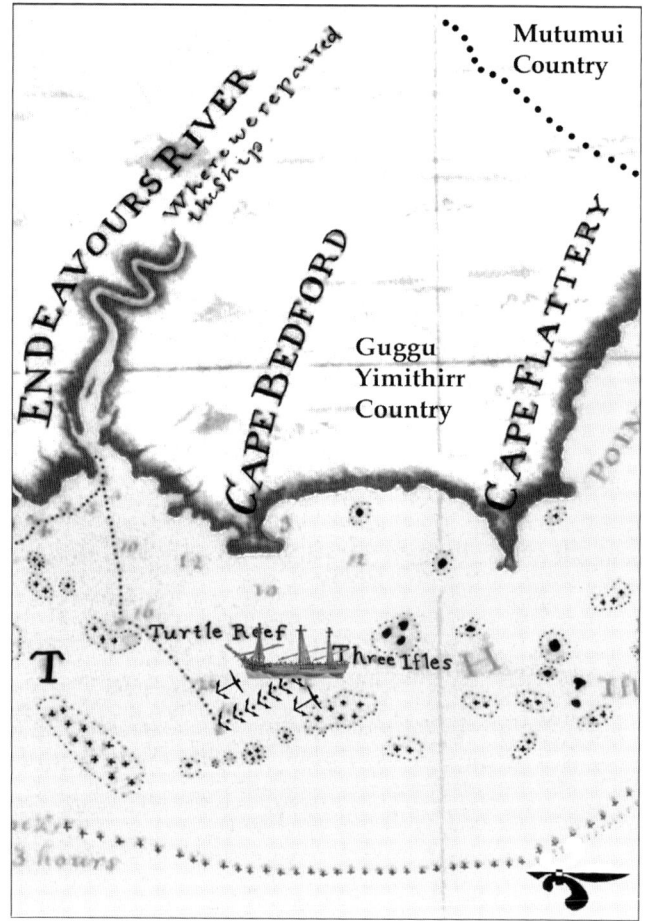

BANKS: During last night the gale had freshened much and in the morn we found that we had Drove above a League. Fortunately no shoal had in that distance taken us up but one was in sight astern and the ship drove fast towards it, on this another **anchor** was let go [the small bower anchor] and much cable veered out [1,500 feet/457 metres] but even this would not stop her. Our prospect was now more melancholy than ever: the shoal was plainly to be seen and the ship still driving gently down towards it, a sea running at the same time which would make it impossible ever to get off if we should be unfortunate enough to get on. Yards and Topmasts were therefore got down and everything done which could be thought of to make the ship snug, without any effect: she still drove and the shoal we dreaded came nearer and nearer to us. The sheet anchor our last resource was now thought of and prepared, but fortunately for us before we were drove to the making use of that expedient the ship stopped and held fast, to our great joy.

SHIP'S LOG: After this she rid fast; but least she should start again we got the Sheet and Coasting anchors [stream and kedge anchors] over the side ready for letting go for there were breakers not far to leeward which made our situation none the best. [For particulars about *Endeavour*'s anchors see glossary 'Anchors'.]

BANKS: During the time of its blowing yesterday and today we became certain that between us and the open sea was a ledge of rocks or reef just the same as we had seen at the Islands, no very agreeable discovery, for should that at any time join in with the main land we must wait for another season when different winds from the present ones prevailed; in which case we must infallibly be short of provisions or, if the turtle should fail us, Salt provisions without bread was all we had to trust to.

8th August

COOK: Strong gales at SSE all this day in so much that I dirst not get up yards and Topmasts.

BANKS: The night Dark as pitch passed over not without much anxiety: whether our anchors held or not we could not tell and maybe might when we least thought of it be upon the very brink of destruction. Day light however releivd us shewd us that the anchors had held and also brought us rather more moderate weather.

9th August

COOK: In the p.m. the weather being something moderater we got up the Topmast, but kept the Lower Yards down. At 6 o'clock in the Morning we began to heave in the Cable thinking to get under sail, but it blowed so fresh together with a head sea that we could hardly heave the Ship ahead and at last was obliged to desist.

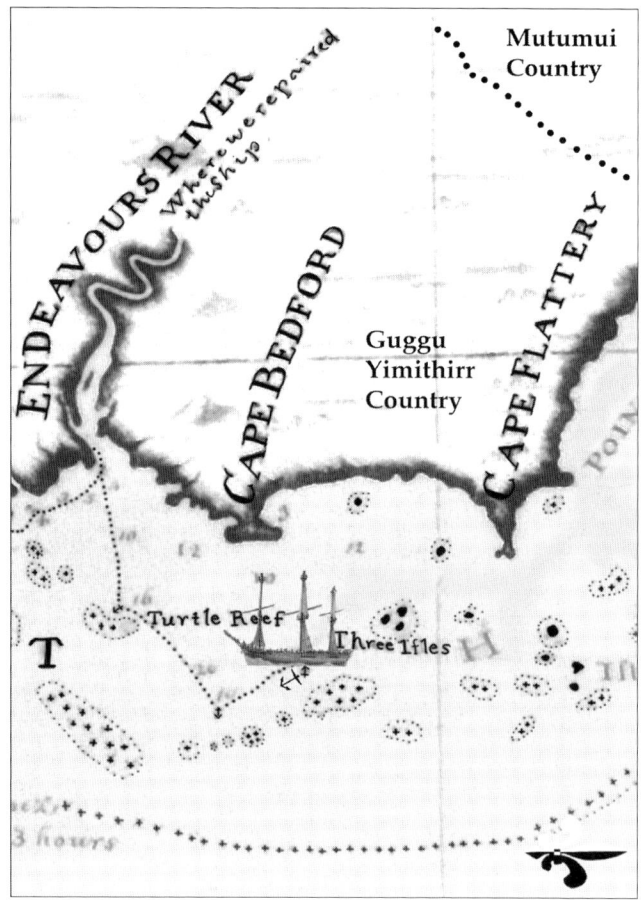

10th August

COOK: In the p.m. the wind fell so as we got up the small bower Anchor and hove in to a whole Cable on the best bower.

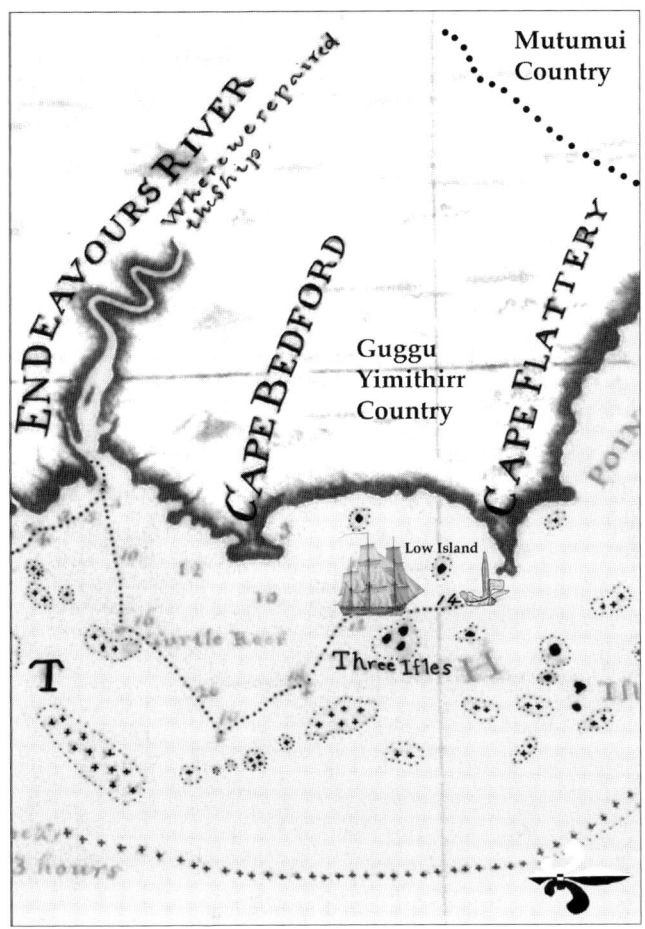

COOK: At 3 o'clock in the morning we got up the lower yards and at 7 weighed and stood in for the land intending to seek a passage along shore to the northward having a Boat ahead sounding.

After standing in an hour we edged away for 3 small Islands [Three Isles] that lay NNE ½ E 3 Leagues from Cape Bedford, to these Islands the Master had been in the Pinnace when the ship was in Port. [Marked 'Low Islands' on Pickersgill's chart. See page 203]

At 9 o'clock we were abreast of them [Three Isles] and between them and the Main having another low Island between us and the latter. In this Channel had **14** fathom water: the northermost Point of the Main we had in sight bore from us NNW ½ W distant 2 Leagues [Cape Flattery].

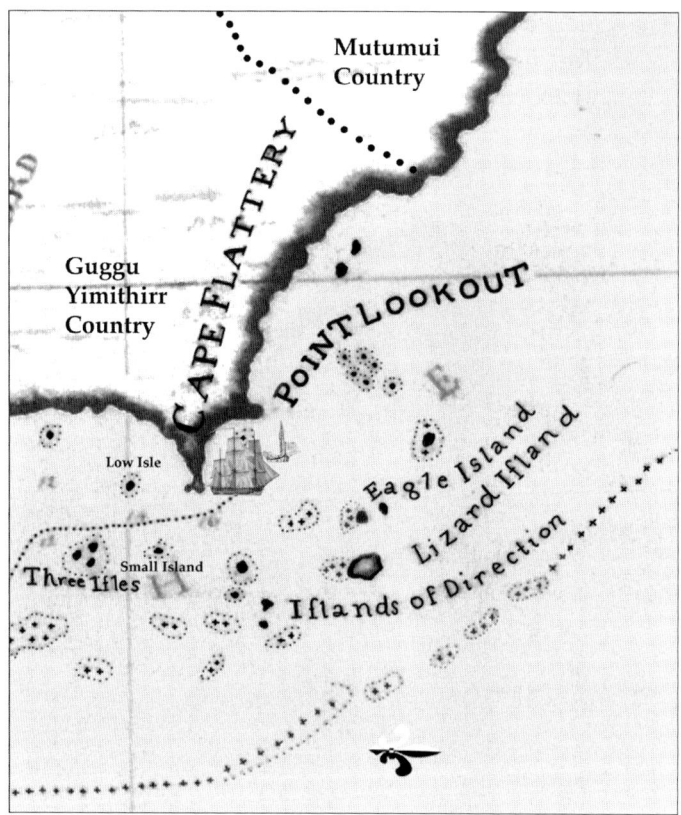

COOK: Four or 5 Leagues to the NE of this headland [Cape Flattery] appeared three high Islands [Lizard Group] with some smaller ones near them [Isles of Direction] and the Shoals and reefs without us we could see extending to the northward as far as these Islands: we directed our Course between them [the shoals] and the above headland leaving a small Island to the eastward of us, all the while a boat ahead sounding.

At Noon we were got betwixt the headland [Cape Flattery] and the 3 High Islands [Lizard Group].

A VIEW of CAPE FLATTERY bearing West 1 League

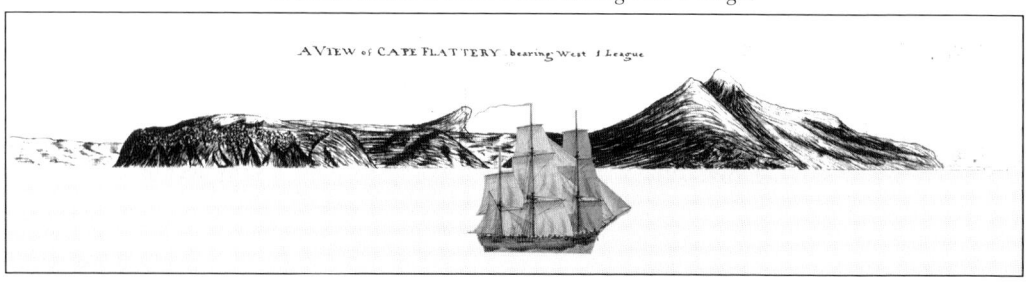

Charles Praval – copied from a lost original by Parkinson or Spöring.[3]

COOK: We now judged ourselves to be clear of all danger having as we thought a clear open sea before us, but this we soon found otherwise and occasioned my calling the headland above mentioned <u>Cape Flattery</u>.

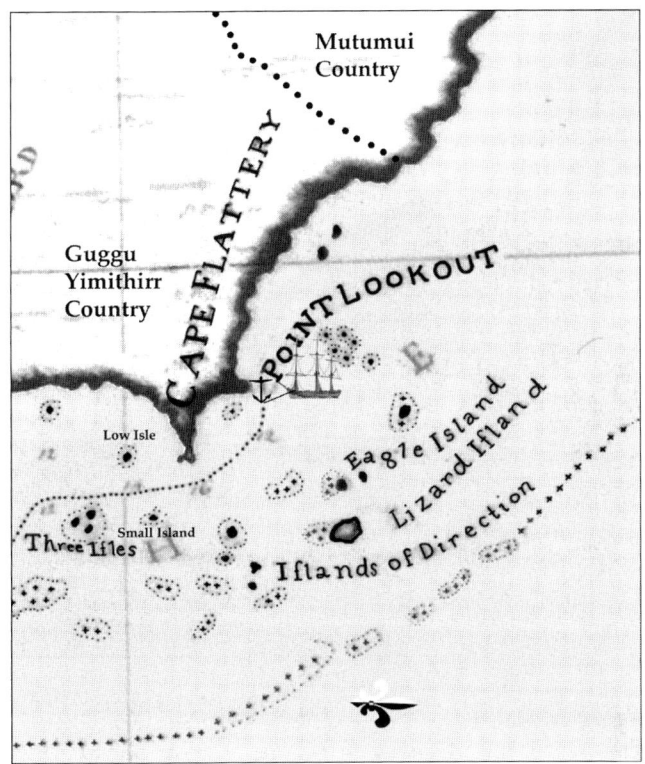

11th August

COOK: At 1 o'clock a Petty officer at the mast head called out that he saw land ahead extending quite round to the Islands without us and a large Reef between us and them. Upon this I went to the mast head myself, the Reef I saw very plain which was now so far to wind ward that we could not weather it, but what he took for main land ahead were only small Islands, for such they appeared to me, but before I had well got from mast head the Master and some others went up who all asserted that it was a continuation of the main land and to make it still more alarming they said they saw breakers in a manner all round us. We immediately hauled upon a wind in for the land and soon after came too an **Anchor** [anchorage marked on chart] under a point of the Main about a Mile from the Shore.

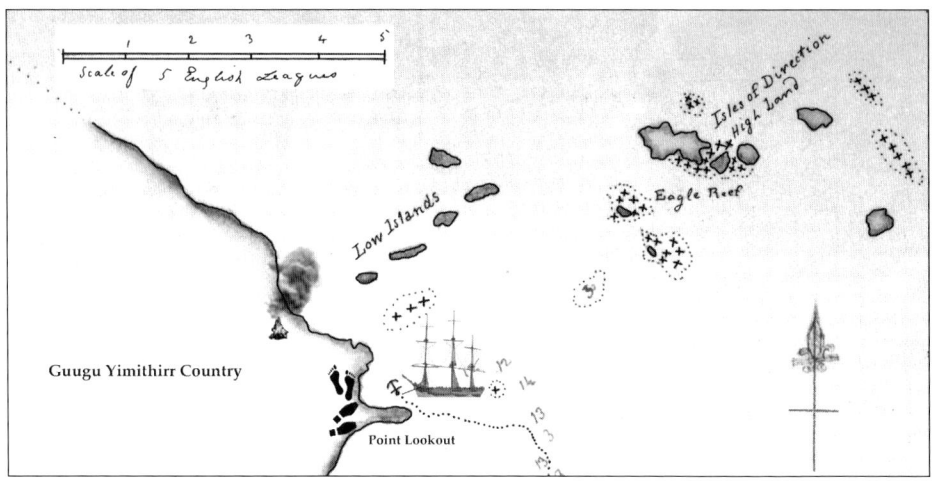

Portion of Master's Mate, Richard Pickersgill's chart 'A draft of part of the shoals...' showing their anchorage at Point Lookout.[4]

COOK: After this I landed and went upon the point which is pretty high from which I had a view. I saw 9 or 10 small low Islands and some shoals laying off the Coast and some large shoals between the Main and the three high Islands without ['Isles of Direction' – Lizard Group] which I was now well assured were Islands and not a part of the Main land as some had taken them to be. The Point I have named Point Lookout.

BANKS: The point we went upon was sandy and very Barren so it afforded very few plants. [*Dillenia alata* was among the plants collected here.] (Colour Plate No. 61)

The Sand itself indeed with which the whole country in a manner was covered was infinitely fine and white, but till a glass house was built here that would turn to no account.

COOK: We saw the footsteps of People upon the Sand [Guugu Yimithirr people] and smoke and fire up in the Country.

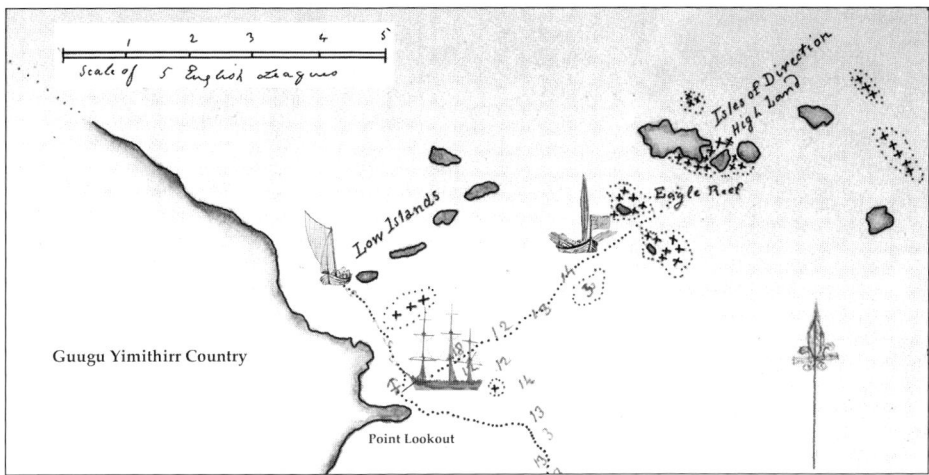

'A draft of part of the shoals …'

COOK: In the evening returned on board where I came to a resolution to Visit one of the high Islands in the offing in my Boat as they lay at least 5 Leagues out to sea and seemed to be of such a height that from the top of one of them I hoped to see and find a Passage out to sea clear of the shoals; accordingly in the morning I set out in the Pinnace for the northermost and largest of the three [Lizard Island – 'Isles of Direction' marked on chart] accompanied by Mr Banks, at the same time I sent the Master in the Yawl to lee-ward to sound between the low Islands and the Main.

BANKS: In going out we passed over 2 very large shoals on which we saw great plenty of Turtle but we had too much wind to strike any. The great Reef was a plenty of Turtle hardly to be credited, every shoal swarmed with them.

COOK: The finest Green Turtle in the world.

Chapter 16

The Indians Had Been Here

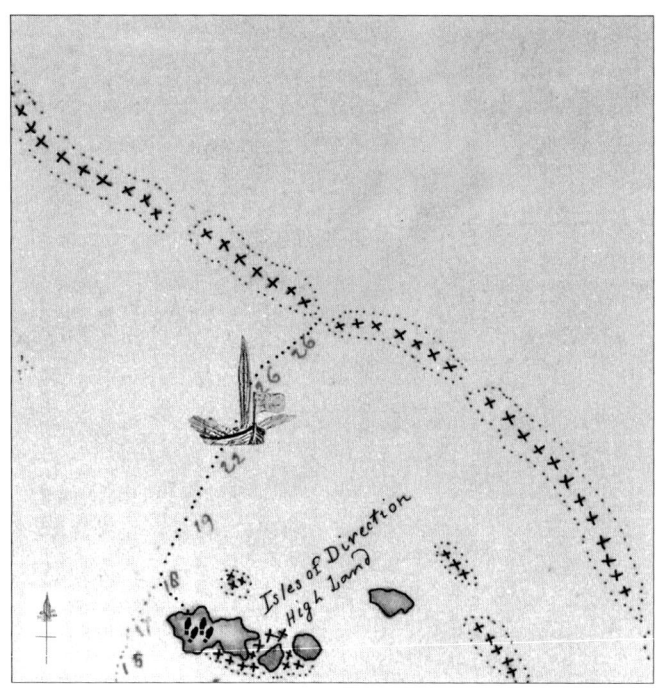

Detail of Master's Mate, Richard Pickersgill's chart 'A draft of part of the shoals…'.[1]

12th August

COOK: I did not reach the island until half an hour after 1 o'clock in the pm.

BANKS: The Island itself was high; we ascended the hill and when we were at the top saw plainly the Grand Reef still extending itself Parallel with the shore at about the distance of 3 leagues from us or 8 from the main; through it were several channels exactly similar to those we had seen in the Islands. Through one of these we determined to go.

COOK: I stayed upon this hill until near sun set but the weather continued so hazy all the time that I could not see above 4 or 5 Leagues round me so that I came down much disappointed in the prospect I expected to have had, but being in hopes the morning might prove clearer and give me a better View of the Shoals. With this view I stayed all night upon the Island.

BANKS: We slept under the shade of a Bush that grew on the beach very comfortably.

COOK: At 3 in the Morning sent the Pinnace with one of the Mates I had with me to sound between the Island and the reefs and to examine one of the breaks or Channels, and in the mean time I went again upon the hill where I arrived by sun rise but found it much hazier than in the evening.

BANKS: Great part of yesterday and all this morn till the boat returned I employed in searching the Island. On it I found some few plants which I had not before seen. [*Josephinia imperatricis* was one of several plants collected here.] (Colour Plate No. 62)

BANKS: The Island itself was small and Barren; on it was however one small tract of woodland which abounded very much with large Lizards some of which I took [Yellow-spotted monitor – *Varanus panoptes*]. (Colour Plate No. 63)

COOK: These seemed to be pretty plenty which occasioned my naming the Island <u>Lizard Island</u>.

Here is also fresh water in two places, the one is a running stream the water a little brackish where I tasted it which was close to the sea, the other is a standing Pool close behind the sandy beach of good sweet water as I dare say the other is a little way from the Sea beach. We were surprised to find Houses &c. upon

Lizard Island which lies 5 Leagues from the nearest part of the Main, a distance we before thought that they could not have gone in their Canoes.

BANKS: Distant as this Isle was from the main, the Indians had been here in their poor embarkations, sure sign that some part of the year must have very settled fine weather; we saw 7 or 8 frames of their huts and vast piles of shells the fish of which had I suppose been their food. All the houses were built upon the tops of eminences exposed entirely to the SE, contrary to those of the main which are commonly placed under the shelter of some bushes or hill side to break off the wind.

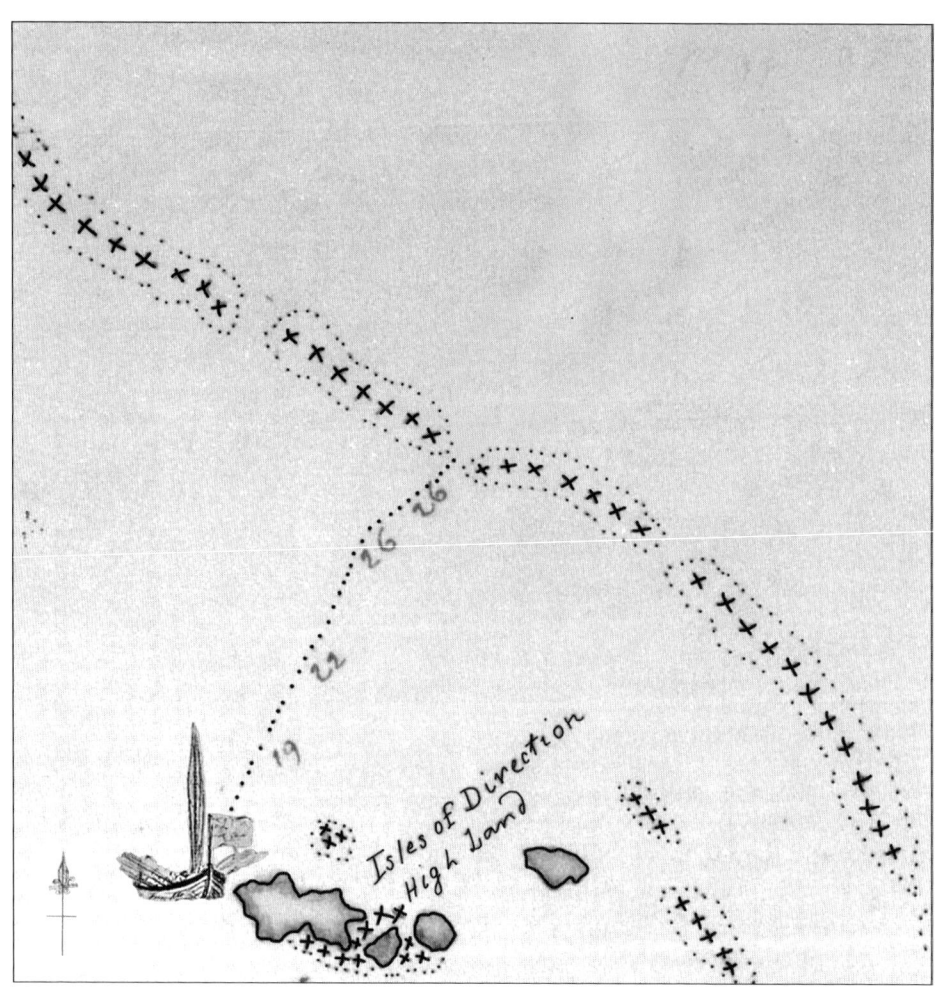

COOK: About noon the Pinnace returned having been out as far as the reef and found from 15 to 28 fathom water. It blowed so hard that they durst not venture into one of the Channels which the Mate said seemed to him to be very narrow but this did not discourage me for I thought from the place he was at he must have seen it at a disadvantage.

BANKS: The officer returned with an account that the sea broke vastly high upon the reef and the swell was so great in the opening that he could not go into it to sound. This was sufficient to assure us of a safe passage out, so we got into the boat to return to the ship in high spirits, thinking our danger now at an end as we had a passage open for us to the main Sea.

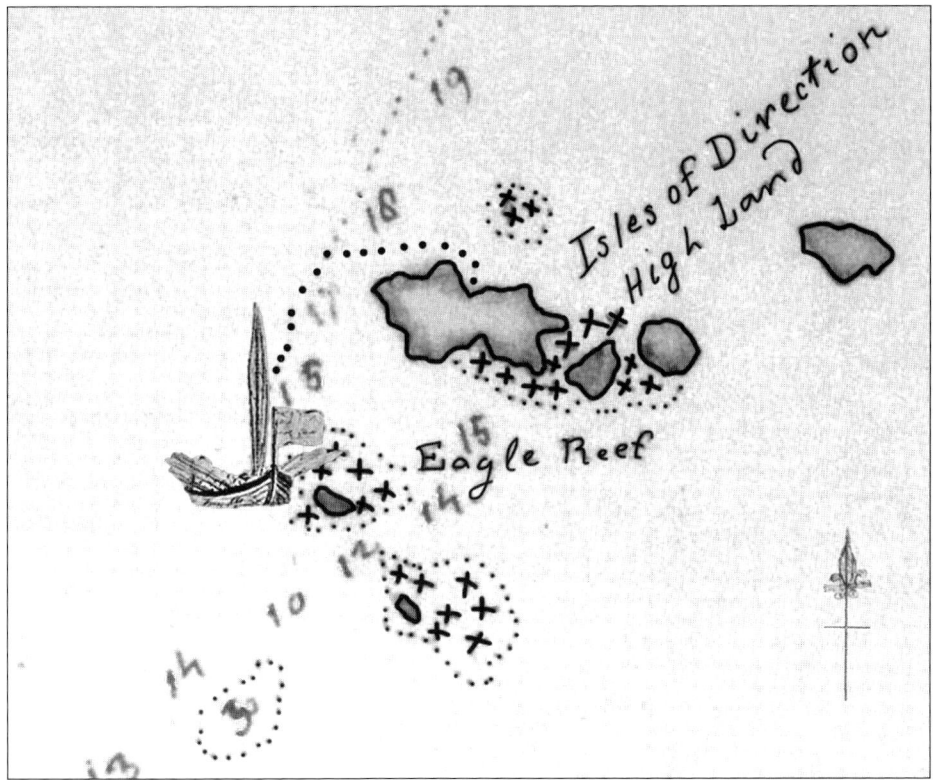

13th August

BANKS: In our return we went ashore upon a low Island where we shot many birds [Eagle Reef marked on chart]; on it was an Eagles nest the young ones of which we killed [Osprey – *Pandion haliaetus*] (Colour Plate No. 64) and another built on the ground by I know not what bird, of a most enormous magnitude – it was in circumference 26 feet [8 metres] and in height 2 feet 8 [81 cm] built of sticks; the only Bird I have seen in this country capable of building such a nest seems to be the Pelican [more probably the Orange-footed scrub fowl – *Megapodius reinwardt*]. (Colour Plate No. 65) The Indians had been here likewise and livd upon turtle, as we could plainly see by the heaps of callipashes [carapaces] which were pild up in several parts of the island.

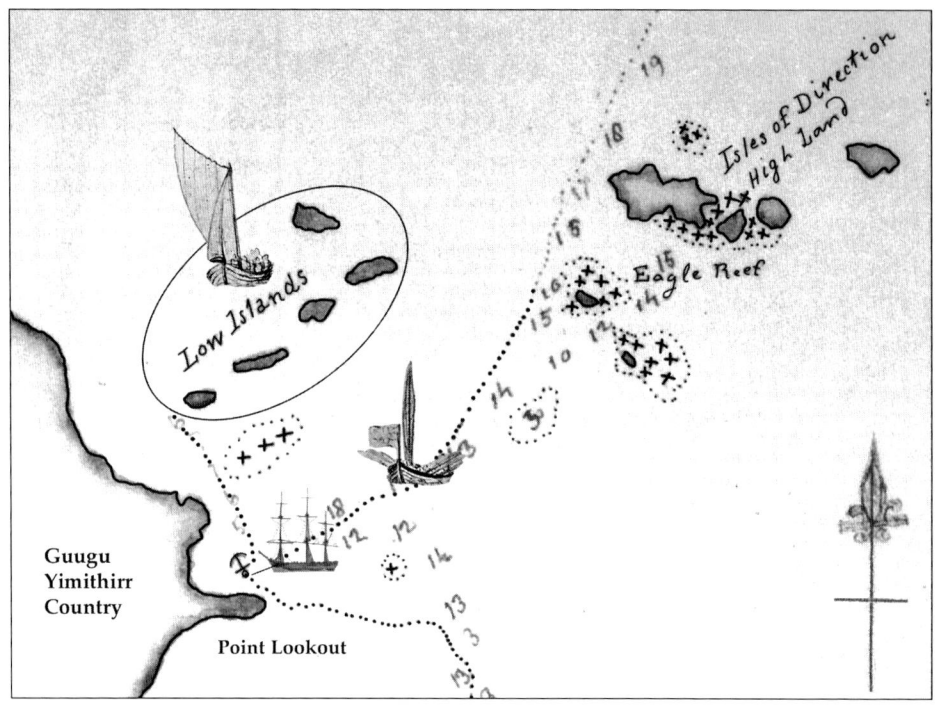

COOK: After leaving Eagle Island we stood SW directly for the Ship.

BANKS: Our Master who had been sent to leeward to examine that Passage went ashore upon a <u>low Island</u> where he slept. Here he saw vast plenty of turtle shells, and so great plenty had the Indians had when there that they had hung up the fins with the meat left on them in trees, where the sun had dried them so well that our seamen eat them heartily.

He saw also two spots clear of grass which had lately been dug up; they were about 7 feet long and shaped like a grave, for which indeed he took them.

COOK: After well considering both what I had seen myself and the report of the Master, who was of opinion that the Passage to Leeward would prove dangers; this I was pretty well convinced of myself that by keeping in with the main land we should be in continual danger besides the risk we should run in being locked in within the Main reef at last and have to return back to seek a passage out, an accident of this kind or any other that might happen to the Ship would infallibly loose our passage to the East Indies this season and might prove the ruin of the Voyage as we have now little more than 3 Months provisions on board and that at short allowance in many articles. These reasons had the weight with all the officers. I therefore resolved to weigh in the morning and endeavour to quit the coast altogether until we could approach it with less danger.

Chapter 17

Great Dangers Swallow Up Lesser Ones

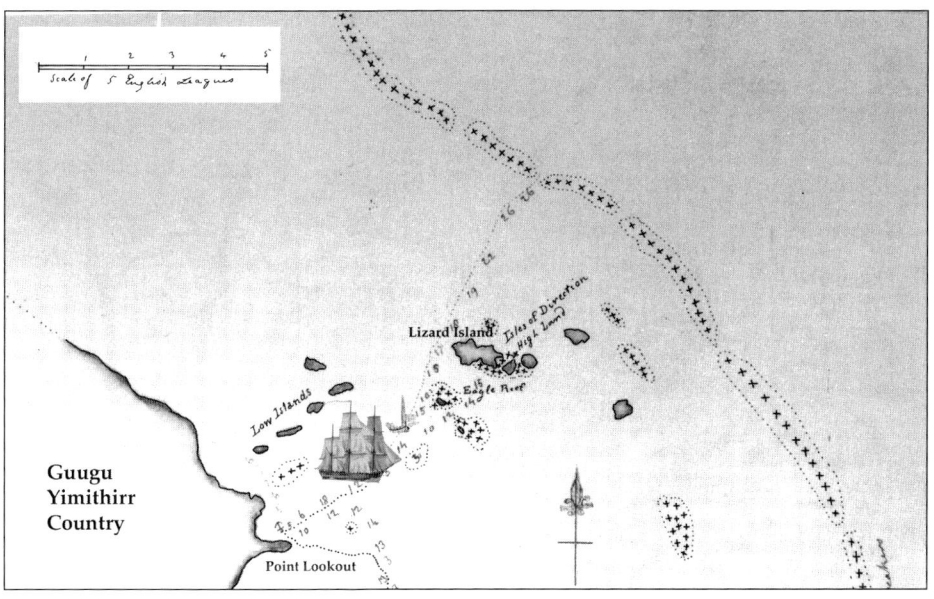

Detail of chart 'A draft of part of the shoals...' by Richard Pickersgill.[1]

13th August (cont'd)

COOK: With this view we got under sail at day light and stood out NE for the NW end of Lizard Island.

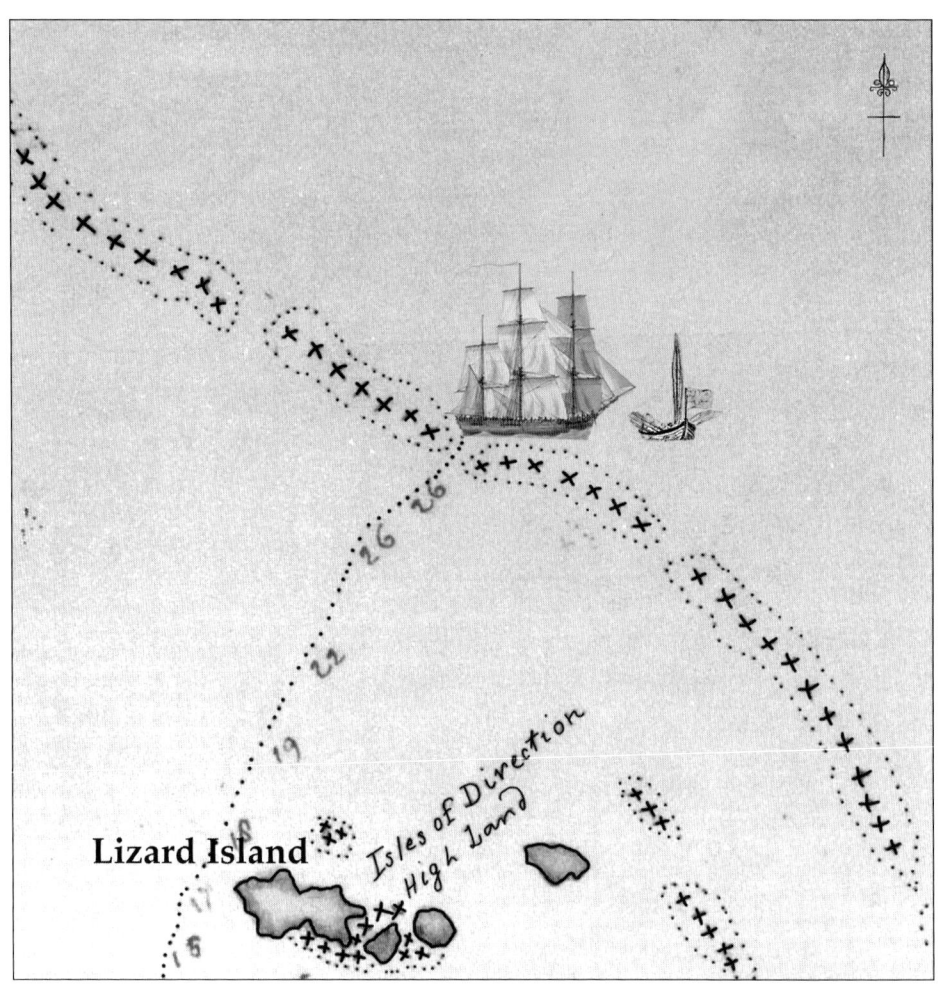

14th August

COOK: By 2 o'clock we just fetched to windward of one of the Channels in the outer Reef I had seen from the Island, we now tacked and made a short trip to the SW while the Master in the Pinnace examined the channel, he soon made the Signal for the Ship to follow which we accordingly did and in a short time got safe out. The Channel [now Cook's Passage] is about one third of a Mile broad and not more in length [passage marked on chart].

Great Dangers Swallow Up Lesser Ones

A VIEW of the ISLANDS of DIRECTION taken at the entrance of the Channel without the Reef

Lizard Island SW by distant 10 Miles

Charles Praval – copied from a lost original by Parkinson or Spöring.[2]

COOK: It may always be found by three high islands within it which I have called <u>Islands of Direction</u> because by them a safe passage may be found, even by strangers.

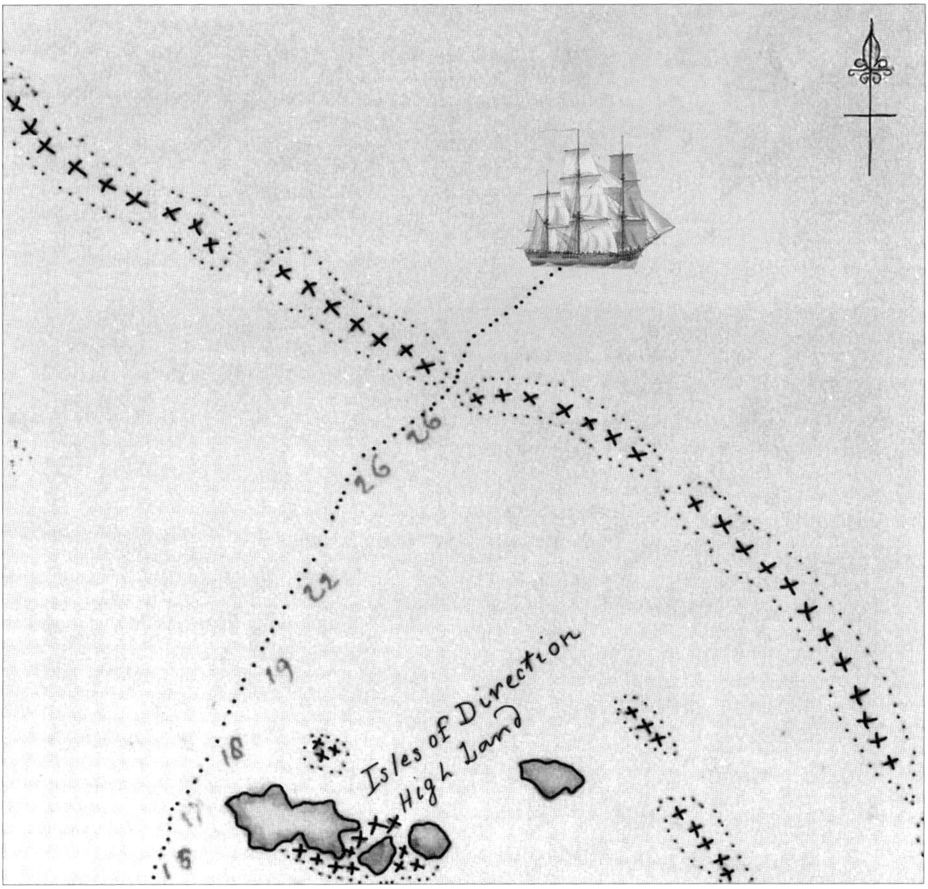

PARKINSON: This reef ran farther than the eye could reach, on the outermost side of all the rest, like a wall, and the sea broke very high upon it: We found no sounding in the passage.

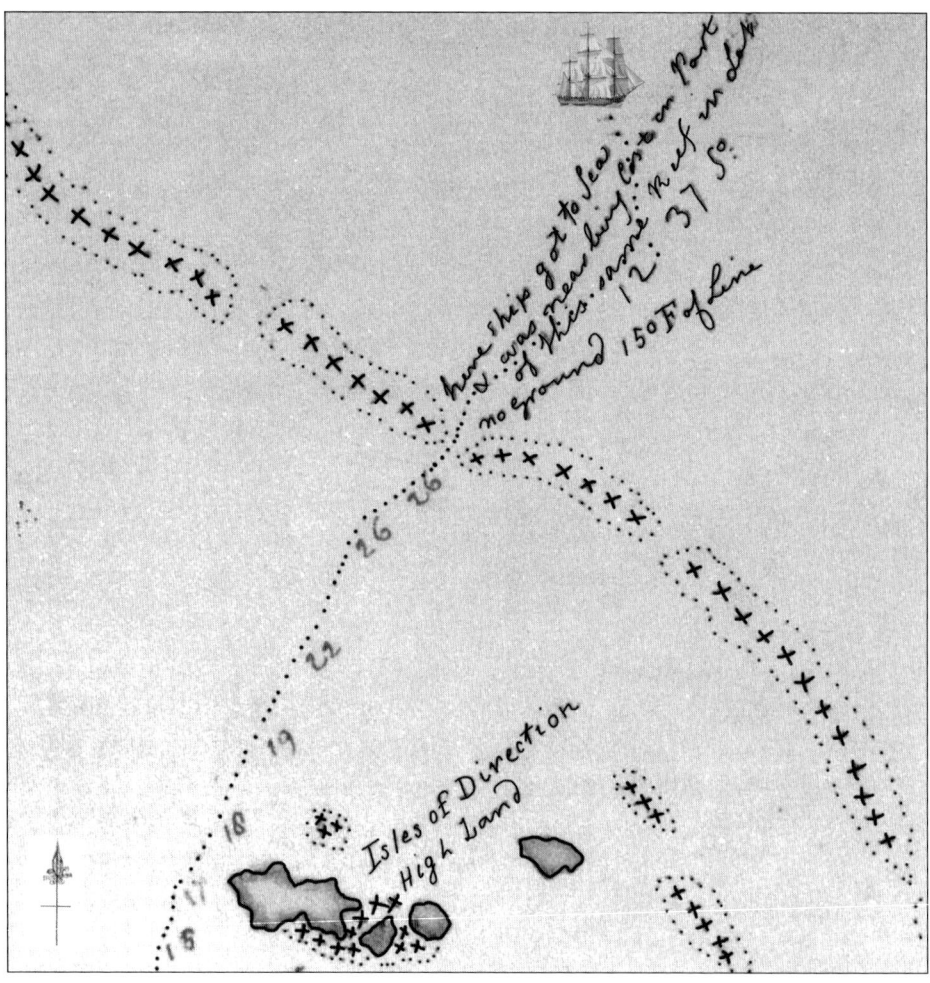

COOK: We had no sooner got without the breakers than we had no ground with 150 fathom of line and found a well grown sea rowling in from the SE certain signs that neither land nor shoals were in our neighbourhood in that direction which made us quite easy at being freed from fears of Shoals, after having been entangled among them more or less ever since the 26th of May, in which time we have sailed 360 Leagues without ever having a Man out of the chains heaving the Lead when the Ship was under way, a circumstance that I dare say never happened to any ship before, and yet it was absolutely necessary.

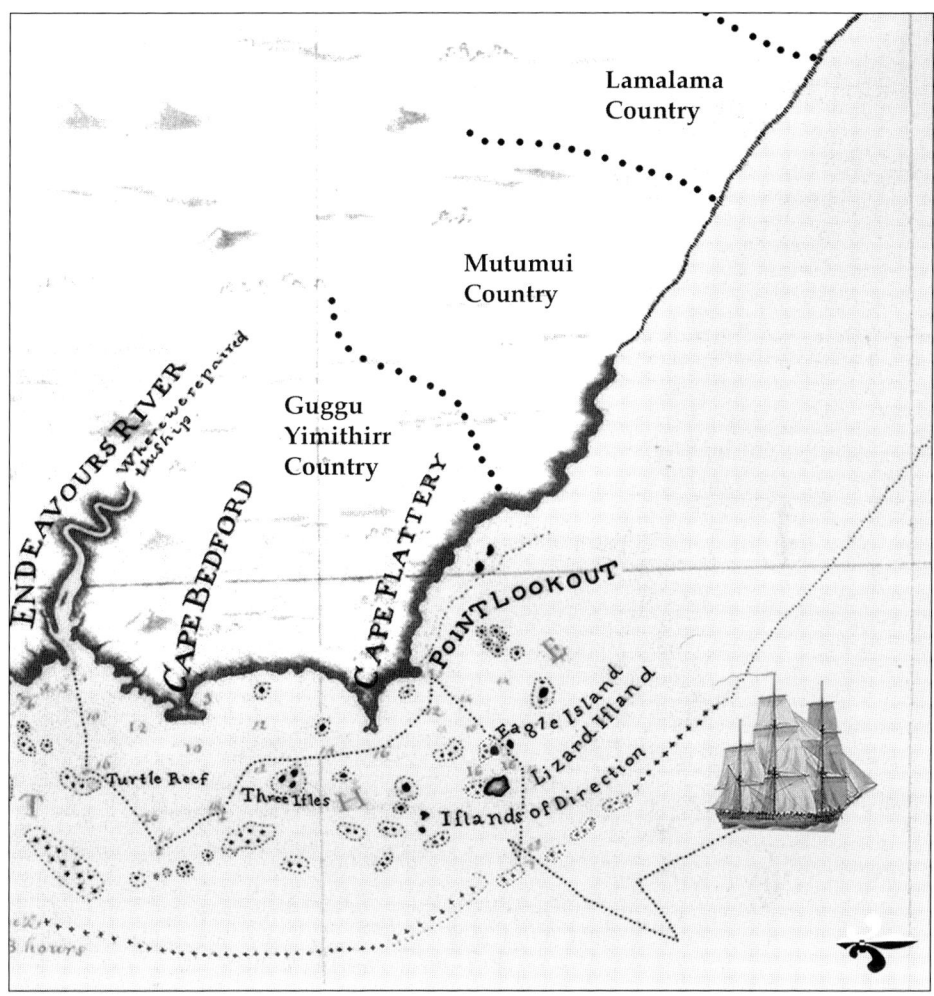

Portion of Cook's chart 'Cape Tribulation to *Endeavour*s Streights'.³

BANKS: For the first time these three months we were this day out of sight of Land to our no small satisfaction: that very Ocean which had formerly been look'd upon with terror by (maybe) all of us was now the Assylum we had long wishd for and at last found. Satisfaction was clearly painted in every mans face: The day was fine and the trade wind brisk before which we steered to the Northward; the well grown waves which followed the ship, sure sign of no land being in our neighbourhood, were contemplated with the greatest satisfaction, notwithstanding we plainly felt the effect of the blows they gave to our crazy ship, increasing her leaks considerably so that she made now 9 inches water every hour. This however was looked upon as a light evil in comparison to those we had so lately made our escape from.

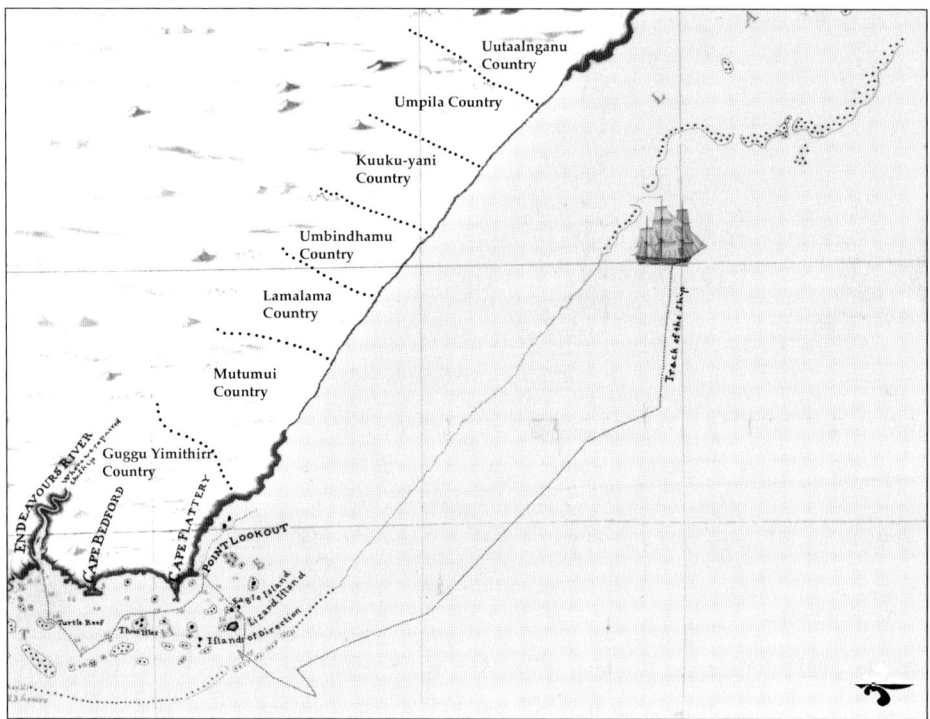

The unexplored coast.

COOK: It was with great regret I was obliged to quit this coast unexplored to its Northern extremity which I think we were not far off, for I firmly believe that it doth not join to New Guinea, however this I hope yet to clear up being resolved to get in with the land again as soon as I can do it with safety.

15th August

BANKS: Fine weather and moderate trade. The Captain fearful of going too far from the Land, least he should miss an opportunity of examining whether or not the passage which is laid down in some charts between New Holland and New Guinea really existed or not, steered the ship west right in for the land; about 12 o'clock it was seen from the Mast head and about one the Reef laying without it in just the same manner as when we left it.

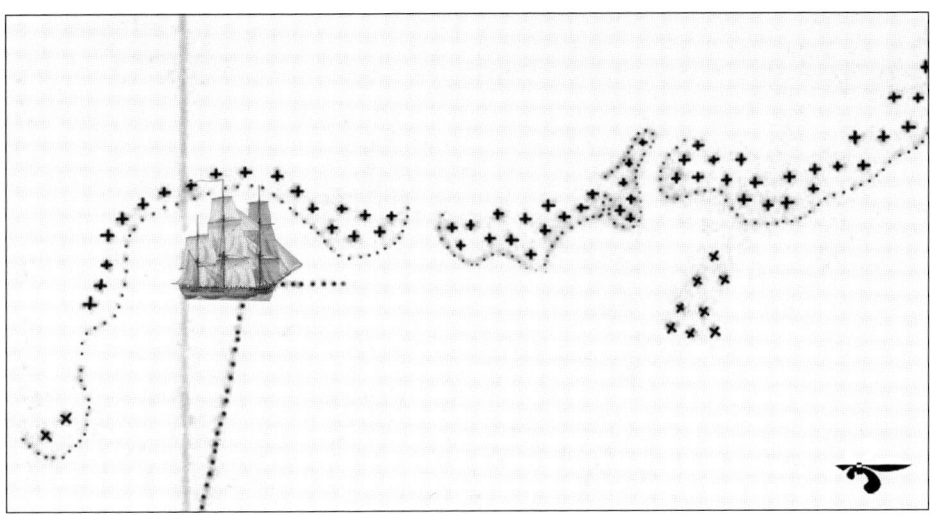

Detail of Cook's chart 'Cape Tribulation to *Endeavour*s Streights'.[4]

16th August

BANKS: He stood on however resolving to stand off at night after having taken a nearer view, but just at night fall found himself in a manner embayd in the reef so that it was a moot point whether or not he could weather it on either tack.

We stood however to the Northward [change of direction from west to north shown on chart] and at dark it was concluded that she would go clear of everything we could see.

The night however was not the most agreeable: all the dangers we had escaped were little in comparison of being thrown upon this reef if that should be our lot. A Reef such a one as I now speak of is a thing scarcely known in Europe or indeed anywhere but in these seas: it is a wall of Coral rock rising almost perpendicularly out of the unfathomable ocean, always overflown at high water commonly 7 or 8 feet, and generally bare at low water; the large waves of the vast ocean meeting with so sudden a resistance make here a most terrible surf breaking mountain high, especially when as in our case the general trade wind blows directly upon it.

At 3 o'clock this morn it dropped calm on a sudden which did not at all better our situation: we judged ourselves not more than 4 or 5 leagues from the reef, maybe much less, and the swell of the sea which drove right in upon it carried the ship towards it fast. We tried the lead often in hopes to find ground that we might anchor but in vain; before 5 the roaring of the Surf was plainly heard and as day broke the vast foaming billows were plainly enough to be seen scarce a mile from us and towards which we found the ship carried by the waves surprisingly fast, so that by 6 o'clock we were within a Cables length of them, driving on as fast as ever and still no ground with 100 fathom of line.

Every method had been taken since we first saw our danger to get the boats out in hopes that they might tow us off but it was not yet accomplished; the Pinnace had had a plank stripped off her for repair and the longboat under the Booms was lashed and fastened so well from our supposed security that she was not yet got out. Two large oars or sweeps were got out at the stern ports to pull the ships head round the other way in hopes that might delay till the boats were out. All this while we were approaching and came I believe before this could be effected within 40 yards of the breaker; the same sea that washed the side of the ship rose in a breaker enormously high the very next time it did rise, so between us and it was only a dismal valley the breadth of one wave; even now the lead was hove 3 or 4 lines fastened together but no ground could be felt with above 150 fathom.

Now was our case truly desperate, no man I believe but who gave himself entirely over, a speedy death was all we had to hope for and that from the vastness of the Breakers which must quickly dash the ship all to pieces was scarce to be doubted. Other hopes we had none: the boats were in the ship and must be dashed in pieces with her and the nearest dry land was 8 or 10 Leagues distant. We did not however cease our endeavours to get out the long boat which was by this time almost accomplished.

At this critical juncture, at this I must say terrible moment, when all assistance seemed too little to save even our miserable lives, a small air of wind sprang up, so small that at any other time in a calm we should not have observed it. We however plainly saw that it instantly checked our progress; every sail was therefore put in a proper direction to catch it and we just observed the ship to move in a slanting direction off from the breakers.

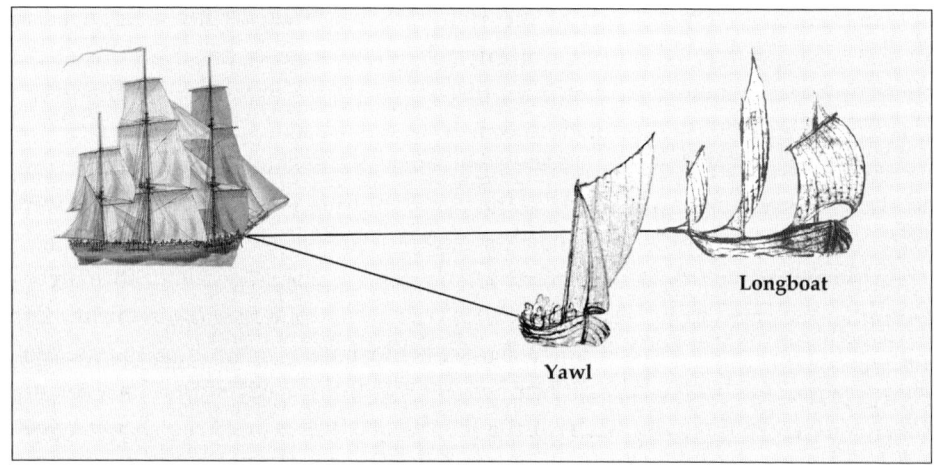

BANKS: This at least gave us time and redoubling our efforts we at last got out the long boat and manning her sent her a head. The ship still moved a little off but in less than 10 minutes our little Breeze died away into as flat a calm as ever. Now was our anxiety again renewed: innumerable small pieces of paper &c. were thrown over the ships side to find whither the boats really moved her ahead or not and so little did she move that it remained almost every other time a matter of dispute. Our little friendly Breeze now visited us again and lasted about as long as before, thrusting us possibly 100 yards farther from the breakers: we were still however in the very jaws of destruction.

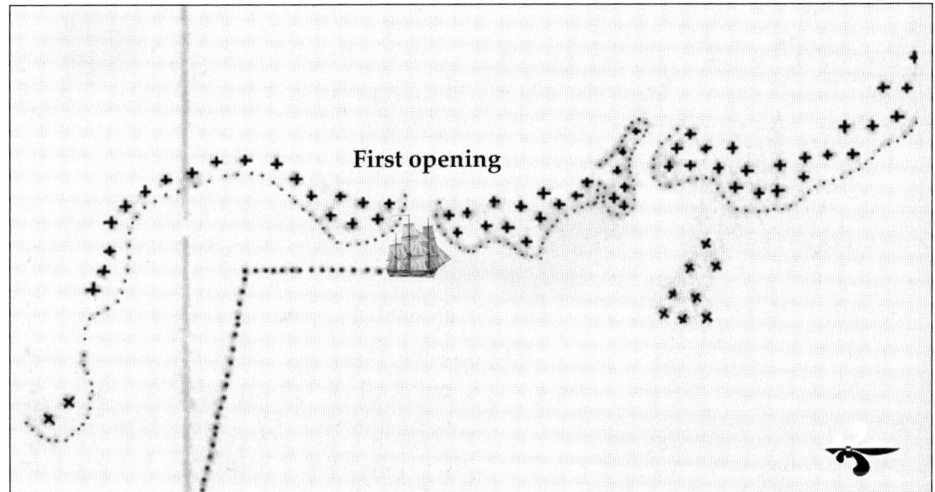

BANKS: A small opening had been seen in the reef about a furlong from us, its breadth was scarce the length of the ship, into this however it was resolved to push her if possible. Within was no surf, therefore we might save our lives: the doubt was only whether we could get the ship so far: our little breeze however a third time visited us and pushed us almost there. The fear of Death is Bitter: the prospect we now had before us of saving our lives tho at the expense of everything we had made my heart set much lighter on its throne, and I suppose there were none but what felt the same sensations. At length we arrived off the mouth of the wished for opening and found to our surprize what had with the little breeze been the real cause of our Escape, a thing that we had not before dreamt of. The tide of flood it was that had hurried us so unaccountably fast towards the reef, in the near neighbourhood of which we arrived just at high water, consequently its ceasing to drive us any farther gave us the opportunity we had of getting off. Now however the tide of Ebb made strong and gushed out of our little opening like a mill stream, so that it was impossible to get in; of this stream however we took the advantage as much as possible and it Carried us out near a quarter of a mile from the reef.

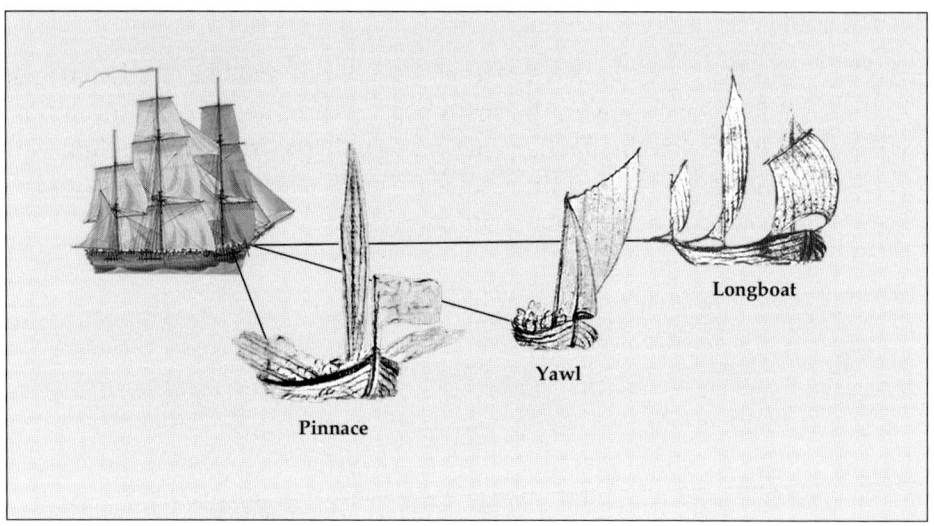

BANKS: We well knew that we were to take all the advantage possible of the Ebb so continued towing with all our might and with all our boats, the Pinnace being now repaired, till we had got an offing of 1 and a half or 2 miles. By this time the tide began to turn and our suspense began again: as we had gained so little while the ebb was in our favour we had some reason to imagine that the flood would hurry us back upon the reef in spite of our utmost endeavours. It was still as calm as ever so no likely hood of any wind today; indeed had wind sprung up we could only have searched for another opening, for we were so embayed by the reef that with the general trade wind it was impossible to get out.

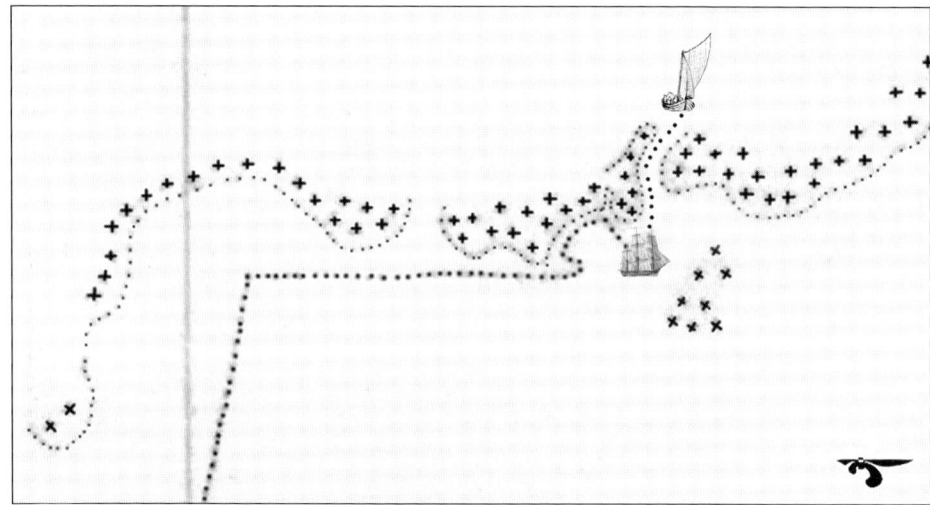

17th August

BANKS: Another opening was however seen ahead and the 1st Lieutenant [Zachary Hickes] went away in the small boat to examine it. In the mean time we struggled hard with the flood, sometimes gaining a little then holding only our own and at others losing a little, so that our situation was almost as bad as ever, as the flood had not yet come to its strength.

BANKS: At 2 however the Lieutenant arrived with news that the opening was very narrow: in it was good anchorage and a passage quite in free from shoals. The ships head was immediately put towards it and with the tide she towed fast so that by three we entered and were hurried in by a stream almost like a mill race, which kept us from even a fear of the sides tho it was not above a quarter of a mile in breadth.

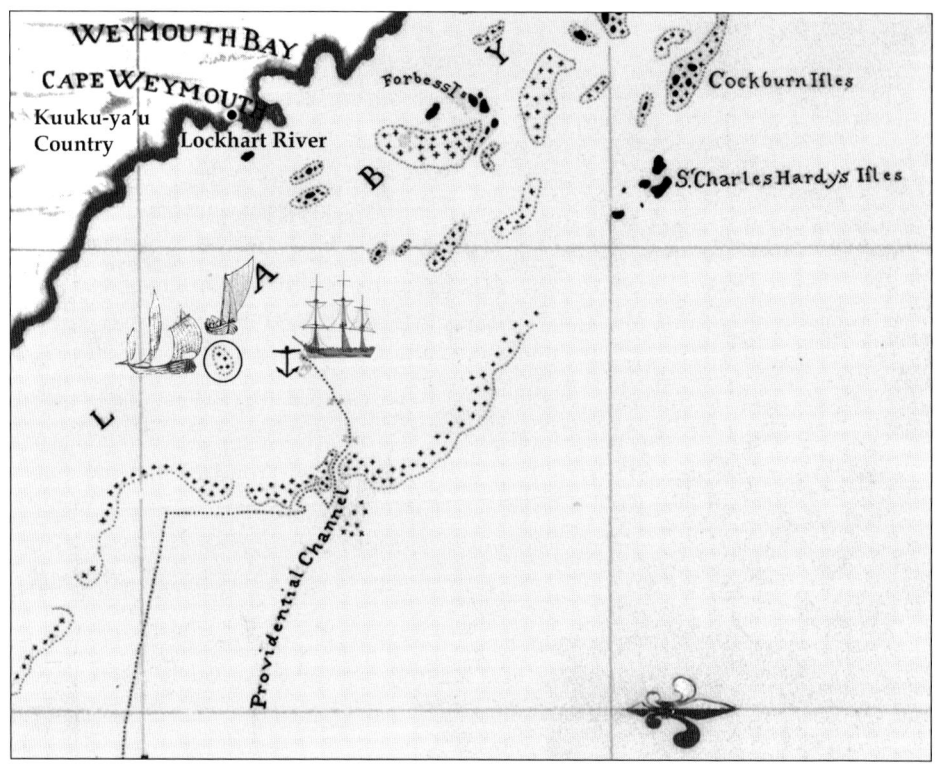

Portion of Cook's chart 'Cape Tribulation to *Endeavour*s Streights'.[5]

BANKS: By 4 [p.m.] we came to an **anchor** happy once more to encounter those shoals which but two days before we thought ourselves supremely happy to have escaped from. How little do men know what is for their real advantage: two days ago our utmost wishes were crowned by getting without the reef and today we were made again happy by getting within it.

COOK: Such are the vicissitudes attending this kind of service and must always attend an unknown Navigation: Was it not from the pleasure which naturally results to a Man from being the first discoverer, even was it nothing more than sands and Shoals, this service would be insupportable, especially in far distant parts, like this, short of Provisions and almost every other necessary. The world will hardly admit of an excuse for a man leaving a Coast unexplored he has once discovered, if dangers are his excuse, he is then charged with Timorousness and want of Perseverance and at once pronounced the unfittest man in the world to be employed as a discoverer. If on the other hand he boldly encounters all the dangers and obstacles he meets and is unfortunate enough not to succeed he is than charged with Temerity and want of conduct. The former of these aspersions cannot with Justice be laid to my charge and if I am fortunate enough to surmount all the dangers we may meet the latter will never be brought in question. I must own I have engaged more among the Islands and shoals upon this Coast than may be thought with prudence I ought to have done with a single Ship and every other thing considered, but if I had not I should not have been able to give any better account of the one half of it than if we had never seen it, that is we should not have been able to say whether it consisted of main land or Islands and as to its produce; but it is time I should have done with this subject which at best is disagreeable & which I was led into on reflecting on our late danger.

I now came to a fixed resolution to keep the Main land on board in our route to the northward let the consequence be what it will, indeed now it was not advisable to go without the reef, for by it we might be carried so far from the coast as not to be able to determine whether or no New Guinea joins to or makes a part of this land; this doubtful point I had from my first coming upon the Coast determined if possible to clear up.

On the Main land within us was a pretty high Promontory which I called <u>Cape Weymouth</u>, and the channel we came in by, I have named <u>Providential Channel</u>.

BANKS: As we were now safe at an anchor it was resolved to send the boats upon the nearest shoal to search for shell fish, turtle or whatever else they could get. They accordingly went and Dr Solander and myself accompanied them in my small boat. In our way we met with two water snakes one 5 the other 6 feet long; we took them both; they much resembled Land snakes only their tails were flatted sideways, I suppose for the convenience of swimming, and were not venomous.

[Now classified *Aipysurus duboisi*. It is one of the top three most venomous snakes in the world, together with the inland taipan and the eastern brown snake.] (Colour Plate No. 66)

PARKINSON: The reefs were covered with a numberless variety of beautiful corallines of all colours and figures, having here and there interstices of very white sand.

BANKS: Among which was the *Tubipora musica*. (Colour Plate No. 67)

PARKINSON: These made a pleasing appearance under water, which was smooth on the inside of the reef, while it broke all along the outside, and may be aptly compared to a grove of shrubs growing under water. Numbers of beautiful coloured fishes make their residence amongst these rocks, and may be caught by hand on the high part of the reef at low water. There are also crabs, molusca of various sorts, and a great variety of curious shell-fish, which adhere to the old dead coral that forms the reef.

BANKS: I have often lamented that we had not time to make proper observations upon this curious tribe of animals but we were so intirely taken up with the more conspicuous links of the chain of creation as fish, Plants, Birds &c. &c. that it was impossible.

18th August

COOK: At 4 o'clock in the p.m. the boats returned from the reef with about 240 pounds of the meat of shell fish most of Cockles, some of which are as large as 2 Men can move and contain about 20lbs [9 kg] of good meat.

Chapter 18

More Happier Than We Europeans

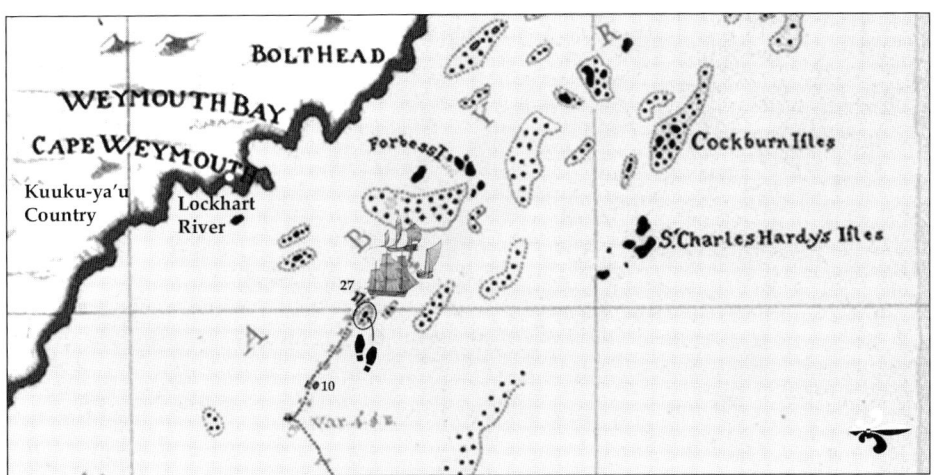

Portion of Cook's chart 'Cape Tribulation to *Endeavour*s Streights'.[1]

18th August (cont'd)

COOK: At 6 o'clock in the Morning we got under sail and Stood away to the NW. Sent the yawl ahead. Our soundings were very irregular from **10** to **27** fathom varying 5 or 6 fathom almost every cast of the lead. A little before noon we passed <u>a low small Sandy Isle</u> [marked on chart above] which we left on our starboard side at the distance of 2 Miles. Mr Banks landed upon it and shott several small birds, called Nodies [Common Noddy – *Anous stolidus pileatus*]. (Colour Plate No. 68)

At Noon some Small Islands [Forbes Isles] extending from N. 40° W to 54° W between us and the Main.

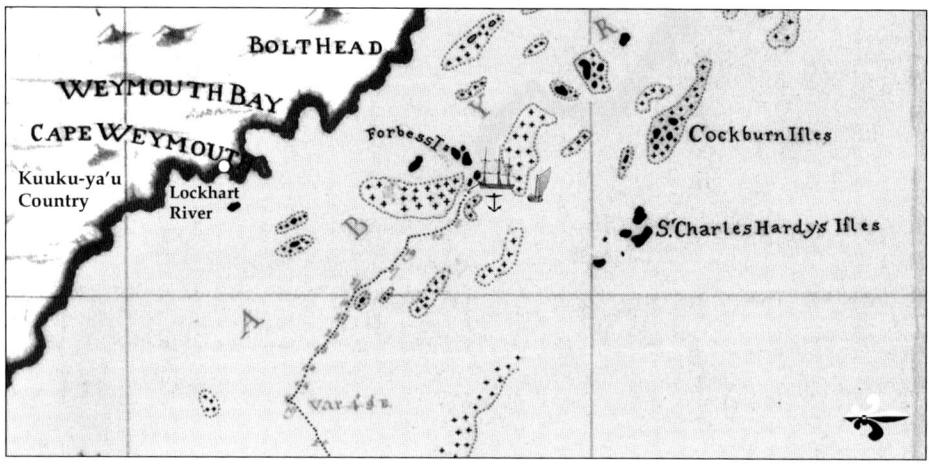

19th August

COOK: At 2 o'clock saw a large shoal right ahead extending 2 or 3 miles on each bow. At half past 6 we **anchored**. These islands which are known in the chart as <u>Forbes's Isles</u> lay about 5 leagues from the main which we called <u>Bolt Head</u>.

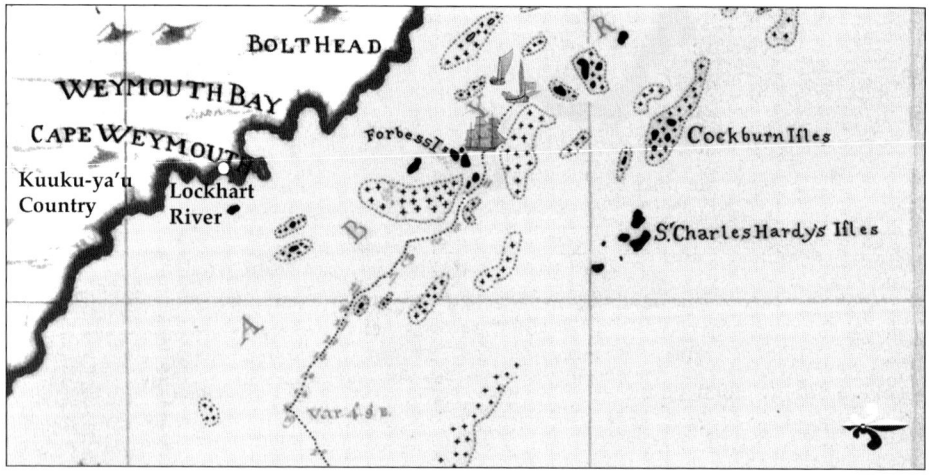

COOK: At 6 o'clock in the a.m. we again got under sail. At half past 8 got the pinnace out. The yawl making the best of her way between the shoals and the main. We followed with the ship.

More Happier Than We Europeans   247

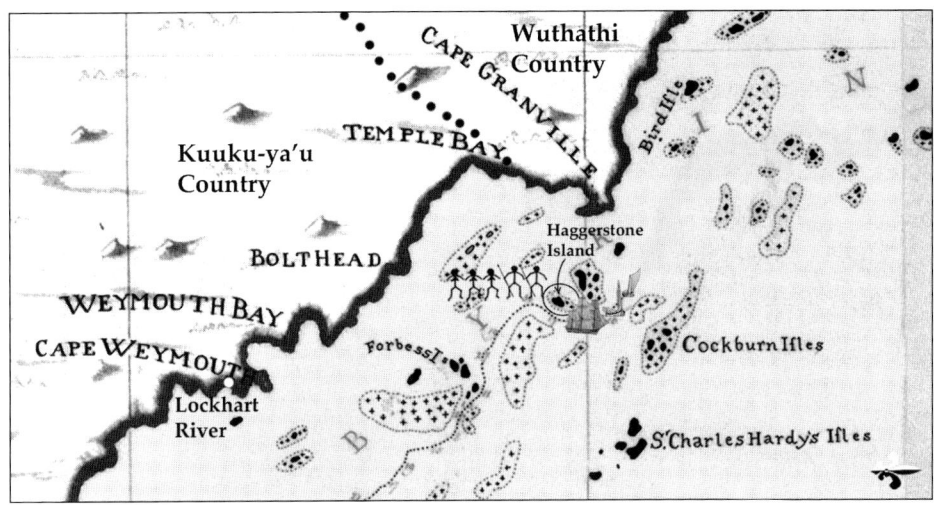

BANKS: All morn were much entangled with Shoals, but so much do great dangers swallow up lesser ones that these once so much dreaded shoals were now looked at with much less concern than formerly.

COOK: By the help of two boats ahead and a good lookout at the mast head we got at last into a fair channel which led us down to an island. We hauled round the NE side of the island. This island is inhabited [now Haggerstone Island – Wuthathi country].

BANKS: We passed very near to were 5 natives [Wuthathi people], 2 of whom carried their Lances in their hands; came down upon a point and looked at the ship for a little while and then retired.

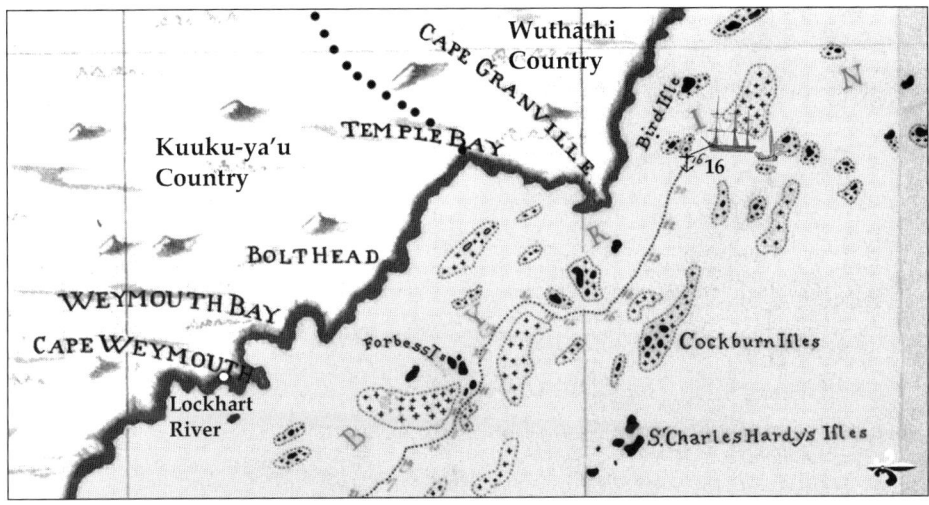

COOK: The main land within these islands forms a point which I call <u>Cape Granville</u>. Between this cape and Bolt Head is a bay which I named <u>Temple Bay</u>. E ½ N 9 leagues from Cape Granville lay some tolerable high islands which I called <u>Sr Charles Hardys Isles</u>, those which lay off the Cape I named <u>Cockburns Isles</u>.

20th August

COOK: About 1 o'clock the pinnace having got ahead and the yawl we took in tow, we filled our sails and steered NbW for some small islands [Cockburn Isles].

At 4 o'clock we discovered some low Islands and rocks bearing WNW [Bird Isles] which we stood directly for. At half past 6 o'clock we **Anchored** on the NE side of the northernmost in **16** fathom. On these Islands we saw a good many birds which occasioned my calling them <u>Bird Isles</u> [marked on chart].

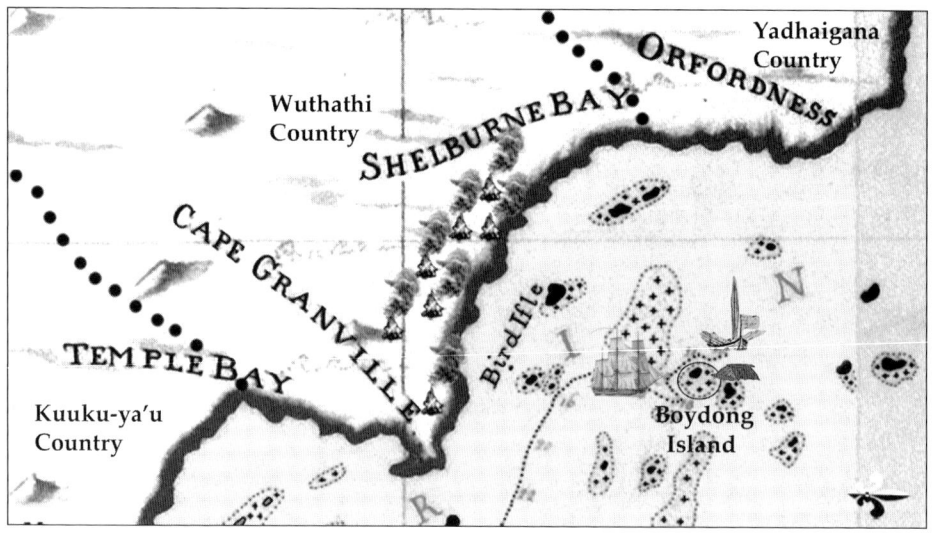

COOK: At 6 o'clock in the Morning we got again under sail; the pinnace ahead.

BANKS: Steering along shore as usual among many shoals, luffing up for some and bearing away for others. We are now pretty well experienced in their appearances so as seldom to be deceived and easily to know asunder a bottom coloured by white sand from a coral rock, the former of which, tho generally in 12 or 14 fathom water, some time ago gave us much trouble. The reef was still supposed to be without us from the smoothness of our water. The mainland appeared very low and sandy and had many fires upon it, more than we had usually observed [Wuthathi country].

COOK: Upon an Island [Boydong Island] which is only a small spot of Sand with some trees upon it, we saw a good many huts or habitations of the Natives [Wuthathi people] which we supposed comes over from the Main to these Islands from which they are distant about 5 Leagues to Catch Turtle at the time these animals come a Shore to lay their Eggs.

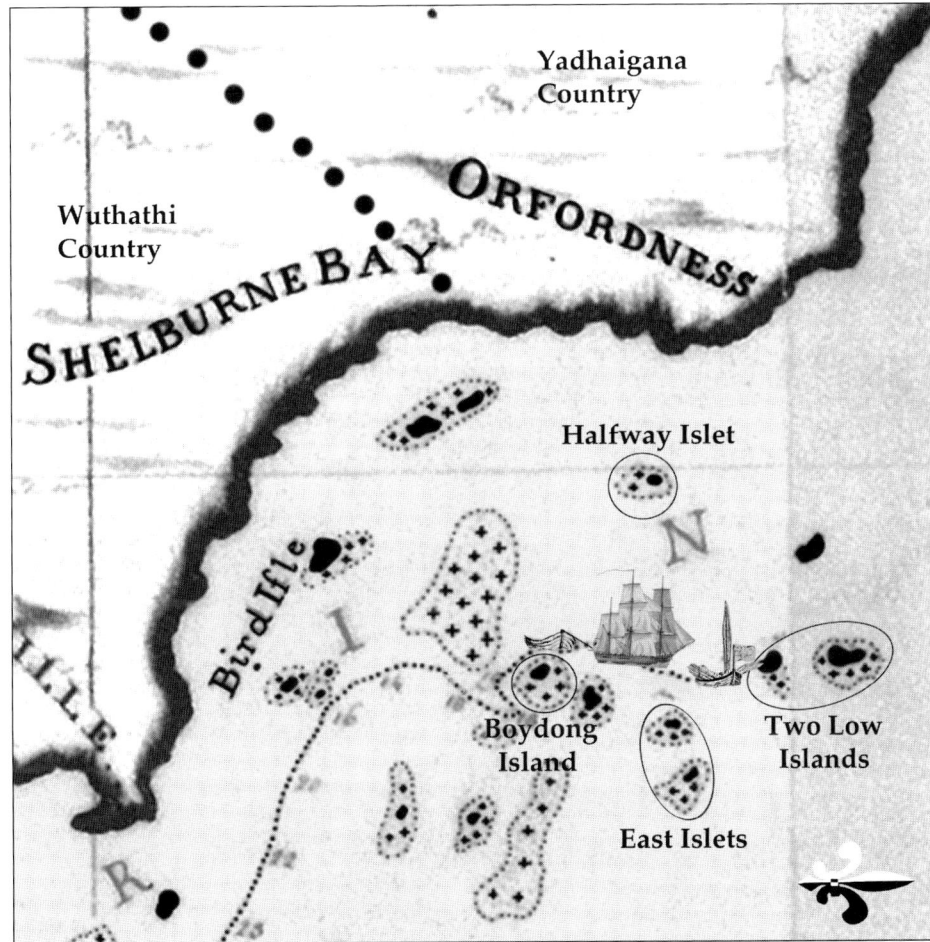

COOK: Having got the yawl in tow we stood away after the pinnace NNE and NBE to <u>two other low islands,</u> having two shoals which we could see without us [East Islets] and one between us and the main [Halfway Islet].

BANKS: We have constantly found the best passage to lie near the main, and the farther that you go near the reef the more numerous are the shoals.

250   The Endeavour Journals

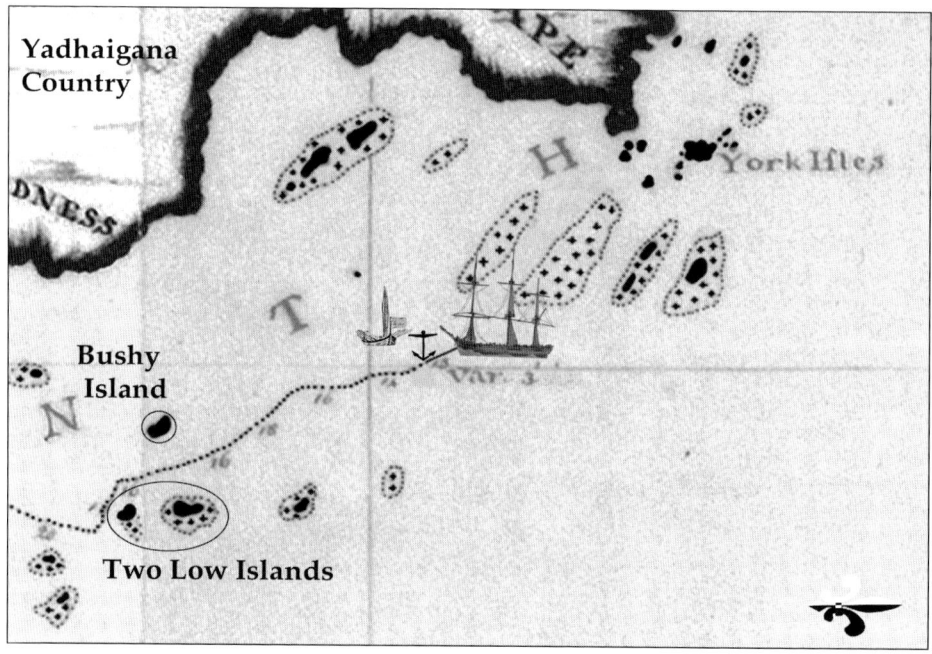

21st August

COOK: By 1 o'clock we had run nearly the length of the southernmost of the two islands. We steered NbW in a parallel direction with the main land leaving a small island [Bushy Island] between us and it, and some <u>low sandy isles</u> and shoals without us. At 7 in the p.m. we **anchored**.

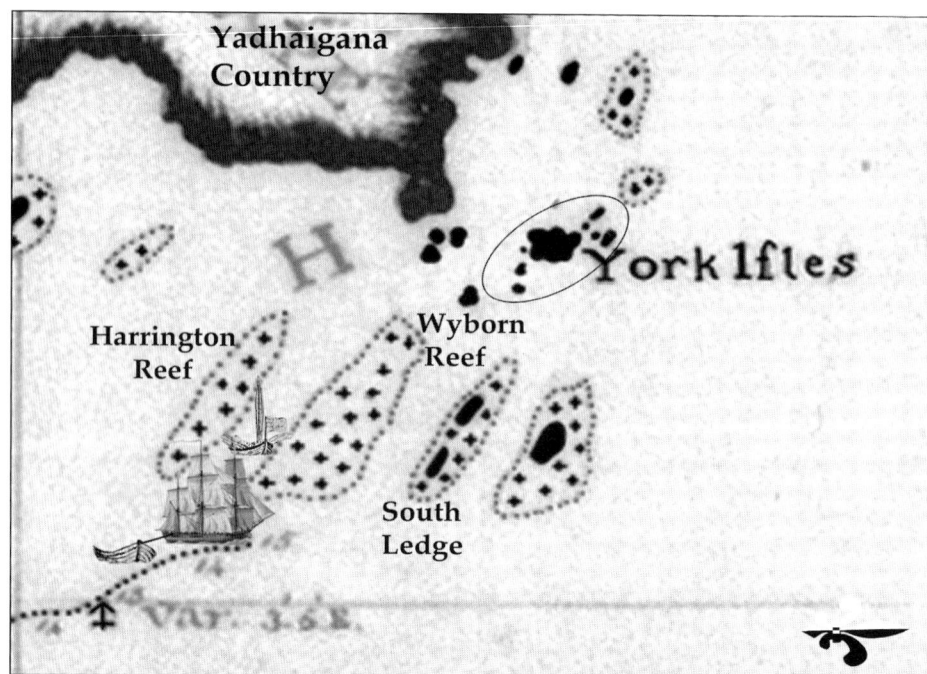

More Happier Than We Europeans   251

COOK: At daylight we got under sail again. Seeing no danger in our way we took the yawl in tow and made all sail we could until 8 o'clock [a.m.] at which time we discovered shoals ahead [South Ledge and Wyborn Reef] on our larboard bow and saw that the northernmost land, which we had taken to be part of the main, was an island or islands [York Isles] between which and the main there appeared to be a good passage, through which we might pass by running to leeward of the shoals on our larboard bow [Wyborn Reef]; upon which we wore and brought too.

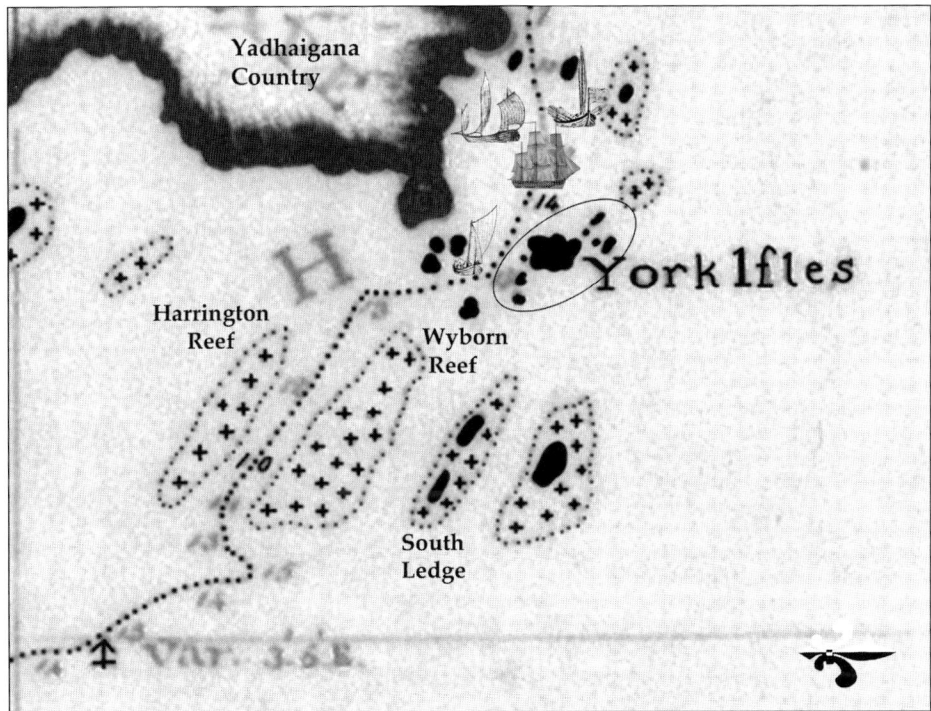

COOK: We sent away the pinnace and yawl to direct us clear of the shoals and then stood after them. After having got round the SE point of the shoal [Wyborn Reef] we steered NW along the SW or inside of it keeping a good lookout from the mast head, having another shoal on our larboard side [Harrington Reef]. But we found a good channel of a mile broad between them wherein were from **10** to **14** fathom water. At 11 o'clock being nearly the length of the islands before mentioned. These islands are known in the chart by the name of York Isles.

BANKS: We began to look out for the Passage we expected to find between New Holland and New Guinea. At noon one was seen very narrow but appearing to widen: we resolved to try it so stood in.

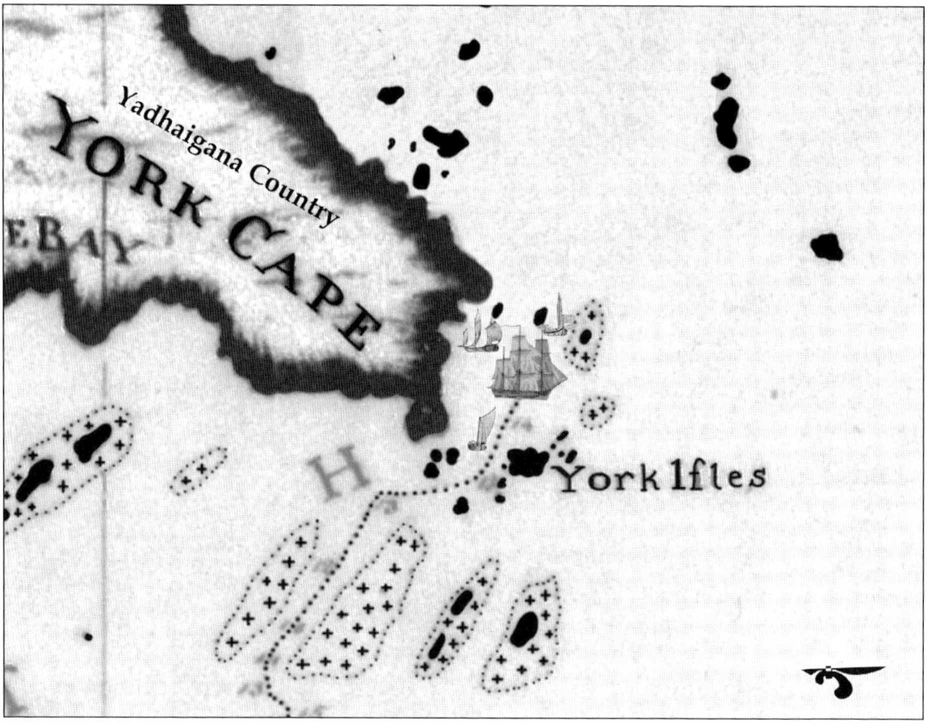

COOK: As soon as the boats were ahead we stood after them and got through by noon [between York Isles – present-day Adolphus Islands – and York Cape as on chart above]. The point of the Main which forms one side of the Passage which is the Northern Promontory of this Country I have named York Cape in honour of His late Royal Highness the Duke of York.

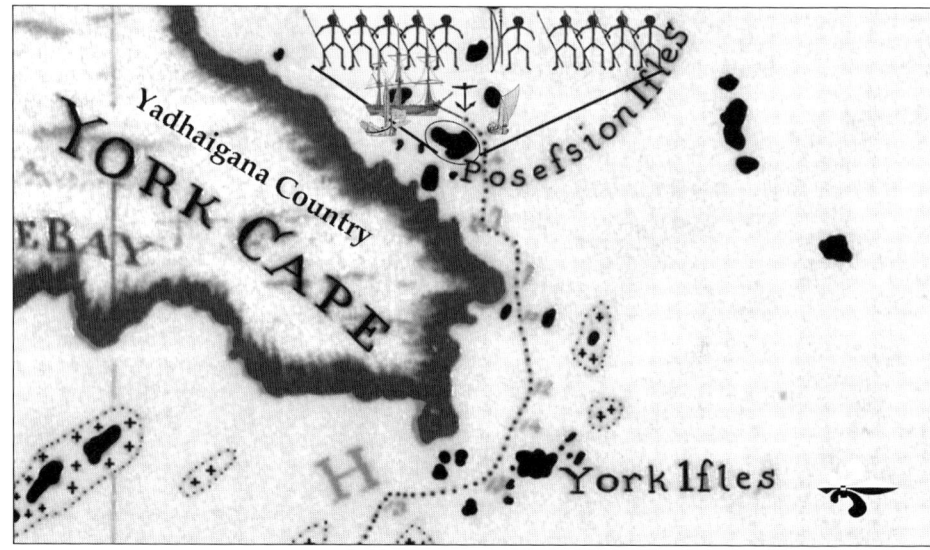

22nd August

COOK: We had not stood above 3 or 4 Miles along shore to the westward before we discovered the Land ahead to be Islands [Possession Isles] detached by several channels from the Main land. I made the signal for the boats to lead through the channel to the northward laying between the islands, which they accordingly did we following with the ship. The channel was about a mile and a half broad from island to island.

BANKS: In passing through, for it was not more than a mile in length before it widened very much, we saw 10 Indians standing on a hill [on Possession Island, marked on chart – Kaurareg or Ankamuti people]; 9 were armed with lances as we had been used to see them, the tenth had a bow and arrows; 2 had also large ornaments of mother of Pearl shell hung round their necks. After the ship had passed by 3 followed her, one of whom was the bow man.

COOK: At 4 o'clock we **anchored** about a Mile and a half or 2 Miles distant from the Islands on each side of us [anchorage marked on chart]. We were in great hopes that we had at last found out a Passage into the Indian seas.

BANKS: We concluded we might have a much better view than from our mast head, so we prepared ourselves to go ashore [on Possession Island] to examine whether the place we stood into was a bay or a passage.

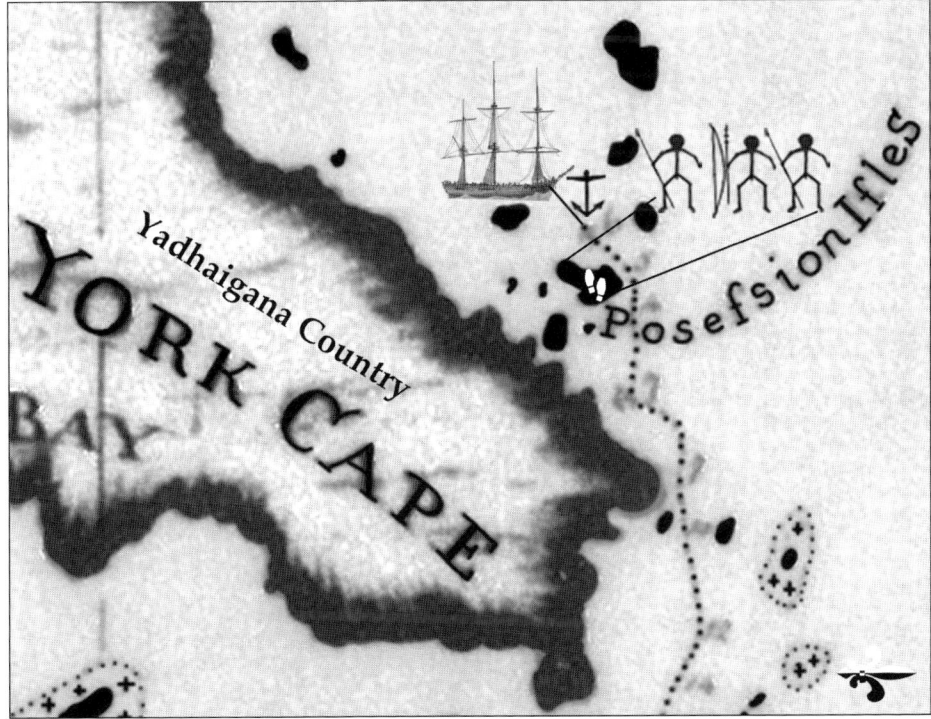

BANKS: The 3 Indians placed themselves upon the beach opposite to us as if resolved either to oppose or assist our landing; when however we came about Musquet shot from them they all walked leisurely away.

COOK: And left us in peaceable possession of as much of the Island as served our purpose. I landed with a party of Men accompanied by Mr Banks and Dr Solander upon the Island which lies at the SE point of the Passage [Possession Island]. After landing I went upon the highest hill.

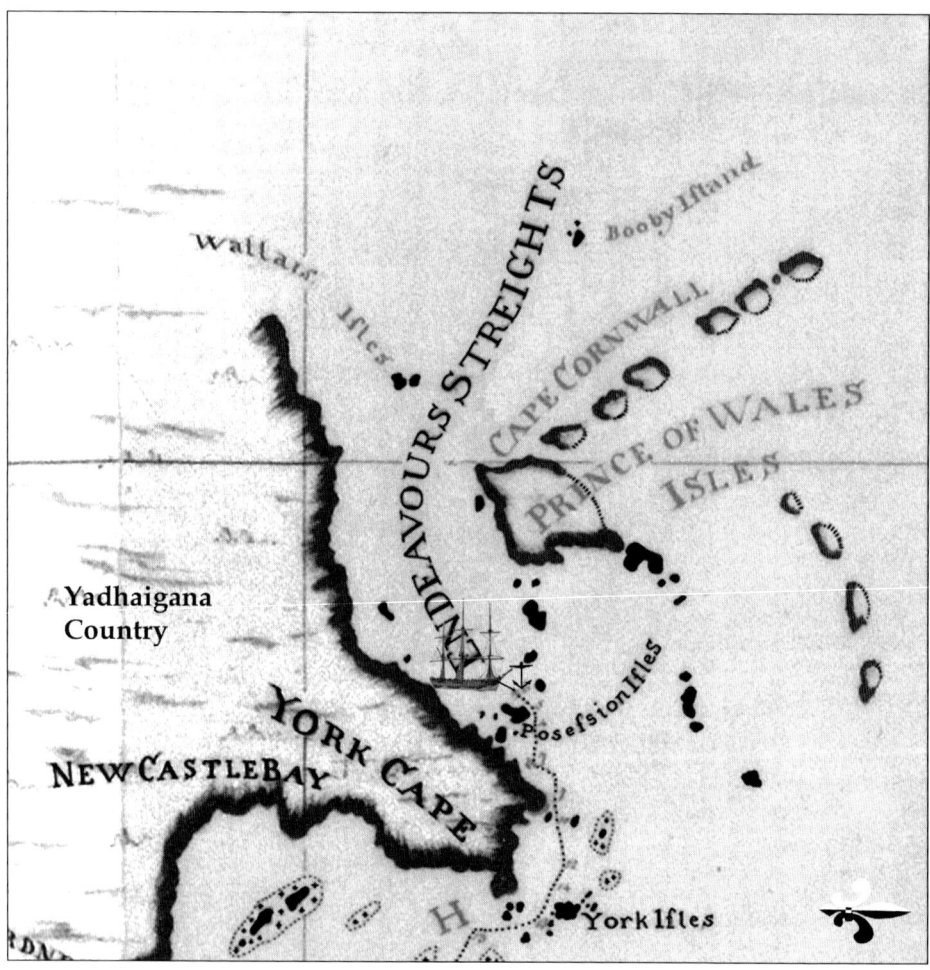

BANKS: The hill we were upon was by much the most barren we had been upon; it however gave us the satisfaction of seeing a straight [Endeavours Streights], at least as far as we could see, without any obstruction.

COOK: I could see no land between SW and WSW so that I did not doubt but there was a passage. Having satisfied myself of the great probability of a passage, thro' which I intend going with the Ship and therefore may land no more upon this eastern coast of New Holland, and on the western side I can make no new discovery, the honour of which belongs to the Dutch Navigators, but the eastern coast from the Latitude of 38° South down to this place I am confident was never seen or visited by any European before us …

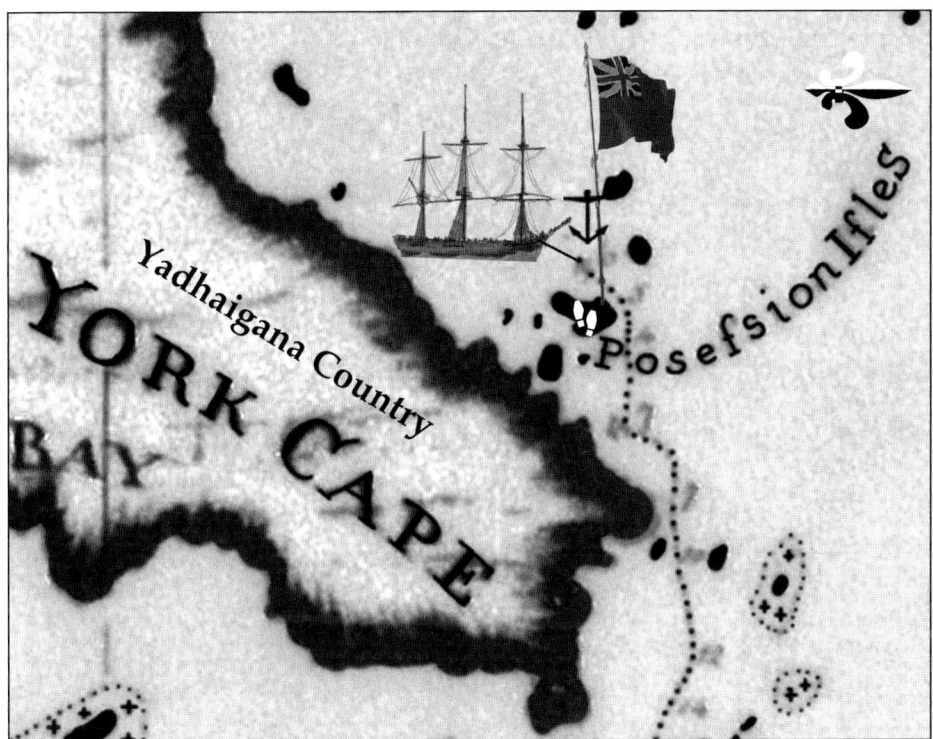

COOK: … and not withstand I had in the Name of his Majesty taken possession of several places upon this coast I now once more hoisted English Colours and in the Name of His Majesty King George the Third took possession of the whole Eastern Coast … (Colour Plate No. 69)

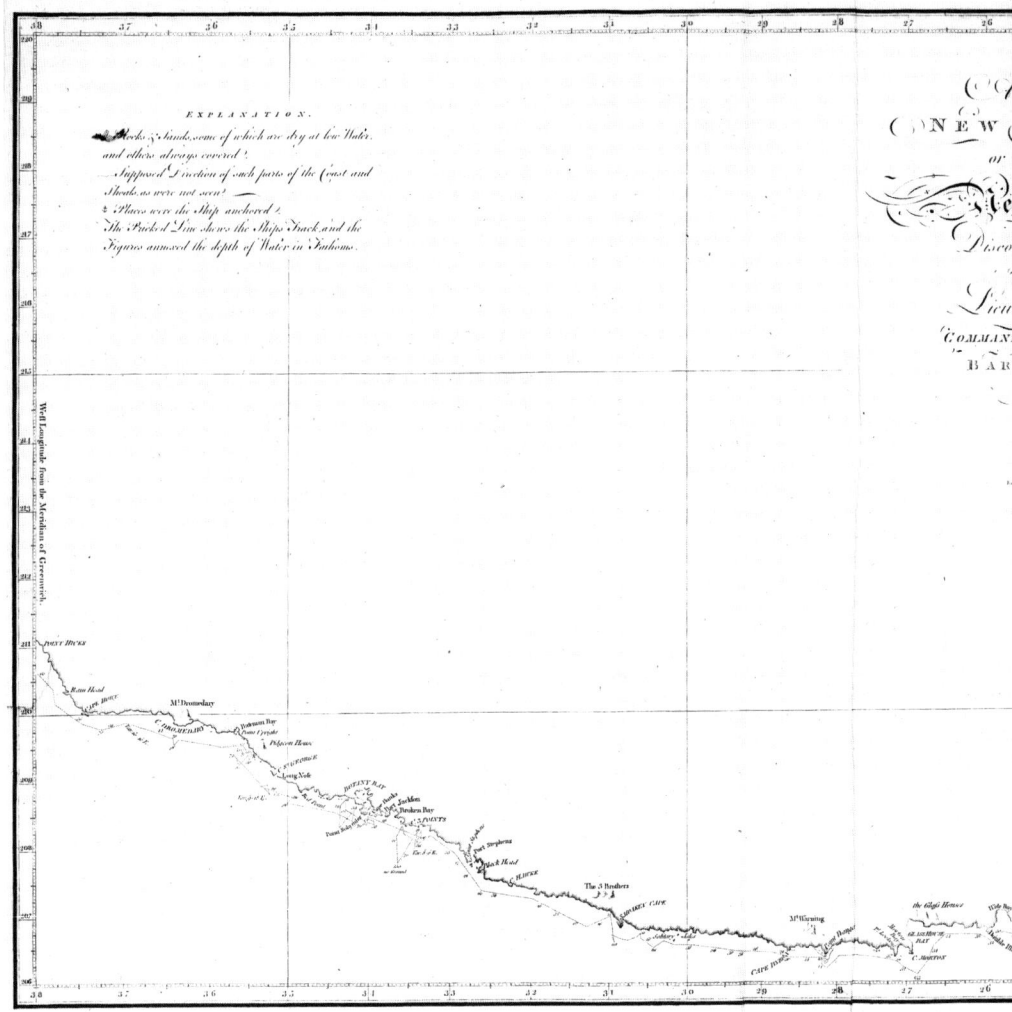

38 degrees South

'A Chart of NEW SOUTH WALES or the East Coast of New Holland. Discover'd and Explored by Lieutenant J. Cook COMMANDER of his MAJESTY'S BARK *ENDEAVOUR* in the Year MDCCLXX'.[2]

COOK: … from the above Latitude [38° South] down to this place [Possession Island] by the Name of New South Wales, together with all the Bays, Harbours Rivers and Islands situate upon the said coast, after which we fired three Volleys of small arms.

PARKINSON: Which was answered by the marines below, and the marines by three vollies from the ship, and three cheers from the main shrouds.

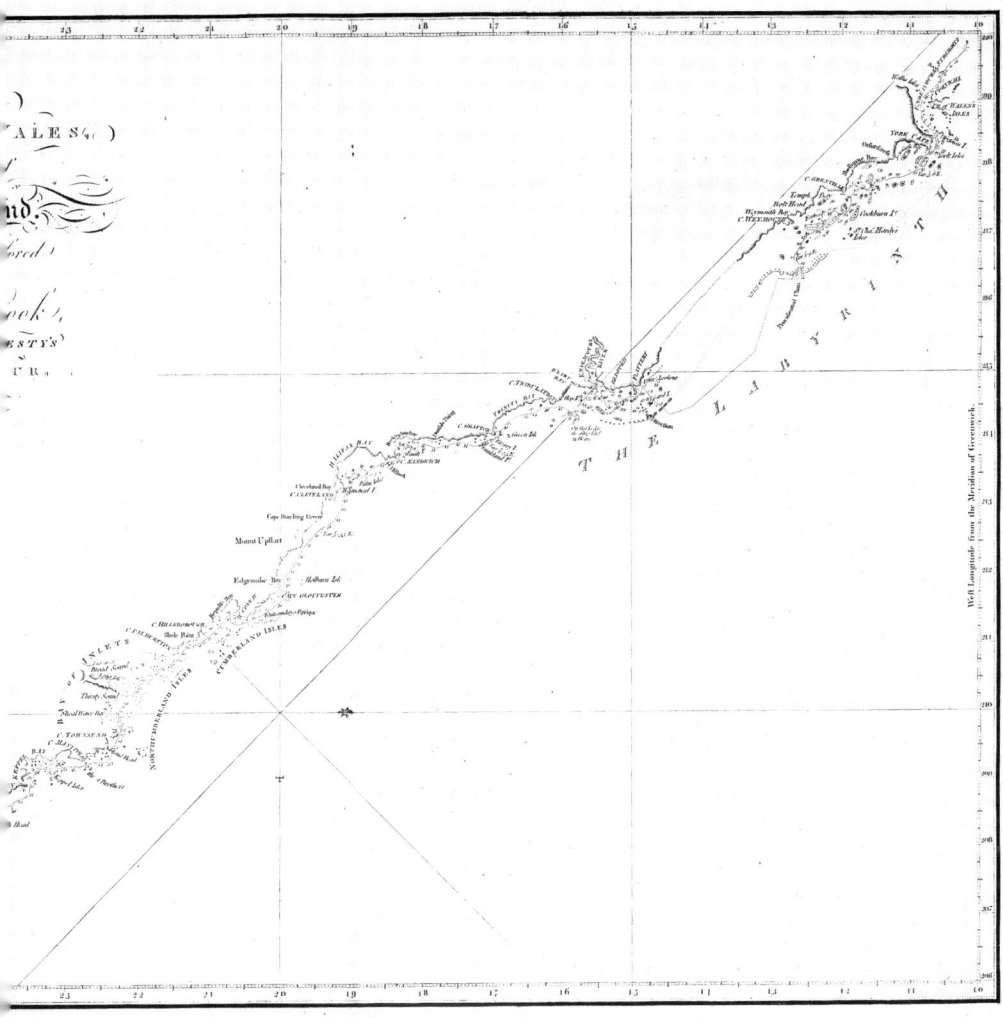

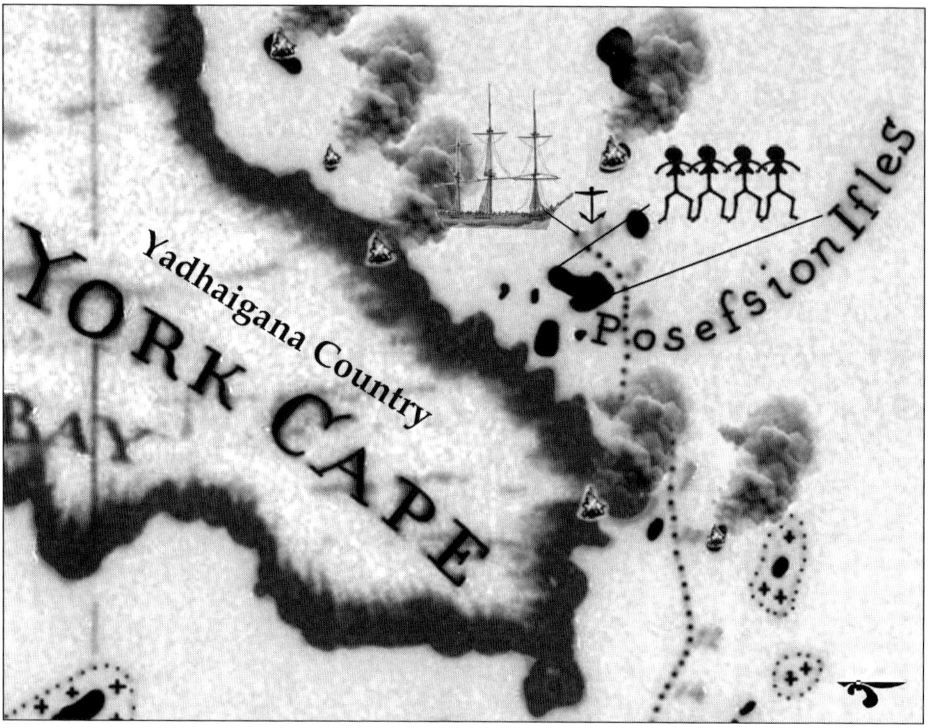

![Map detail showing York Cape, Yadhaigana Country, and Posession Isles]

COOK: We saw on all the Adjacent Lands and Islands a great number of smokes a certain sign that they are Inhabited, and we have daily seen smokes on every part of the coast we have lately been upon.

BANKS: In the morn 3 or 4 women [Kaurareg or Ankamuti people] appeared upon the beach gathering shellfish: we looked with our glasses and to us they appeared as they always did more naked than our mother Eve.

COOK: They had not a single rag of any kind of Clothing upon them and both these and those we saw yesterday were in every respect the Same sort of people we have seen everywhere upon the Coast.

COOK: From what I have said of the Natives of New-Holland they may appear to some to be the most wretched people upon Earth, but in reality they are far more happier than we Europeans; being wholly unacquainted not only with the superfluous but the necessary conveniences so much sought after in Europe, they are happy in not knowing the use of them. They live in a Tranquillity which is not disturbed by the Inequality of Condition: The Earth and sea of their own accord furnishes them with all things necessary for life; they covet not Magnificent Houses, Household-stuff &c. They live in a warm and fine Climate and enjoy a very wholesome Air, so that they have very little need of Clothing and this they seem to be fully sensible of, for many to whom we gave Cloth &c. to, left it carelessly upon the Sea beach and in the woods as a thing they had no manner of use for. In short they seemed to set no Value upon any thing we gave them, nor would they ever part with anything of their own for any one article we could offer them; this, in my opinion argues that they think themselves provided with all the necessaries of Life and that they have no Superfluities.

23rd August 1770

Extract from Cook's original journal, 23rd August 1770, page 753.[3]

BANKS: Thus live these I had almost said happy people, content with little nay almost nothing, Far enough removed from the anxieties attending upon riches, or even the possession of what we Europeans call common necessaries: anxieties intended maybe by Providence to counterbalance the pleasure arising from the Possession of wished for attainments, consequently increasing with increasing wealth, and in some measure keeping up the balance of happiness between the rich and the poor ...

23rd August 1770

BANKS: From them appear how small are the real wants of human nature, which we Europeans have increased to an excess which would certainly appear incredible to these people could they be told it. Nor shall we cease to increase them as long as Luxuries can be invented and riches found for the purchase of them; and how soon these Luxuries degenerate into necessaries may be sufficiently evinced by the universal use of strong liquors, Tobacco, spices, Tea &c. &c. In this instance again providence seems to act the part of a leveller, doing much towards putting all ranks into an equal state of wants and consequently of real poverty: the Great and Magnificent want as much and may be more than the middling: they again in proportion more than the inferior: each rank still looking higher than his station but confining itself to a certain point above which it knows not how to wish, not knowing at least perfectly what is there enjoyed.

Extract from Banks's original journal, 23rd August 1770, page 291.[4]

> & the poor. From them appear how small are the real wants of human nature which we Europeans have increasd to an excess which would certainly appear incredible to these people. Could they be told it nor shall we cease to increase them as long as Luxuries can be invented & riches found for the purchace of them & how soon these Luxuries degenerate into necessaries may be sufficiently evined by the universal use of strong liquers, Tobacco, spices, Tea &c&c. in this instance again providence seems to act the part of a Leveler doing much towards putting all ranks into an equal state of wants & consequently of real poverty the great & magnificent want as much & may be more than the midling they again in proportion more than the inferior each rank still looking higher than his station but confining itself to a certain point above which it knows not how to wish not knowing at least perfectly what is there enjoyd.

Extract from Banks's original journal, Page 292.

COOK: I do not look upon them to be a warlike People, on the Contrary I think them a timorous and inoffensive race, no ways inclinable to cruelty, as appeared from their behaviour to one of our people in Endeavour River which I have before mentioned.

At low-water about 10 o'clock we got under sail and stood to the SW with a light breeze.

Epilogue

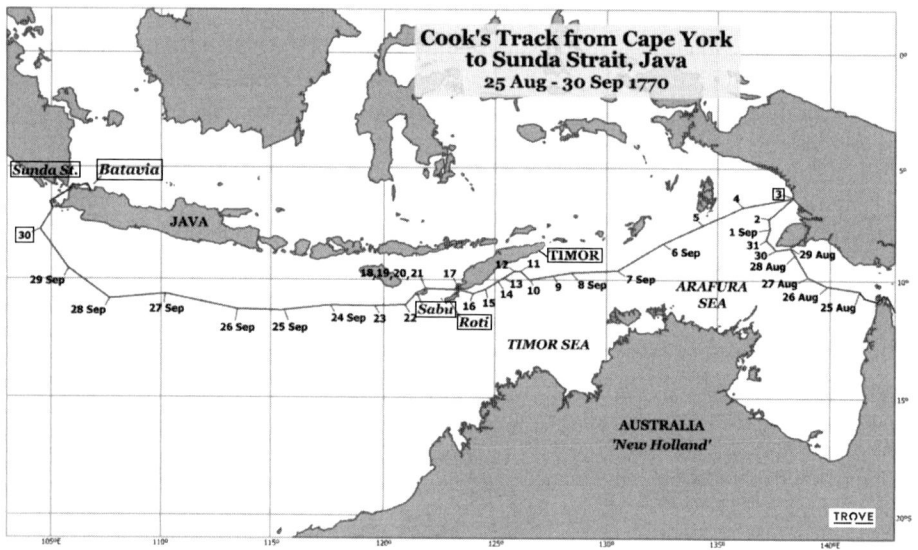

Cook's route from Cape York to Sunda Strait, Java.¹

Their route along Endeavours Streights took them through difficult shallow water, and the shoals and sandbanks continued to endanger them:

> 23rd August – We had now an open Sea to the westward, which gave me no small satisfaction being able to prove that New-Holland and New-Guinea are 2 Separate Lands or Islands, which until this day hath been a doubtful point with Geographers. **(Cook)**

In a leaking ship they steered north-west, along the coast of New Guinea.

> 28th August – Vast quantities of the little substances floating upon the water in large lines a mile or more long and 50 or 100 yards wide. The seamen who are now convinced that it was not as they had thought the spawn of fish began to call it Sea sawdust, a name certainly not ill adapted to its appearance. [*Trichodesmium* is a genus of filamentous cyanobacteria. It fixes atmospheric nitrogen into ammonium, a nutrient used by other organisms.] Some of our people observed they smelt an uncommon stink which they supposed to proceed from it. **(Banks)**

> 3rd Sept – A party of our people went, in the pinnace, to examine the country [New Guinea] while we stood off and on. They soon returned with an account that a great number of the natives threatened them on the beach, who had pieces of bamboo, or canes, in their hands, out of which they puffed some smoke, and then threw some darts at them about a fathom long, made of reeds, and pointed of Etoa wood, which were barbed, but very blunt. Our people fired upon them, but they did not appear to be intimidated; our men, therefore, thought proper to embark. They observed that these people were not negroes, as has been reported, but are much like the natives of New Holland, having shock hair, and being entirely naked. **(Parkinson)**

Leaving New Guinea, they turned west and the water deepened. Their destination now Batavia, the capital city of the Dutch East Indies, the outskirts of European civilization:

> As soon as ever the boat was hoisted in we made sail and steered away from this land to the no small satisfaction of I believe three fourths of our company the sick became well and the melancholy looked gay. The greatest part of them were now pretty far gone with the longing for home which the Physicians have gone so far as to esteem a disease under the name of Nostalgia; indeed I can find hardly anybody in the ship clear of its effects but the Captain Dr Solander and myself, indeed we three have pretty constant employment for our minds which I believe to be the best if not the only remedy for it. **(Banks)**

They passed several islands and the land of Timor. Then the island Roti. Then Sabu, where they landed, met the local Raja and his Dutch master, who dined on board and promised to supply them with provisions.

> At setting down however the King [the local Raja] excusd himself, saying that he did not imagine that we who were white men would suffer him who was black to set down in our company. A complement however removd his scruples and he and his prime minister sat down and eat sparingly. During all dinner time we receivd many professions of freindship from both the King and the European who was a native of Saxony by name Johan Christopr Lange. Mutton was our fare: the King expressd a desire of having an English sheep; we had one left which was presented to him. An English dog was then askd for and my greyhound presented to him. **(Banks)**

> After much shuffling on their part, we made shift to obtain a large number of fowls, eight bullocks, several goats, hogs, a great quantity of syrup, and a few fruits. **(Parkinson)**

Here also Parkinson recorded many words of the native language, as he had previously done in all the different countries they had visited.

After a stay of two or three days, they left Sabu and continued out of sight of land until 30th September.

> In the a.m. I took into my possession the Officers, Petty officers and Seamens Log Books & Journals, at least all that I could find and enjoined every one not to divulge where they had been. (**Cook**)

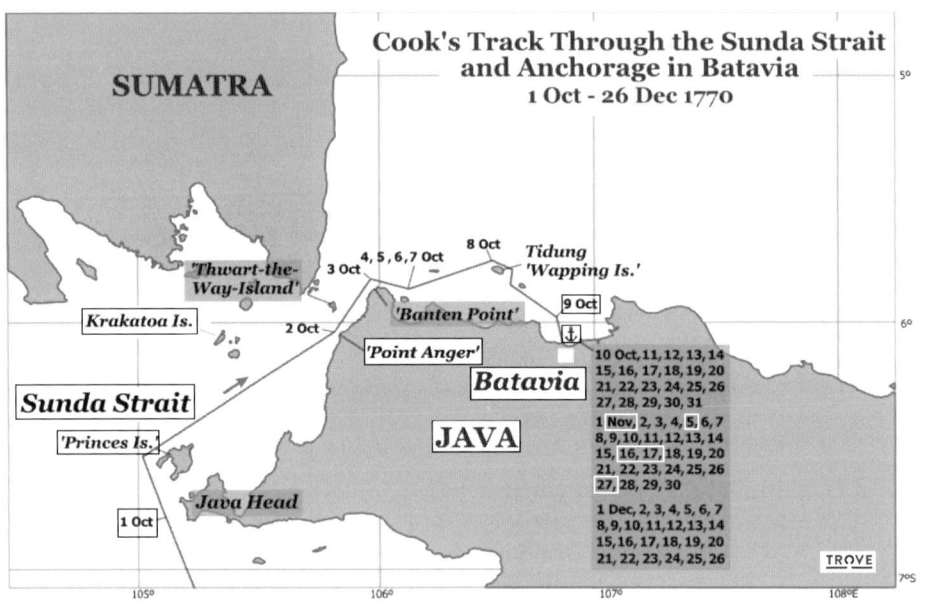

On 1st October, in the morning, they sighted Java and Prince's Island. They made the run up the Sunda Strait, passing the volcanic island of Krakatoa on the way:

> The breeze was fresh and tolerably favourable so that at night we had passed Crocata [Krakatoa] and stood on by very clear Moonlight, tho the clouds about the Horizon threatened and it lightened a good deal. (**Banks**)

On 2nd October they **anchored** behind Point Anger where they received disturbing news from two Indiamen laying at anchor there:

> Their Captains received our officer very politely and told him some European news, as that the government in England were in the utmost disorder, the people crying up and down the streets Down with King George, King Wilkes for ever; that the Americans had refused to pay taxes of any kind in consequence of which was a large force being sent there both of sea and land forces. (**Banks**)

On the 9th of October, we **anchored** in the road of Batavia, – in which we found sixteen large ships, three of which were British. **(Parkinson)**

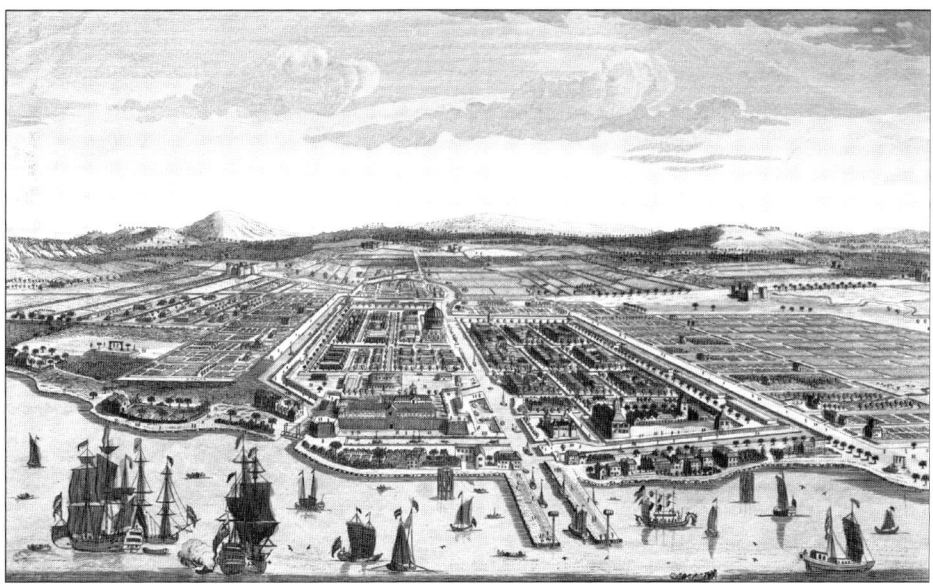

Batavia around 1780 – Present-day Jakarta.[2]

When they arrived, Banks reported:

> A boat came immediately on board us. The officer on board her enquired who we were. Both himself and his people were almost as spectres, no good omen of the healthiness of the country we were arrived at; our people however who truly might be called rosy and plump, for we had not a sick man among us, jeered and flouted much at their brother seamen's white faces.

Banks hired a small house ashore, and Cook began negotiations with the Dutch authorities for repairs to the ship.

Lieutenant Hickes, Mr Green and Tupia were the only people who had any complaints, and Hickes was sent ashore for the recovery of his health.

When Tupia came ashore:

> his spirits were instantly raised by the sights which he saw, and his boy Taiyota who had always been perfectly well was almost ready to run mad. Houses, carriages, streets, in short everything were to him sights which he had often heard described but never well understood, so he looked upon them all with more than wonder, almost made with the numberless novelties which diverted his attention from one to the other he danced about the streets examining everything to the best of his abilities. **(Banks)**

However, Tupia's broken constitution grew worse and worse every day. Then Taiyota was attacked by a cold and inflammation on his lungs. Banks's servants, Peter Brisco and James Roberts, and Banks himself had intermitting fevers, and Dr Solander a constant nervous one; in short everyone on shore and many on board were taken ill.

The Dutch-built canals of the city harboured disease and bred malarial mosquitoes. To make matters worse, the damage to the ship proved more serious than they thought, and delays occurred.

Tupia and Taiyota grew worse every day ashore, and Tupia asked to be returned to the ship where he said he should breathe a freer air, clear of the numerous houses that he believed to be the cause of his disease by stopping the free draught:

> All the ships, which are careened and hove-down here, go to a small island in the bay, called Onrust, about seven miles from Batavia. The whole island is one dockyard. Near Onrust is another island, called Coopers Island. There are many other islands in the bay, named Amsterdam, Rotterdam, and Eadam. **(Parkinson)**

Cook set up the tents for the reception of the sick among the ship's company on Coopers Island. Tupia had a tent pitched for himself there too, in a place he chose where both sea breeze and land breeze blew right over him, a situation in which he expressed great satisfaction. The seamen were now falling sick so fast that the tents ashore were always full of ill men.

After a stay of two days minding Tupia, Banks left him and returned to town where:

> I was immediately seized with a tertian [occurring every third day], the fits of which were so violent as to deprive me entirely of my senses and leave me so weak as scarcely to be able to crawl downstairs. My servants Peter and James were as bad as myself, and Dr Solander now felt the first attacks of his fever. But worst of all was Mr Monkhouse the ships surgeon; he was now confined to his bed by a violent fever which grew worse and worse notwithstanding all the efforts of the physician. **(Banks)**

> <u>5th November</u> – In the afternoon of this day poor Mr Monkhouse departed, the first sacrifice to the climate and the next day was buried. Dr Solander attended his funeral, and I should certainly have done the same had I not been confined to my bed by my fever. **(Banks)**

His loss will be the more severely felt. He was succeeded by Mr Perry, his Mate who is equally well if not better skilled in his profession. **(Cook)**

The ship was transported over to Onrust, alongside one of the careening wharfs:

> We found that many of her planks; and her keel, were much damaged; one part of her not being above one-eighth of an inch thick, which was luckily before one of the timbers, or, in all probability, she would have sunk long before we reached the bay of Batavia. **(Parkinson)**

Surgeon William Brougham Munkhouse (1732–1770).

It was a matter of surprise to everyone who saw her bottom how we had kept her above water and yet in this condition we had sailed some hundreds of leagues in as dangerous a navigation as is in any part of the world happy in being ignorant of the continual danger we were in. **(Cook)**

While our ship was repairing at Onrust, most of the crew were at Cooper's island, where they were taken with a putrid dysentery; three of whom, the steward of the gunroom, one of the seamen, and a boy, died. The disorder also carried off Tupia, and the lad Taiyota. **(Parkinson)**

Taiyota – portrait by Sydney Parkinson.

When Taiyota was seized with the fatal disorder, as if certain of his approaching dissolution, he frequently said to those of us who were his intimates, Tyau mate oee, 'my friends, I am dying'. He took any medicines that were offered him; but Tupia, who was ill at the same time, and survived him but a few days, refused everything of that kind, and gave himself up to grief; regretting, in the highest degree, that he had left his own country, and, when he heard of Taiyota's death, he was quite inconsolable, crying out frequently, Taiyota! Taiyota! They were both buried in the island of Eadam. **(Parkinson)**

I had given him quite over ever since his boy died whom I well knew he sincerely loved, though he used to find much fault with him during his lifetime. **(Banks)**

Tupia's death indeed cannot be said to be owing wholly to the unwholesome air of Batavia, the long want of a vegetable diet which he had all his life before been used to had brought upon him all the disorders attending a sea life. He was a shrewd sensible, ingenious man, but proud and obstinate which often made his situation on board both disagreeable to himself and those about him and tended much to promote the deceases which put a period to his life. **(Cook)**

Dr Solander and Banks still grew worse and worse, and the doctor took to his bed. Banks immediately sent for their local physician, Dr Jaggi, who applied sinapisms to his feet (a poultice containing the volatile oil of mustard seed, a powerful irritant) and blisters to the calves of his legs but expressed little or no hopes of even the possibility of his living till morning. Banks, much alarmed, sat by him till morning when Dr Solander appeared visibly better.

Dr Jaggi insisted on the country air as necessary for their recovery. They procured a small house about two miles out of town, supplied with provisions and the use of five slaves. It was situated on the banks of a briskly running river and well open to the sea breeze, 'in a country perfectly resembling the low part of my native Lincolnshire'. **(Banks)**

Cook, on hearing of their plight, sent Mr Spöring, a seaman and the captain's own servant from the ship. That night, Dr Solander had a relapse but seemed recovered by morning, and Banks, who suffered anxiety about Solander's condition, was also improved:

Dr Solander grew better though by very slow degrees; myself soon had a return of my ague which now became quotidian [recurring daily], the Captain also was taken ill on board and of course we sent his servant to him, soon after which both Mr Spöring and our seaman were seized with intermittents, so that we were again reduced to the melancholy necessity of depending entirely upon the Malays for nursing us, all of whom were often sick together. (**Banks**)

<u>16th November</u> The ship was now repaired; and reloading commenced.

<u>17th November</u> We are now become so sickly that we seldom can muster above 12 or 14 hands to do duty. (**Cook**)

<u>27th November</u> Got all the Sick on board and every other thing from the Island. (**Cook**)

The 25th Xmas day by our account being fixed for sailing, we this morn hired a large country Praw, [a horse-drawn vehicle] which came up to the door and took in Dr Solander, now tolerably recovered, and carried him on board the ship where in the evening we all joined him. (**Banks**)

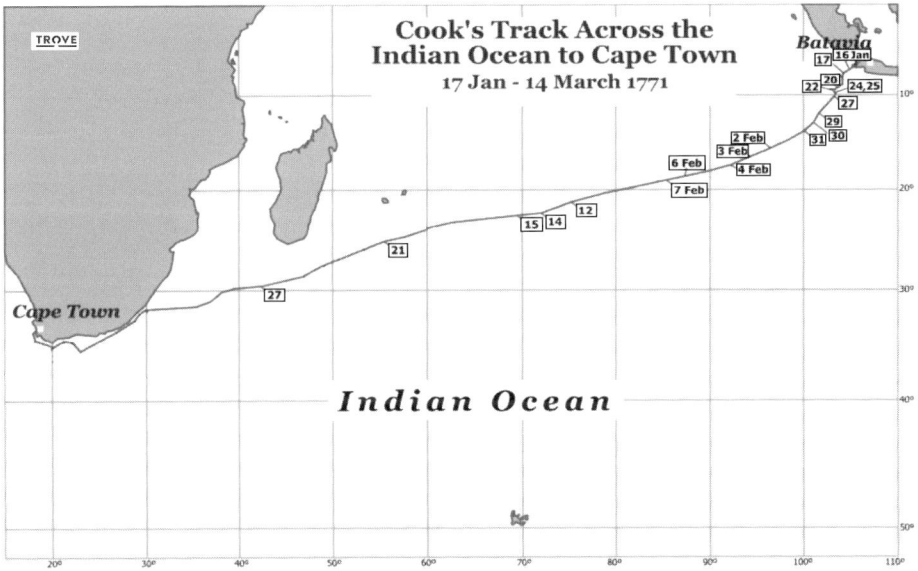

26th December At 6 in the a.m. we weighed and came to sail with a light breeze at SW. The Number sick on board at this time amounts to 40 or up wards and the rest of the Ships company are in a weakly condition having been every one sick except the Sail maker an old man about 70 or 80 years of age and what was still more extraordinary in this man his being generally more or less drunk every day. But notwithstanding this general sickness we lost but seven men in the whole, the Surgeon, three Seamen [Timothy Rearden, John Woodworth and another], Mr Greens Servant [John Reynolds] and Tupia and his servant both of which fell a sacrifice to this unwholesome climate before they had reached the Object of their wishes. (**Cook**)

16th January 1771 This Morn we waked in the open Ocean, nothing in sight but sea and sky. The winds though fair continued yet so gentle that we hardly knew whether we went on or stood still. At night a booby made us a visit and slept his last sleep in the stomachs of some of our men, not induced quite to forsake the old trade of Booby Eating even by the present abundance of victuals. (**Banks**)

17th January Calms and light breezes still detained us till eve when a pleasant breeze sprung up and gave us hopes of soon gaining the trade wind, which we impatiently longed for, especially myself who had my fever every day; nor was I the only sick man, many began to complain of purgings. (**Banks**)

20th January Myself who had begun with the Bark Yesterday missed my fever today, the people however in general grew worse and many had now the dysentery or bloody flux. (**Banks**)

22nd January Almost all the Ships Company were now ill with either fluxes or severe purgings; myself far from well, Mr Spöring very ill and Mr Parkinson very little better, his complaint was a slow fever. (**Banks**)

24th January Myself was too ill today to do anything. (**Banks**)

   In the a.m. died John Truslove Corporal of Marines, a Man much esteemed by every one on board – Many of our people at this time lay dangerously ill of Fevers and fluxes, We are inclinable to attribute this to the water we took in at Princes Island and have put lime into the Casks in order to purify it. (**Cook**)

25th January My distemper this day turned out to be a flux attended (as that disease always is) with excruciating pains in my bowels, on which I took to my bed. (**Banks**)

   Departed this Life Mr Spöring a Gentleman belonging to Mr Banks's retinue. (**Cook**)

   One more of the People died today [Thomas Dunster, Marine Private]. Myself endured the pains of the Dammed almost; at night they became fixed in one point in my bowels on which the surgeon of the ship thought proper to order me the hot bath, into which I went 4 times at the intervals of two hours and felt great relief. (**Banks**)

27th January Departed this Life Mr Sidney Parkinson Natural History Painter to Mr Banks, and soon after John Ravenhill, Sailmaker a Man much advanced in years. (**Cook**)

   Though better than yesterday my pains were still almost intolerable. (**Banks**)

29th January Self something easier but still in great pain. (**Banks**)

In the night died Mr Charles Green who was sent out by the Royal Society to Observe the Transit of Venus; he had long been in a bad state of health which he took no care to repair but on the contrary lived in such a manner as greatly promoted the disorders he had long upon him, this brought on the Flux which put a period to his life. (**Cook**)

30th January Died of the Flux Samuel Moody and Francis Hate, two of the Carpenters Crew. (**Cook**)

For the first time I found myself better and slept some time, which my continual pains had never suffered me to do before notwithstanding the opiates which were constantly administered. So weak were the people in general that, officers and men included, not more than 8 or nine could keep the deck so that 4 in a watch was all they had. (**Banks**)

Sydney Parkinson – Natural history artist (1745–1771).

31st January In the Course of this 24 hours we have had four Men died of the Flux, viz. John Thompson Ships Cook, Benjamin Jordan Carpenters Mate, James Nicholson and Archibald Wolfe Seamen. A Melancholy proof of the Calamitous situation we are at present in, having hardly well men enough to tend the Sails and look after the Sick, many of the latter are so ill that we have not the least hopes of their recovery. I shall mention what effect only the imagery approach of this disorder had upon one man. He had long tended upon the Sick and enjoyed a tolerable good state of health. One morning coming upon deck he found himself a little griped and immediately began to stamp with his feet and exclaim, I have got the Gripes, I have got the Gripes: I shall die, I shall die, in this manner he continued untill he threw himself into a fit and was carried off the deck in a manner dead, however he soon recovered and did very well. (**Cook**)

This day I got out of my bed in good spirits and free from pain but very weak. My recovery had been as rapid as my disease was violent, but to what cause to attribute either the one or the other to we all were equally at a loss. The wind which came to E and SE yesterday blew today in the same direction so we had little reason to doubt its being the true trade, a circumstance which raised the spirits of even those who were most afflicted with the tormenting disease, which now raged with its greatest violence. (**Banks**)

2nd February Departed this Life Daniel Roberts Gunners Servant who died of the flux. Since we have had a fresh Trade wind this fatal disorder hath seemed to be at a stand, yet there are several people which are so far gone and brought so very low by it that we have not the least hopes of their recovery. (**Cook**)

3rd February Departed this Life John Thurman Sailmakers assistant. (**Cook**)

4th February In the night died of the Flux Mr John Bootie Midshipman and Mr John Gathrey Boatswain. (**Cook**)

6th February In the night died Mr Jonathan Monkhouse Midshipman and Brother to the late Surgeon. (**Cook**)

7th February Through the whole course of this distemper Medicine has been of little use, the Sick generally proceeding gradually to their end without a favourable symptom, till the change of weather stopped in a manner instantaneously the Malignant quality of the disease. (**Banks**)

12th February At 7 in the a.m. died of the flux after a long and painful illness, Mr John Satterly, Carpenter, a Man much Esteemed by me and every Gentleman on board, in his room I appoint George Knowel one of the Carpenters Crew, having only him and one More left. (**Cook**)

14th February Departed this Life Alexander Lindsey Seaman, this man was one of those we got at Batavia and had been some time in India. (**Cook**)

15th February Died of the flux Daniel Preston Marine. (**Cook**)

21st February In the night died of the flux Alexander Simpson a very good Seaman. In the Morning Punished Thomas Rossiter with Twelve Lashes for getting Drunk, grossly Assaulting the Officer of the Watch and beating some of the Sick. (**Cook**)

27th February In the a.m. Died of the Flux Henry Jeffs, Emanuel Pharah [Manuel Pereira/Parreyra] and Peter Morgan Seamen, the last came Sick on board at Batavia of which he never recovered and the other two had long been past all hopes of recovery, so that the death of these three men in one day did not in the least alarm us. On the contrary we are in hopes that they will be the last that will fall a Sacrifice to this fatal disorder, for such as are now ill of it are in a fair way of recovering. (**Cook**)

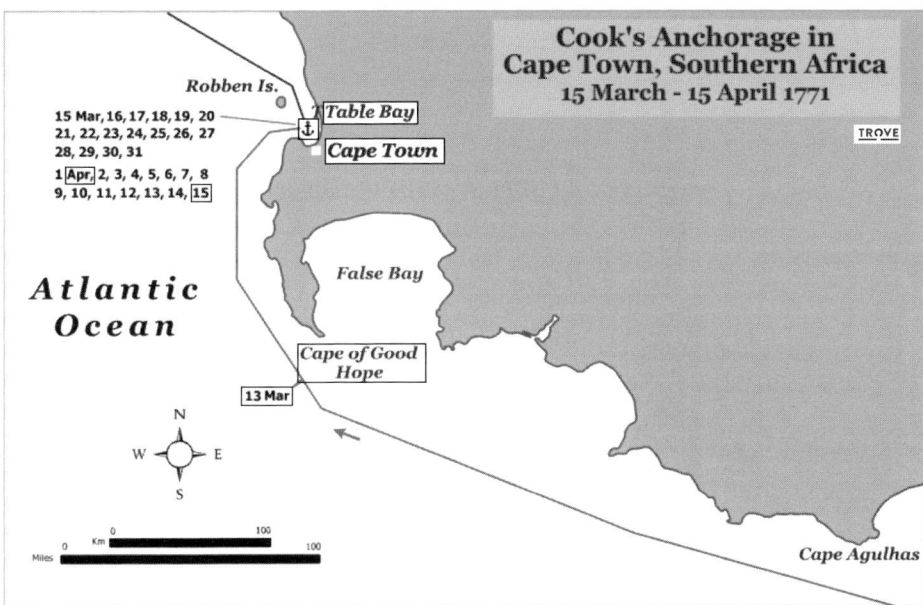

On 13th March they sighted the Cape of Good Hope and two days later anchored in Table Bay, Cape Town. Cook waited upon the Governor, 'who was pleased to tell me that I should have everything I wanted that the place afforded', and a place was found ashore to accommodate the sick.

Here they encountered sixteen other ships, one of which was an English East Indiaman that saluted them with eleven guns, which compliment they returned. She was the *Admiral Pocock*, under the command of Captain Riddell, and homeward bound from Bombay, so Cook delivered his letter for the Admiralty and Royal Society to be taken with her.

Dr Solander suffered a relapse, a consequence of Batavia fevers, and a country doctor was sent for. He was confined to his bed and emerged two weeks later, 'very much emaciated by his tedious Illness'. **(Banks)**

The month they spent in Cape Town saw the ship resupplied with provisions and repairs carried out. During their stay, more of their numbers had perished. 'Three died here [Able Seamen Richard Thomas, John Lorrain and John Dozey] but this loss was made up by the opportunity we had of completing our full Compliment. In the Morning unmoored and got ready for sailing.' **(Cook)**

15th April Weighed and stood out of the Bay – Saluted with 13 Guns which compliment was returned both by the Castle and Dutch Commodore, the Europa saluted us as we passed her, which we returned. **(Cook)**

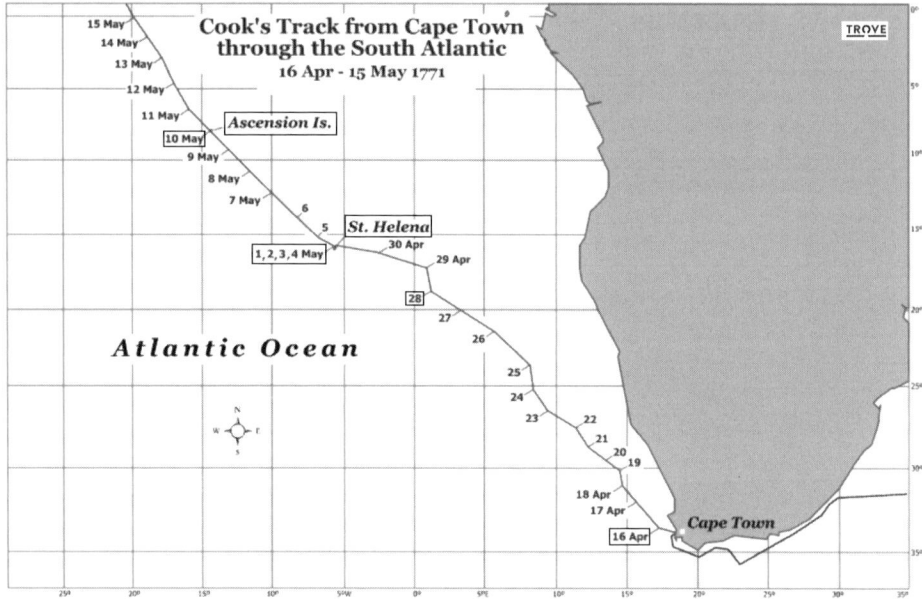

16th April At 3 Weighed with a light breeze at SE and put to Sea – At 4 Departed this Life Mr Robert Molineux Master, a young man of good parts but had unfortunately given himself up to extravagancy and intemperance which brought on disorders that put a period to his life. (**Cook**)

28th April This day we crossed our first meridian and Completed the Circumnavigation of the Globe. (**Banks**)

1st May Saw the Island of St Helena. (**Cook**)

2nd May At Anchor in St Helena Road. (**Cook**)

Ship's Master – Robert Molineux (1746–1771).

3rd May Spent this day in Botanizing on the Ridge where the Cabbage trees grow, visiting Cuckolds point and Dianas peak, the Highest land in the Island as settled by the Observations of Mr Maskelyne, who was sent out to this Island by the Royal Society for the Purpose of Observing the transit of Venus in the Year (1761). (**Banks**)

4th May At 1 p.m. weighed and Stood out of the Road in Company with the Portland and 12 Sail of Indiamen resolved to steer homewards with all expedition in Order (if possible) to bring home the first news of our voyage, as we found that many Particulars of it has transpired and particularly that a copy of the Latitudes and Longitudes of most or all the principal places we had been at had been taken by the Captain's Clerk [Richard Orton] from the Captain's own Journals and Given or Sold to one of the India Captains. (**Banks**)

10th May At 6 in the a.m. saw the Island of Ascention bearing NNW distant 7 Leagues. (**Cook**)

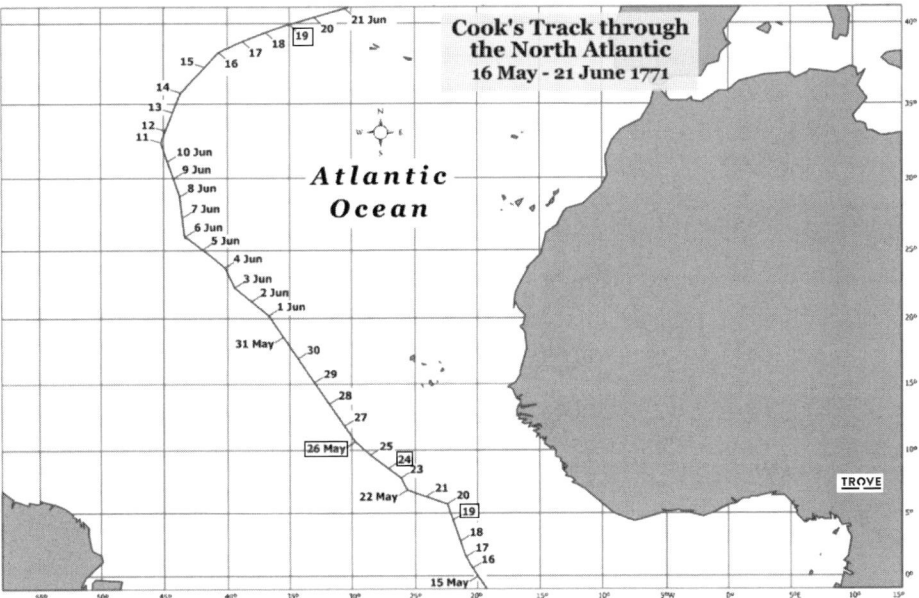

19th May Hoisted a boat out and sent on board the Houghton for the Surgeon Mr Carret in order to look at Mr Hicks who is so far gone in a Consumption that his Life is despaired of. (**Cook**)

24th May It became again hazy and we lost sight of them [the fleet] and notwithstanding we kept close upon a wind all night with as much sail out as we could bear there was not one sail in sight in the Morning. (**Cook**)

26th May About 1 o'clock in the p.m. departed this life Lieutenant Hicks and in the evening his body was committed to the Sea with the usual ceremonies, he died of a Consumption which he was not free from when we sailed from England so that it may be truly said that he hath been dying ever since, though he held out tolerable well until we got to Batavia. (**Cook**)

<u>27th May</u> This day I gave Mr **Charles Clerk** an order to Act as Lieutenant in the Room of Mr Hicks deceased, he being a young Man extremely well qualified for that station. (**Cook**)

Charles Clerke was 27 when he joined *Endeavour* as master's mate. He was to sail with Cook on all three of Cook's voyages. After Hickes's death, John Gore became second lieutenant. The two men had a long history together.

Clerke was a midshipman and was on board when *Bellona* captured the *Courageux* on 13th August 1761, where he possibly first encountered John Gore with whom he developed a close friendship. Clerke was in the mizen-top when it was shot away, and fell with it into the sea but saved himself by climbing up the fore chains. In 1764, as midshipman he sailed under John Byron in HMS *Dolphin* during that ship's voyage round the world. John Gore was also on board as master's mate.

On Cook's third voyage, after Cook was killed on 14th February 1779, Clerke assumed overall command of the expedition and transferred from *Discovery* to *Resolution* and Gore moved across to *Discovery* as its new commander. When Clerke died on 23rd August 1779, Gore re-joined *Resolution* as overall leader of the expedition.

Clerke's first action when Cook was killed was to prevent reprisals being taken against the Hawaiians so that when the ships left, a few days later, a semblance of peace had been restored. Clerke developed an ability in astronomy during the *Endeavour* voyage and took over the duties of the deceased Charles Green.

<u>19th June</u> At 10 o'clock in the a.m. Saw a Sail ahead which we soon came up with and sent a boat on board. She was a Schooner from Road Island out upon the Whale fishery – from her we learnt that all was peace in Europe and that the America disputes were made up, to Confirm this the Master said that the Coat on his back was made in Old England. (**Cook**)

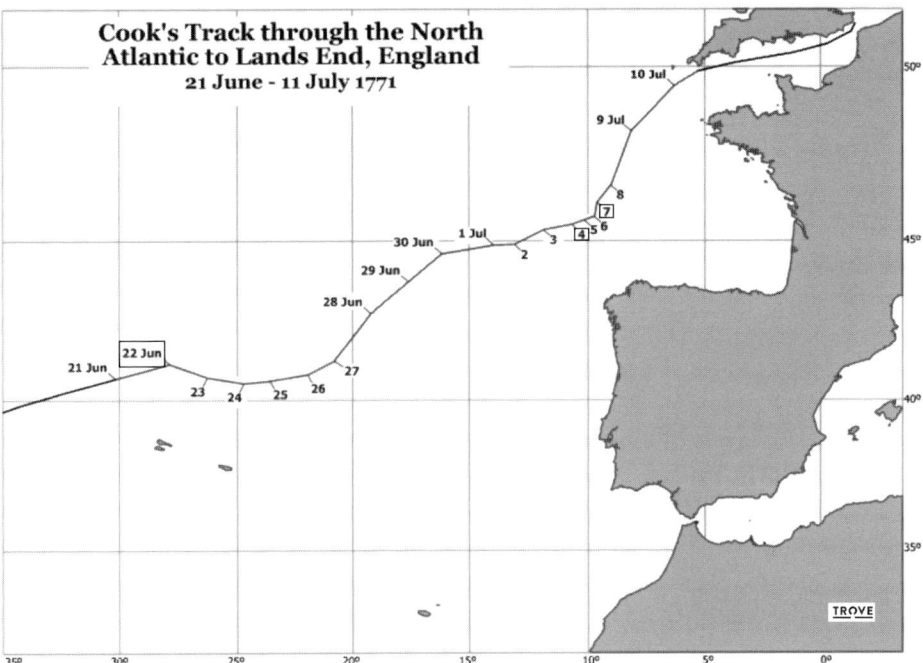

22nd June Our Rigging and Sails are now so bad that some thing or another is giving way every day. (**Cook**)

4th July My Bitch Lady was found dead in my Cabin laying upon a stool on which she generally slept. She had been remarkably well for some days; in the night she shrieked out very loud so that we who slept in the great Cabin heard her, but becoming quiet immediately no one regarded it. Whatever disease was the cause of her death it was the most sudden that ever came under my Observation. (**Banks**)

7th July At 9 a.m. spoke to a Brig from London bound to the Grenades. We learnt from this Vessel that no accounts had been received in England from us and that Wagers were held that we were lost. (**Cook**)

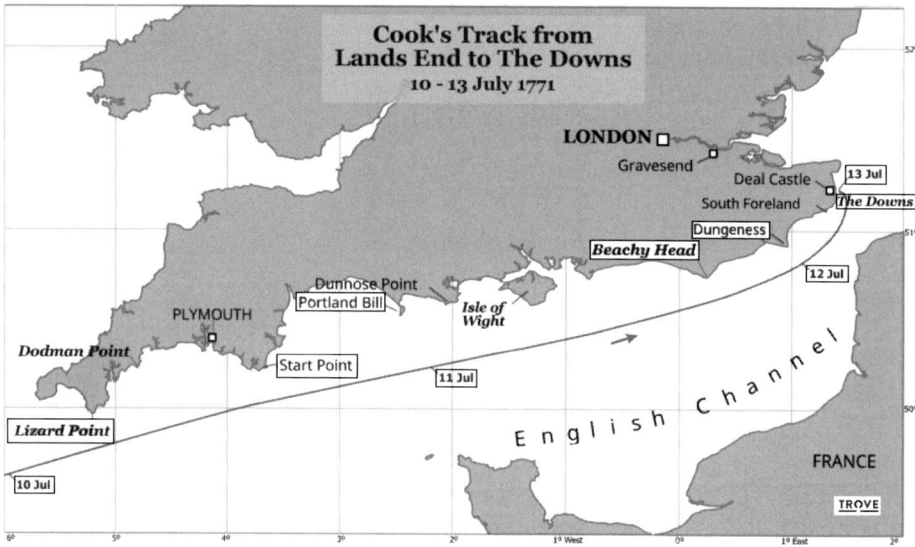

10th July This Morning the land was discovered by Young Nick the same boy who first saw New Zealand: it proved to be the Lizzard. (**Banks**)

11th July At 7 o'clock in the Morning the Start Point bore NWBN distant 3 Leagues. (**Cook**)

12th July p.m. passed the Bill of Portland and at 7 Peverell point. At 6 a.m. passed Beachy head at the distance of 4 or 5 Miles, at 10 Dungenness at the distance of 2 Miles and at Noon we were abreast of Dover. (**Cook**)

Saturday, 13th July 1771 At 3 o'clock in the p.m. Anchored in the Downs, & soon after I landed in order to repair to London. (**Cook**)

Francis Wilkinson, Master's Mate, died a few days after *Endeavour* had tied up in the Thames in August 1771.

Glossary of Terms

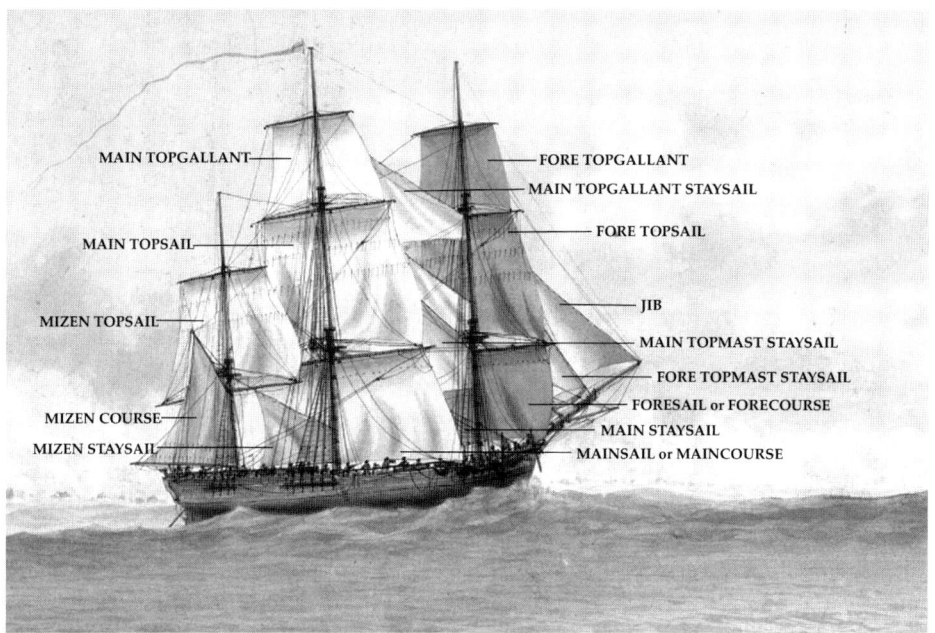

Illustration 1.

Abaft – A location toward the stern of the vessel. Sometimes with reference to another point, as abaft the beam.

Anchors – *Endeavour* had two bow anchors (best bower and small bower), a sheet anchor, two stream anchors, and one kedge anchor.

Bower anchors were the two main anchors of the ship. The small bower weighed 17cwt 3qr 14lbs. The best bower weighed much the same. The bower anchors were carried permanently attached to their cables on each side of the bow, always ready to be let go in case of an emergency.

The sheet anchor, another large anchor of similar weight to the bower anchors, was stowed on the starboard abaft the best bower, to back up the other anchors if needed. As well as these three main anchors, there were two spares in the hold. A stream anchor is a light anchor for use in narrow waterways. A kedge anchor is also a light anchor but lighter than the stream anchor. Both the stream and kedge anchors could be used as 'coasting' anchors, for anchoring the ship in light condition, to save the heavy work of weighing the bower anchors. Otherwise, they were normally used to carry out astern when the bower anchor was down, with a hawser (a strong thick rope) to prevent the ship swinging. They were commonly used when warping the ship, being placed ahead of the ship on which she could haul herself. [We are indebted to Ray Parkin for this information.]

Arrack – A name in the East Indies and the Indian islands for all ardent spirits. Arrack is often distilled from a fermented mixture of rice, molasses, and palm wine of the cocoanut tree or the date palm.

Auger – A hand tool, typically having a threaded shank and cross handle, used for boring holes in wood.

Ball – A lethal charge of a single lead ball fired from a musket.

Ballast – Heavy material that is placed in a position low in the hull to provide stability.

Beam – The breadth of the ship at the widest point.

Bear away – To turn or steer a vessel away from the wind.

Bent – The sails were attached, or 'bent', to long horizontal spars of wood called 'yards' suspended above the deck through a complex system of ropes.

Berth – A sleeping space on a ship. Also, the space allotted to a vessel at anchor or at a wharf.

Bilge – The compartment at the very bottom of the ship's hull where water collects and must be pumped out of the vessel.

Blacklist – A list of people who are to be punished.

Block – A pulley with one or more sheaves or grooves over which a line is roved. It can be used to change the direction of the line, or in pairs used to form a lifting tackle.

Boat hook – A pole with a blunt tip and a hook on the end. Typically used to assist in docking and undocking a boat, with its hook used to pull a boat towards a dock and the blunt end to push it away from a dock, as well as to reach into the water to help people catch buoys or other floating objects or to reach people in the water.

Boatswain – A non-commissioned officer usually referred to as Bosun, who had responsibility for rigging, cables, anchors, sails and boats, who issues 'piped' commands to seamen. The sailmaker and boatswain's mates were under the command of the boatswain.

Booms – Masts or yards, lying on board in reserve.

Bow – The forward part of the hull of a ship.

Brig – A vessel with two square-rigged masts.

Bring to – To cause the ship to be stationary by arranging the sails.

Broadside – One side of the vessel.

Broiling – Cooking by exposing food to direct radiant heat from the fire.

Buoy – A floating conical cask used to show where the anchor lies.

Butt – Cask.

Cable – An especially large or thick rope. For the two bower anchors and sheet anchor *Endeavour* was supplied with seven cables, each 13½ inches in circumference by 120 fathoms (220 metres). Each weighed 2 tons 5cwt 2qr. and had a breaking strain of around 55½ tons. For the kedge and stream anchors, which were smaller, there were 5- and 6-inch hawsers (small cables). [We are indebted to Ray Parkin for this information.]

Cag – A small cask or barrel.

Careen – To incline the ship on one side so that her bottom on the other side may be examined.

Cask – A barrel.

Caulking – Oakum or other fibres driven into planking seams to make them watertight.

Chains – Platforms projecting from each side of the hull to increase the width where shrouds join the chain plates. This is where the Leadsmen stood, swinging the lead to take soundings.

Chart – A map showing coastlines, water depths, or other information of use to navigators.

Civil time – The new day begins at midnight.

Cockswain – The helmsman or crew member in command of a boat.

Collier – A collier is a bulk cargo ship designed or used to carry coal.

Compass –

Illustration 2.

Compass abbreviations – NBE (North by East), NWBW (Northwest by West) etc.

Consumption – Tuberculosis.

Course – The point of a compass to which the ship steers.

Courses – The lowest square sails on square-rigged masts [see illustration 1].

Court-martial – A court consisting of commissioned officers for the trial of members of the armed forces.

Crossing the Line – Crossing the Equator.

Dampier – William Dampier (baptised 5th September 1651, died March 1715) was an English explorer, pirate, privateer, navigator and naturalist who became the first Englishman to explore parts of what is today Western Australia.

Draw'd (also draft) – The depth of a ship's keel below the waterline.

Dredge – An implement consisting of a net on a frame, used for gathering shellfish.

Dysentery – A general term for a group of gastrointestinal disorders characterised by inflammation of the intestines, particularly the colon. Characteristic features include abdominal pain and cramps, straining at stool (tenesmus), and frequent passage of watery diarrhea or stools containing blood and mucus.

Equator – An imaginary circle around the earth's surface, equidistant from the poles, that divides the earth into the Northern Hemisphere and the Southern Hemisphere.

False Keel – A sacrificial length of timber fastened under the keel to take the wear caused by the vessel bottoming out or being hauled up a beach.

Fathom – A unit of length equal to 6 feet (1.8m).

Flood – The incoming rising tide.

Flux – A flowing of fluid from the body, such as diarrhoea.

Fore and aft – From the bow of a ship to the stern; lengthwise.

Forecastle – The upper deck of the ship forward of the foremast at the head of the vessel.

Fore foot – The lower part of the stem of a ship.

Fothering – Covering the leak in the ship with a sail containing rope fibres or other similar material to stop the leak. Tarred wool was substituted for rope fibres when the *Endeavour* was fothered after she got off Endeavour Reef.

Foul wind – A term for the wind when unfavourable to the ship's course.

Four pounder – A cannon capable of firing an iron cannon ball weighing 4lb (1.8kg).

Gig – A pronged spear for fishing.

Glasses – Telescopes.

Grapeshot – Small balls of lead fired from a cannon, analogous to shotgun shot but on a larger scale.

Great Cabin – A relatively spacious cabin at the rear of the *Endeavour* designed as a workroom for Cook and the Royal Society.

Gymp – A narrow flat braid or rounded cord of fabric used for trimming dresses, furniture, etc.

Haul the wind – To turn the head of the ship nearer to the point from which the wind blows.

Hawser – A small kind of cable.

Heaving her down – To turn the vessel on its side.

Helm – The ship's steering mechanism.

Hoist – To raise or haul up.

Hold – The space between the lower deck and the bottom of the ship where her stores are held.

Hoops – Metal parts around the barrel that hold the staves together.

Hove – Heaved.

Keel – The main structural longitudinal member or backbone at the lowest point of the ship's hull.

Larboard (port side) – An obsolete term for the left side of a ship, looking from the stern forward to the bow.

Lay to – To bring a vessel into the wind and hold her stationary. A vessel doing this is said to be 'laying to'.

League – A unit of length used to measure distances, normally equal to 3 nautical miles.

Lee-shore – That shore upon which the wind blows.

Leeward – The direction toward which the wind is blowing.

Lieutenant – A commissioned officer who assumed command when the captain was absent. Lieutenants were responsible for standing watches, i.e. taking routine command of the deck when the ship was at sea with responsibility for maintaining discipline and navigation, and had overall command of a particular mast during setting and taking in sail. They would also oversee particular evolutions such as taking in stores or weighing anchor. An officer was required to have at least six years' service at sea before passing the examination for promotion to lieutenant and 'appear' to be of the age of 18.

Log – A navigation tool consisting of a wooden board attached to a line (the log line) lowered into the sea. Used to estimate the speed of a vessel through water. The log line is wound on a reel so the user can easily pay it out and has a number of knots at uniform intervals. Counting the knots as the line pays out as the ship moves forward against an instrument for measuring time (sandglass) gives the ship's speed.

Long boat – The largest of the *Endeavour*'s three boats, propelled by sails or oars, three masted and rowed by eight oars.

Luff – To point a sailing vessel closer to the wind.

Made sail – To set sail.

Main chains – The chain platform abreast the main mast.

Main yard – The lower yard on a mainmast.

Man at the lead – Also called a leadsman, who stands in the chains of the ship, up against the shrouds where he swings a lead plummet attached to a line of thin rope (lead line) to sound the depth of water.

Marine – A soldier trained for service afloat with many and varied duties including providing guard to ship's officers should there be a mutiny aboard.

Masthead – A small platform partway up the mast, just above the height of the mast's main yard. A lookout is stationed here, and men who are working on the main yard will embark from here.

Midshipman – This rank was usually filled by a 'young gentleman' with aspirations to become a commissioned officer who joined from the age of 9 onwards. They were responsible for overseeing tasks under the overall direction of a lieutenant, such as going aloft to supervise sail handling or casting the log. They might be put in command of a ship's boat.

Miss stays – A ship is said to miss stays, when her head will not fly up into the direction of the wind, in order to get her on the other tack.

Moor – To secure the vessel with a cable or an anchor.

Morasses – An area of low-lying, soggy ground.

Musket – A muzzle-loading smoothbore flintlock long-barrelled weapon. The late eighteenth-century British military issue musket was nicknamed the 'Brown Bess'. It could be loaded with a lethal single lead ball, or with a less lethal charge of many smaller lead balls called 'small shot'.

Nautical Mile – Nautical miles are used to measure the distance travelled through the water. A nautical mile is slightly longer than a mile on land, equalling 1.1508 land measured (or statute) miles.

Oakum – Oakum was recycled from old tarry ropes and cordage used for packing the joints of timbers in wooden vessels.

Officer of the Watch – The officer responsible for the navigation of the ship, in the absence of the captain, during a certain watch.

Opiate – A substance derived from opium.

Pinnace – The second largest of the *Endeavour*'s three boats, propelled by sails or oars, double masted and rowed by six oars.

Poop – The rear deck of a ship is often called the afterdeck or poop deck.

Portable soup – A kind of dehydrated food of English origin used in the eighteenth century. It was a precursor of meat extract and bouillon cubes.

Privateer – A privately owned ship authorised by a national power to conduct hostilities against an enemy.

Punishment – Floggings were administered with a cat o' nine tails. A total of twelve lashes per incident was most common, sometimes twenty-four, with six as the least number. A total of twenty men are recorded as being punished on *Endeavour*. Stealing offences account for most. Other offences included disobedience, assault, desertion, leaving duty ashore, using abusive language to the officer of the watch, drunkenness, beating the sick and for refusing to eat their allowance of fresh meat. Some were punished more than once. Boatswain's mate John Reading was punished for not doing his duty in correctly administering punishment to two others he was assigned to flog.

Purging – Vomiting.

Quadrant – An instrument for measuring the altitude of the sun or a star above the horizon to find geographic position at sea.

Quarter deck – The aftermost or stern deck of a ship. During the Age of Sail, the quarterdeck was the preserve of the ship's officers.

Quartermaster – Responsible for the steering of a ship.

Queirós – Pedro Fernandes de Queirós (1563–1614) was a Portuguese navigator in the service of Spain, who in May 1606 reached the islands later called the New Hebrides and now known as the independent nation of Vanuatu. Queirós landed on a large island that he took to be part of the mythical 'great south land', and named it Australia del Espíritu Santo – 'The Great South Land of The Holy Spirit'.

Quires – A set of four sheets of parchment or paper folded so as to make eight leaves: the ordinary unit of construction for early manuscripts and books.

Reefing the sails – Reefing reduces the area of a sail, usually by folding or rolling one edge of the canvas in on itself and attaching the unused portion to a spar or a stay. Or by means of lengths of rope attached to the sail itself (at reef-points). This was the primary measure to preserve a sailing vessel's stability in strong winds, or to slow the vessel.

Relative bearings –

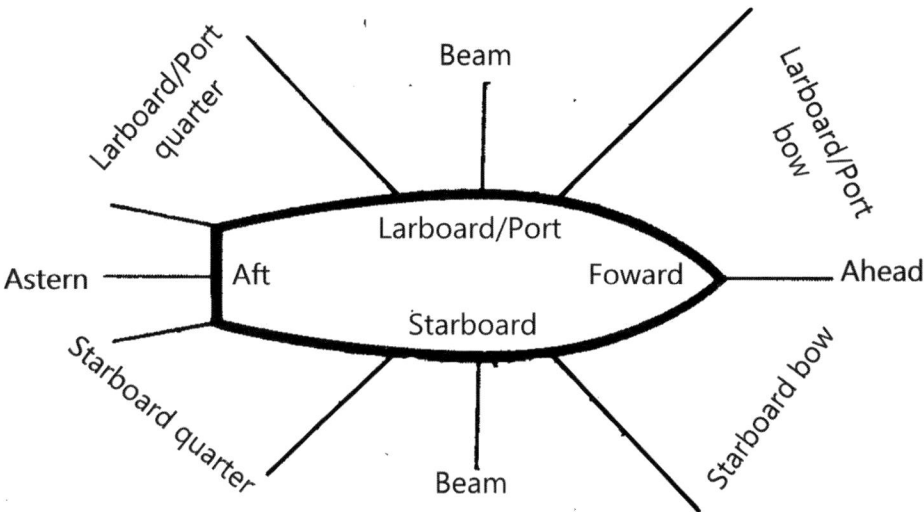

Illustration 3.

Roundshot – Cannonball.

Scurvy – A disease caused by deficiency of Vitamin C, characterised by spongy and bleeding gums, bleeding under the skin, and extreme weakness.

Scurvy grass – *Cochlearia* is a genus of about thirty species of annual and perennial herbs in the family *Brassicaceae*. Most commonly found in coastal regions, on cliff-tops and salt marshes. Scurvy grass gets its name from the fact that sailors used to take it to prevent the disease.

Scuttle – A small hatchway or opening in the deck, with a lid for covering it.

Seine – A long fishing net that hangs vertically in the water, having floats at the upper edge and sinkers at the lower. It is used to enclose and catch fish when its ends are pulled together or are drawn ashore.

Sextant – A navigational instrument used to measure a ship's latitude.

Sheathing board – A secondary layer of timber laid over the hull surface.

Ship's time – The new day begins at noon.

Ship's Master – The master was the senior warrant rank and reported directly to the captain. Those classed as sea officers had equal status as commissioned officers and could stand on the quarterdeck. The master's main duty was navigation, taking the ship's position daily and setting the sails as appropriate for the required course. He was responsible for ensuring the maintenance of the rope rigging and sails. Other duties included the stowing of the hold, inspecting provisions, taking in or moving stores so that the ship was not badly trimmed and reporting defects to the captain. He was also responsible for the security and issue of drink on board and supervised entry of parts of the official log such as weather, position, and expenditure.

Shoal – A stretch of shallow water, a sandbank, rocky area or coral reef, especially one that is visible at low water, but hidden at high water.

Shorten sail – To reduce sail by taking it in.

Shrouds – Heavy rope rigging giving lateral support to the masts.

Small shot – A less than lethal charge of many small lead balls fired from a musket, analogous to shotgun shot.

Sounding – Measuring the depth of the water.

Starboard – That side of a vessel which is on the right hand of a person who stands on board facing the bow.

Starboard Quarter – The starboard surface of the vessel's hull that is located behind the beam (the widest part of the ship).

Stave – A narrow length of wood with a slightly bevelled edge to form the sides of barrels.

Stern Ports – Openings in the stern of the vessel for admitting light and air, fitted with strong shutters (deadlights).

Surgeon – The Surgeon was warranted to the ship by the Navy Board. They were the only medical officer on the ship and were assisted by Surgeon's Mates (inferior warrant officers). They were responsible for the sick and injured, performing surgical operations as necessary and dispensing medicine. They were required to keep a journal of treatment and advise the captain on health matters.

Swivel gun – A small cannon, mounted on a swivelling stand or fork that allows a very wide arc of movement.

Tacking – A sailing manoeuvre for a vessel sailing into the wind by which it turns its bow toward and through the wind, thereby allowing progress in the desired direction. A series of tacking moves, effectively 'zig-zagging' back and forth across the wind, is called beating, and allows the vessel to sail directly upwind, which would otherwise be impossible.

Tackle – A pair of blocks through which is rove a rope to provide an advantageous purchase. Used for lifting heavy loads and to raise and trim sails.

Glossary of Terms 289

Taffrail – The handrail around the open deck area toward the stern of the ship.

The Northwest Passage – A sea route between the Atlantic and Pacific oceans through the Arctic Ocean.

The Royal Society – The Royal Society of London, originated in 1660 for improving Natural Knowledge, the oldest national scientific society in the world.

Thwart – A part of an undecked boat (or canoe) that provides structural rigidity for the hull. A thwart goes from one side of the hull to the other.

Topsails – The second sail (counting from the bottom) up a mast (see illustration 1 above).

Torrid zone – The region, between the Tropic of Cancer and the Tropic of Capricorn, characterised by a hot and humid climate; the tropics.

Touchwood – Rotten wood used to catch the fire struck from a flint.

Transit of Venus – Transit of Venus is a phenomenon in which the disc of the planet Venus passes like a small shadow across the face of the sun.

Trim – To modify the angle of a vessel to the water by shifting cargo or ballast.

Tropic – The tropics are the regions of Earth surrounding the Equator. They are defined in latitude by the Tropic of Cancer in the Northern Hemisphere and the Tropic of Capricorn in the Southern Hemisphere.

Under an easy sail – To sail slowly.

Variation – A magnetic interference on the compass from magnetic fields, or from the vessel's own equipment.

Victuals – Food fit for human consumption.

Viscera – The intestines.

Warp – To move a vessel by hauling on a line or cable that is fastened to an anchor.

Watches and Bells – Most of the crew of a ship would be divided into two to four groups, called watches. Each watch would take its turn with the essential activities of manning the helm, navigating, trimming sails, and keeping a lookout. The hours between 16:00 and 20:00 are so arranged because that watch (the 'dog watch') was divided in two. The odd number of watches aimed to give each man a different watch each day; it also allowed the entire crew of a vessel to eat an evening meal, the normal time being at 17:00 with first dog watchmen eating at 18:00.

Number of bells	Bell pattern	Watch				Dog		
		Middle	Morning	Fore-noon	After-noon	First	Last	First
One bell	1	00:30	04:30	08:30	12:30	16:30	18:30[a]	20:30
Two bells	2	01:00	05:00	09:00	13:00	17:00	19:00[a]	21:00
Three bells	2 1	01:30	05:30	09:30	13:30	17:30	19:30[a]	21:30
Four bells	2 2	02:00	06:00	10:00	14:00	18:00		22:00
Five bells	2 2 1	02:30	06:30	10:30	14:30		18:30	22:30
Six bells	2 2 2	03:00	07:00	11:00	15:00		19:00	23:00
Seven bells	2 2 2 1	03:30	07:30	11:30	15:30		19:30	23:30
Eight bells	2 2 2 2	04:00	08:00	12:00[b]	16:00		20:00	24:00

Illustration 4.

Waterspout – A funnel-shaped or tubular column of rotating cloud-filled wind usually extending from the underside of a cumulus or cumulonimbus cloud down to a cloud of spray torn up by the whirling winds from the surface of an ocean.

Weigh – To heave up an anchor preparatory to sailing.

Well – A place in the ship's hold for pumps.

Windward – The direction that the wind is coming from.

Wore/wear ship – To turn away from the wind.

Yard – A horizontal spar from which a square sail is suspended.

Yawl – The smallest of *Endeavour*'s boats, propelled by sails or oars, single-masted and rowed by four oars.

Ship's Muster on *Endeavour*

We are indebted to Ray Parkin and John Robson (Captain Cook Society) for this information.

NAME	RATING	BIOGRAPHY
COOK, James. 39	1st Lieutenant (Captain and Purser*).	Born Marton-In-Cleveland, Yorkshire.
HICKES, Zachary. 29	2nd Lieutenant.	Born Stepney. Died consumption 26 May 1771 North Atlantic Ocean.
GORE, John. 38	3rd Lieutenant.	Born Virginia. Ex *Dolphin* under Captain Bryon and Captain Wallis. Sailed on third voyage *Resolution*.
MOLINEUX, Robert. 22	Master.	Born Hale near Liverpool. Ex *Dolphin* under Captain Wallis. Died 16 April 1771 Nr Robben Island.
SMITH, Isaac. 16	AB* then Midshipman then Master's Mate.	Born London. Ex *Grenville*. Sailed on second voyage in *Resolution*.
BOOTIE, John. (18) ?	Midshipman.	Died 4 February 1771 Indian Ocean.
MUNKHOUSE, Jonathan.	Midshipman.	Born Penrith, Cumberland. Died 6 February 1771 Indian Ocean.
MAGRA, James Mario (Maria). 24	AB then Midshipman.	Born New York. 1746 (?) Dismissed 24 May 1770 from the quarter deck and suspended from doing any duty on board. Restored to duty 13 June 1770.
SAUNDERS, Patrick.	Midshipman then AB.	Deserted in Batavia.
MUNKHOUSE, William Brougham.	Surgeon.	Born Penrith, Cumberland. Died 5 November 1770 Batavia.
PERRY, William. 21	Surgeon's Mate then Surgeon.	Born Chiswick.

NAME	RATING	BIOGRAPHY
SATTERLEY, John.	Carpenter.	Died 12 February 1771 Indian Ocean.
FORWOOD, Stephen. 33 (?)	Gunner.	Born Sandwich, Kent.
GATHREY, John.	Boatswain.	Died 4 February 1771 Indian Ocean.
TAYLOR, Robert.	Armourer. Acted as Ship's Blacksmith.	Thought to be Scottish.
RAVENHILL, John. 49	Sailmaker.	Born Hull. '*An old Man about 70 or 80 Years of age, and what is still more extraordinary in this man his being generally more or less drunk every day. (Cook)* Died 27 January 1771 Indian Ocean.
THOMSON, John.	Ship's Cook (had lost his right hand).	Died 31 January 1771 Indian Ocean.
ORTON, Richard.	Captain's Clerk.	Often drunk.
PICKERSGILL, Richard. 19	Master's Mate then Master.	Born West Tanfield, Yorkshire. Ex *Dolphin* under Captain Wallis. Punished on 12 October 1768 for disobedience: '*Mates and Midshipmen commanded to scrape and clean between decks, which Mr Pickersgill (only) having the spirit to refuse was order'd before the Mast.*' (Cook) Sailed on second voyage in *Resolution*. In 1779 he drowned in the River Thames.
CLERKE, Charles. 27	Master's Mate then 3rd Lieutenant.	Born Weathersfield, Essex. Ex *Dolphin*. Sailed second voyage in *Resolution* and third voyage in *Discovery*.
WILKINSON, Francis. 21	AB then Master's Mate.	Born Chatham, Kent. Ex *Dolphin* under Captain Wallis. Died August 1771.
WEIR, Alexander. 35	Quartermaster.	Born Fife, Scotland. Accidently drowned when his leg became entangled in a buoy rope 14 September 1768 Maderia.
EVANS, Samuel. 32	Quartermaster (Boatswain's Mate and Coxswain* of the Pinnace) then Boatswain.	Ex *Dolphin* under Captain Wallis.

Ship's Muster on *Endeavour*

NAME	RATING	BIOGRAPHY
ANDERSON, Robert. 28 (?)	AB then Quartermaster.	Born Inverness. Punished on 30 November 1768, with 12 lashes for attempted desertion and on 21 June 1769 for disobedience. Sailed on second and third voyage in *Resolution*.
READING, John. 24	Boatswain's Mate.	Born Kinsale, County Cork. Punished with 12 lashes on 30 November 1768 for neglect of duty. Died 28 August 1769 Pacific Ocean, from '*an excess of rum*'.
HARDMAN, Thomas. 33	Boatswain's Mate then AB then Sailmaker.	Born London. Ex *Grenville*.
HUTCHINS, Richard. 27	AB then Boatswain's Mate.	Born Deptford. Punished on 16 April 1769 with 12 lashes for disobedience.
PARKER, Isaac. 27	AB then Boatswain's Mate then Yeoman of the Sheets*.	Born Ipswich.
HUGHES, Richard. 22	AB then Carpenter's Mate.	Born London.
NORWELL, George.	AB then Carpenter's Crew then Carpenter.	
TERREL, Edward. 19	Carpenter's Servant then AB.	Born Spitalfields, London. Sailed on second voyage in *Resolution*.
HAITE, Francis. 42	AB then Carpenter's Crew.	Born Rochester, Kent. Ex *Dolphin* under Captain Byron and Captain Wallis. Died 30 January 1771 Indian Ocean.
MOODY, Samuel. 40	AB then Carpenter's Crew.	Born Worcester. Died 30 January 1771 Indian Ocean.
THURMAN, John. 20	AB then Sailmaker's Mate. (Pressed at Madeira).	Born New York. Punished on 19 November 1768 with 12 lashes for disobedience; and on 12 June 1769 with 24 lashes for theft. Died 3 February 1771 Indian Ocean.
HOWSON, William. 16	Captain's Servant then AB.	Born London. Ex *Grenville*.
CHARLTON, John. 15	Captain's Servant.	Born London. Ex *Grenville*.
HARVEY, William. 17	2nd Lieutenant Hickes's Servant then AB then Midshipman.	Born London. Sailed on second and third voyage in *Resolution*.
MOREY, Nathaniel.	3rd Lieutenant Gore's Servant.	

NAME	RATING	BIOGRAPHY
MANLEY, Isaac George. 12	Master's Servant then Midshipman.	Born London.
JONES, Thomas.	Surgeon Munkhouse's Servant, then AB, then Astronomer's Servant.	
MATHEWS, Thomas.	Ship's Cook's Servant.	
JORDAN, Thomas.	Boatswain's Servant then AB.	
ROBERTS, Daniel.	Gunner's Servant.	Died 2 February 1771 Indian Ocean.
REARDON, Timothy. 25	AB Boatswain's Yeoman*.	Born Cork, Ireland. Ex *Grenville*. Died 24 December 1770 Batavia.
JOHNSON, Isaac. 26 (?)	AB Cooper*.	Born Knutsford, Cheshire.
GOODJOHN, John.	AB Cooper's Mate.	
DAWSON, William. 19	AB Cooper's Steward.	Born Deptford, London. Sailed on second and third voyage in *Resolution*.
CHILDS, Josiah. 29	AB Captain's Cook then Ship's Cook.	Born Dublin.
KNIGHT, Thomas.	AB (Lieutenant's Cook).	
COLLETT, William. 20	AB Barber.	Born High Wycombe. Sailed on second and third voyage in *Resolution*.
WOLFE, Archibald. 39	AB Tailor.	Born Edinburgh. Punished with 24 lashes on 4 June 1769 for theft. Died 31 January 1771 Indian Ocean.
JEFFS, Henry.	AB Butcher.	Punished on 29 April 1769 with 12 lashes for aggression. Died 27 February 1771 Indian Ocean.
SUTHERLAND, Forby. 29	AB Poulterer*.	Born Orkney, Scotland. Died 1 May 1770 Buried at Kurnell, Botany Bay, Australia.
FLOWER, Peter. 18	AB.	Born Guernsey. Ex *Grenville*. Drowned 2 September 1768 after falling overboard from *Endeavour* in the harbour at Rio De Janeiro.
GRAY, James. 24	AB then Quartermaster.	Born Leith. Ex *Dolphin* under Captain Byron. Sailed on second voyage in *Resolution*.

Ship's Muster on *Endeavour*

NAME	RATING	BIOGRAPHY
RAMSAY, John. 21	AB.	Born Plymouth. Sailed on second voyage in *Resolution*.
JORDAN, Benjamin. 30	AB Carpenter's Mate.	Born Deptford. Died 31 January 1771 Indian Ocean.
JONES, Samuel. 22	AB.	Born London. Punished with 12 lashes for disobedience on two occasions.
NICHOLSON, James. 21	AB.	Born Inverness. Punished with 24 lashes on 12 June 1769 for theft. Died 31 January 1771 Indian Ocean.
SIMMONDS, Thomas. 24	AB.	Born Brentford.
STAINSBY, Robert. 27	AB.	Born Low Dinsdale, County Durham.
COX, Mathew. 22 (?)	AB.	Born Gillingham, Dorset. Punished on 30 November 1769 at the Bay of Islands, New Zealand with 6 lashes for stealing potatoes from a Māori garden. He must have then disputed that sentence as he was given another 6 lashes on 1 December 1769. After the voyage, Cox began an action against Cook for the punishments he received on *Endeavour*.
WILLIAMS, Charles. 38	AB.	Born Bristol.
SIMPSON, Alexander.	AB.	Punished on 2 December 1769 with 12 lashes for theft. Died 21 February 1771 Indian Ocean.
STEPHENS, Henry. 28	AB.	Born Falmouth. Punished 16 September 1768 with 12 lashes for disobedience; and on 30 November 1769, again with 12 lashes, for theft.
JONES, Thomas. 27	Surgeon's Servant then AB.	Born Bangor, North Wales.
PONTO, Antonio. 24	AB.	Born Venice.
DOZEY, John. 20	AB.	Born Brazil. Died 7 April 1771 Cape Town.
FUNLEY, James. 24	AB.	Born Blackwell, East London. Punished on 19 June 1769 with 12 lashes for theft.

LITTLEBOY, Michael. 20	AB.	Born Deptford.
WOODWORTH, John. 48	AB.	Died 24 December 1770 Batavia.
PECKOVER, William. 21	AB.	Born Eyot, Northamptonshire. Sailed on second voyage in *Resolution* and third voyage in *Discovery*.
LITTLEBOY, Richard. 25	AB.	Born Deptford. Punished during the voyage on 2 December 1769 with 12 lashes for theft.
PERIERA, Manoel.	AB entered Muster List at Rio de Janeiro.	Born Portugal (?). Punished on 30 November 1769 with 12 lashes for theft. Died 27 February 1771 Indian Ocean.
YOUNG, Nicholas. 12 (?)	Boy then Surgeon's servant entered Muster List 18 April in Tahiti.	

Crew who joined *Endeavour* post New Holland (Australia)	RATING	BIOGRAPHY
MORGAN, Peter. 29	AB entered Muster List at Batavia.	Born Strabane, Ireland. Died 7 March 1771 Indian Ocean.
THOMAS, Richard. 20	AB entered Muster List at Batavia.	Born Scilly. Died 15 March 1771 Cape Town.
LORRAIN, John	AB entered Muster List at Batavia.	Born Middlesex. Died 4 April 1771 Cape Town.
LINDSAY, Alexander.	AB entered Muster List at Batavia.	Born Mount Holly, northern England. Died 19 February 1771 Indian Ocean.
GOLDSMITH, Thomas. 22	AB entered Muster List at Batavia.	Born Yarmouth.
MARRA, John. 23 (?)	AB entered Muster List at Batavia.	Born Cork. Sailed on second voyage in *Resolution* as gunner's mate where he was punished on five separate occasions.
JOYCE, James. 28	AB entered Muster List at Batavia.	Born Portsmouth.
MOULTON, William.	AB entered Muster List on 1 February 1771, Batavia.	
MATHIAS, Thomas.	AB entered Muster List on 1 February 1771, Batavia.	

Ship's Muster on *Endeavour*

Crew who joined *Endeavour* post-New Holland (Australia)	RATING	BIOGRAPHY
HILL, Richard.	Master's Servant entered Muster List 5 February 1771, Batavia.	
BAPTISA (BAPTISO), John.	AB entered Muster List at Batavia.	
BREWER, John.	AB entered Muster List at Batavia.	
LEGG, John.	AB entered Muster List 5 February 1771, Batavia.	
PRAVAL, Charles.	Entered Muster List 19 December 1770 at Batavia, then 7 February 1771 as AB.	
SMITH, Samuel. 31	Entered Muster List 19 December 1770 at Batavia, then 13 February 1771 as AB.	Born Hull.
BURN, William. 30	Entered Muster List 19 December 1770 at Batavia, then 13 February 1771 as AB	Born Dublin.
SMITH, John. 26	Entered Muster List 19 December 1770 at Batavia, then 17 February 1771 as AB.	Born Ostend.
NICHOLS, Peter. 23	Entered Muster List 19 December 1770 at Batavia, then 20 February 1771 as AB.	Born Bristol.
CAMPBELL, James. 18	Entered Muster List 19 December 1770 at Batavia, then 20 February 1771 as AB.	Born Leith.
Crew who joined the *Endeavour* at CAPE TOWN		
HANSON, Turkel. 24	AB entered Muster List 14 March 1771, Cape Town.	
ROBERTS, Thomas. 23	AB entered Muster List 14 March 1771, Cape Town.	Born Copenhagen.
CLOSS (CLAUS), Fick (Vick). 21	AB entered Muster List 14 March 1771, Cape Town.	Born Hanover.
STALBONE, Christopher. 33	AB entered Muster List 14 March 1771, Cape Town.	Born Hanover.
OLAFSON, Canute. 23	AB entered Muster List 14 March 1771, Cape Town.	Born Christiana, Oslo.
VAN SAKE, Antonia. 20	AB entered Muster List 14 March 1771, Cape Town.	Born Oporto.

Crew who joined *Endeavour* post-New Holland (Australia)	RATING	BIOGRAPHY
BODE, Johan Arnes. 27	Entered Muster List on 16 March 1771 at Cape Town then 1 April 1771 as AB.	Born Frankfurt.
SAUNDERS, Lowrand. 31	Entered Muster List on 16 April 1771 at Cape Town then 1 April 1771 as AB.	Born Copenhagen.
COOPER, Peter. 22	Entered Muster List on 16 March 1771 at Cape Town then 5 April 1771 as AB.	Born Ghent.
VAN CANT, Jacobus. 23	Entered Muster List on 16 March 1771 at Cape Town then 8 April 1771 as AB.	Born Dunkirk.
STILL, John.	Gunner's Servant entered Muster List 1 July 1771.	

MARINES	RATING	BIOGRAPHY
EDGECUMBE, John.	Sergeant.	Sailed on second voyage in *Resolution*.
TRUSLOVE, John.	Corporal.	Died 24 January 1771 Indian Ocean.
ROSSITER, Thomas.	Drummer.	Punished 2 December 1769 with 12 lashes for theft; and on 21 February 1771 with 12 lashes for drunkenness.
JUDGE, William.	Private.	Punished on 30 November 1768 with 12 lashes for insolence.
PAUL, Henry.	Private.	
PRESTON, Daniel.	Private.	Died 15 February 1771 Indian Ocean.
WILTSHIRE, William.	Private.	
GREENSLADE, William.	Private.	Died 26 March 1769 S.E. Pacific Ocean. Suicide.
GIBSON, Samuel.	Private.	Punished on 14 July by being '*close confin'd*' for a while and with 24 lashes for trying to desert at Tahiti. Sailed on second and third voyage in *Resolution*.
DUNSTER, Thomas.	Private.	Punished on 16 September 1768 with 12 lashes for disobedience. Died 25 January 1771 Indian Ocean.

Ship's Muster on *Endeavour*   299

MARINES	RATING	BIOGRAPHY
WEBB, Clement.	Private.	Punished with 24 lashes on 14 July 1769 for having tried to desert at Tahiti with Samuel Gibson.
BOWLES, John.	Private.	Punished on 7 April 1770 with 12 lashes for disobedience.

SUPERNUMERARIES	RATING	BIOGRAPHY
BANKS, Joseph. 25	Naturalist.	Born Westminster, London.
SOLANDER, Daniel Carl. 35	Naturalist.	Born Norrland, Sweden.
GREEN, Charles. 33	Astronomer.	Born Yorkshire. Died 26 January 1771 Indian Ocean.
REYNOLDS, John.	Green's Servant.	Died 18 December 1770 Batavia.
PARKINSON, Sydney. 23	Botanical Artist.	Born Edinburgh, Scotland. Died 27 January 1771 Indian Ocean.
BUCHAN, Alexander.	Landscape Artist.	Born Scotland. Died of epilepsy. Died 17 April 1769 Tahiti.
SPÖRING, Herman Dietrich. 35	Assistant Naturalist.	Born Abo, Sweden. Died 25 January 1771 Indian Ocean.
ROBERTS, James. 16(?)	Banks's Footman.	Born Lincolnshire.
BRISCOE, Peter. 26	Banks's Footman.	
RICHMOND, Thomas.	Banks's Negro Servant.	Died of hypothermia 16 January 1769 Tierra Del Fuego.
DALTON, George.	Banks's Negro Servant.	Died of hypothermia 16 January 1769 Tierra Del Fuego.
TUPIA.	Tahitian.	Died 20 December 1770 Batavia.
TAIYOTA.	Tahitian, Tupia's Boy Servant.	Died 17 December 1770 Batavia.
Alexander.	Entered 6 November 1770 as Banks's servant.	

Legend*

SHIPS UNDER COOK'S COMMAND
Newfoundland and Labrador – *Grenville* (1763–1767)
First Pacific voyage – *Endeavour* (1768–1771)
Second Pacific voyage – *Resolution/Adventure* (1772–1775)
Third Pacific voyage – *Resolution/Discovery* (1776–1779)

Dolphin had made two Pacific circumnavigation voyages before *Endeavour*. The first between June 1764 and May 1766 under Lord Byron. The second under Samuel Wallis, left Plymouth on 21st August 1766, they discovered Tahiti, and arrived back in England in May 1768.

Purser – A ship's purser is the person on a ship principally responsible for the handling of money on board.

AB – Able seaman.

Boatswain's Yeoman – Was responsible for keeping the storerooms for the boatswain.

Yeoman of the Sheets – Petty officers in charge of overseeing the smooth performance of the fore and/or aft sails.

Coxswain – Person in charge of a boat, particularly its navigation and steering.

Cooper – Barrel maker.

Poulterer – Responsible for the poultry on board.

Notes

1. Journal of HMS *Endeavour*, 1768–1771. Courtesy of National Library of Australia. Object – 228958440. (This entry 23rd August 1770.)

CONTENTS
1. A Chart (with alterations) of New South Wales, or the East Coast of New Holland. Discover'd and Explored by Lieutenant J. Cook, Commander of his Majesty's Bark *Endeavour*, in the Year MDCCLXX. Engraved by W. Whitchurch after Cook. Hawkesworth part III, f.p.481 Courtesy of National Library of Australia.

THE AUTHORS
(i) Captain James Cook of the *Endeavour* by William Hodges, painted 1775 (Detail). From Wikimedia Commons, the free media repository. Public Domain in UK. Also, PD-US.
(ii) Sir Joseph Banks by Sir Joshua Reynolds, painted 1773 (Detail). From Wikimedia Commons, the free media repository. Public Domain in UK. Also, PD-US.
(iii) Sydney Parkinson, self-portrait. From Wikimedia Commons, the free media repository. Public Domain in UK. Also, PD-US.

PROLOGUE
1. Painting of the *Earl of Pembroke*, later HMS *Endeavour*, leaving Whitby Harbour in 1768 by Thomas Luny. From Wikimedia Commons, the free media repository. Public Domain in UK.
2. Plans of His Majesty's Bark *Endeavour* as fitted at Deptford in 1768. Courtesy of Royal Museums Greenwich.
3. John Robson (Captain Cook Society) – Information about the men mustered for the HMB *Endeavour* voyage, 1768–1771 (private communication).
4. Portion of a painting *Endeavour off the Coast of New Holland* by Samuel Atkins (1787–1808) – 1794. From Wikimedia Commons, the free media repository. Public Domain.
5. This and the following twelve maps (with minor alterations) are reproduced with permission from the Australian National University and National Library of Australia. https://webarchive.nla.gov.au/awa/20091011115149/http://southseas.nla.gov.au/journals/maps/05_atlantic.html
6. *Hollandia Nova Terre Australe* by Thevenot Melchisedec, Paris 1696. Courtesy of the Mitchell Library, State Library of New South Wales.

CHAPTER 1
1. Captain James Cook of the *Endeavour* by William Hodges, painted 1775 (Detail). From Wikimedia Commons, the free media repository. Public Domain in UK. Also, PD-US.
2. *Journal of HMS Endeavour, 1768–1771.* Courtesy of National Library of Australia. Object – 228958440. (Here and elsewhere in the entire text the words of James Cook are taken from this journal. This entry 18th April 1770.)

3. The *Endeavour* at Sea. Pencil sketch by Sydney Parkinson © British Library Board Add. 9345, f.16v.
4. Sydney Parkinson, self-portrait. From Wikimedia Commons, the free media repository. Public Domain in UK. Also, PD-US.
5. A journal of a voyage to the South Seas: in His Majesty's ship, the *Endeavour* faithfully transcribed from the papers of the late Sydney Parkinson, draughtsman to Joseph Banks, esq. on his late expedition with Dr. Solander around the world / embellished with views and designs delineated by the author. Courtesy of the Mitchell Library, State Library of New South Wales. (Here and elsewhere in the entire text the words of Sydney Parkinson are taken from this journal. This entry 18th April 1770.)
6. Sir Joseph Banks by Sir Joshua Reynolds, painted 1773. (Detail). From Wikimedia Commons, the free media repository. Public Domain in UK. Also, PD-US.
7. Port Egmont Hen, Brown Skua – *Catharacta skua antarctica*. By Winfried Bruenken from Wikimedia Commons, the free media repository. Public Domain in UK. Also, PD-US.
8. Joseph Banks – *Endeavour journal, 15 August 1769–12 July 1771*. Courtesy of the Mitchell Library, State Library of New South Wales. (Here and elsewhere in the entire text the words of Joseph Banks are taken from this journal. This entry 18th April 1770.)
9. A Chart of the East Coast of New Holland from Point Hickes to Smoky Cape by Lieutenant J. Cook Commander of the *Endeavour*. (All following charts in Chapter 1 are details taken from this same source.) © British Library Board Add. 7085, No f. 35.
10. Portion of A Chart of the East Coast of New Holland from Point Hickes to Smoky Cape by Lieutenant J. Cook Commander of the *Endeavour*. © British Library Board Add. 7085, No f. 35. (The image of the ship here and elsewhere in the entire text is a detail of a painting, 1794, by Samuel Atkins (1787–1808) of *Endeavour* off the coast of New Holland. From Wikimedia Commons, the free media repository. Public Domain.
11. Altered photograph of waterspouts courtesy of National Oceanic and Atmospheric Administration USA.
12. TOP Sydney Parkinson. Pencil sketch View of part of the Eastern Coast of New Holland. Lat 35. © British Library Board Add MS 9345, ff. 60^v – 61 (a).
    BOTTOM Charles Praval. A view of the Pigeon House and land adjacent. © British Library Board Add. 7085, No 36.
13. Sydney Parkinson. Pencil drawing depicting the pinnace. © British Library Board Add MS 9345, f. 22
14. Sydney Parkinson. Pencil drawing depicting the yawl. © British Library Board Add MS 9345, f. 21.
15. Sydney Parkinson. The Lad Taiyota, Native of Otaheite, in the Dress of his Country'. Engraving by R.B. Godfrey from 'A journal of a voyage to the South Seas: in His Majesty's ship, the *Endeavour*: faithfully transcribed from the papers of the late Sydney Parkinson, draughtsman to Joseph Banks, esq. on his late expedition with Dr. Solander around the world / embellished with views and designs delineated by the author. (1773) pl. IX (fp.66) Courtesy of the Mitchell Library, State Library of New South Wales.

CHAPTER 2
1. Portion of A Chart of the East Coast of New Holland from Point Hickes to Smoky Cape by Lieutenant J. Cook Commander of the *Endeavour*. © British Library Board Add. 7085, No f. 35.
2. Journal of HMS. *Endeavour*, 1768–1771. Courtesy of National Library of Australia. Object – 228958440. (Here and elsewhere in the entire text the words of James Cook are taken from this journal. This entry 28th April 1770.)

3. Portrait of Robert Molineux, Master on board Cook's ship the *Endeavour* on his first voyage, 1769–71 (ca. 1760s). Painting by an unknown artist. Oil on canvas: 758 x 628mm. Hocken Collections, Uare Taoka o Hākena, University of Otago, A731.
4. Portion of A Sketch of Botany Bay in New South Wales by James Cook and Isaac Smith. (All following charts in Chapter 2 are from this same source.) © British Library Board Add. 7085, f. 40.
5. Joseph Banks, *Endeavour* journal, 15 August 1769–12 July 1771. Courtesy of the Mitchell Library, State Library of New South Wales. (Here and elsewhere in the entire text the words of Joseph Banks are taken from this journal. This entry 29th April 1770.)
6. Sydney Parkinson – Detail of Two Aborigines; one with markings on the chest and across the shoulders, resembling a crucifix. The other Aborigine is launching a spear from a spear-thrower with his right hand, and holding a shield with his left. Other sketches include two shields; a bark hut; four bark canoes, one containing an Aborigine paddling. Sketches probably made at Botany Bay, April 1770. British Library Board Add. 9345, f. 14v.
7. The anonymous journal (A journal of a voyage round the world in his Majesty's ship *Endeavour*) / attributed to James Mario Matra (Magra). Journal entry page 110 (29th April 1770). Courtesy of National Library of Australia.
8. Detail from portrait of James Douglas, 14th Earl of Morton (1702–1768), Scottish astronomer, with his family (1740) by Jeremiah Davison. From Wikimedia Commons, the free media repository. Public Domain in UK. Also, PD-US.
9. Hints offered to the consideration of Captain Cooke, Mr. Banks, Doctor Solander and other gentlemen who go upon the expedition on board the *Endeavour*. Chiswick, 10 August 1768. MS 9 – Papers of Sir Joseph Banks (bulk 1745–1820) Manuscript/Series 3/ Item113–113h. Courtesy of National Library of Australia.
10. Two of the Natives of New Holland, Advancing to Combat. Parkinson del. Engraving by Thomas Chambers. From A journal of a voyage to the South Seas: in His Majesty's ship, the *Endeavour*: faithfully transcribed from the papers of the late Sydney Parkinson, draughtsman to Joseph Banks, esq. on his late expedition with Dr. Solander around the world / embellished with views and designs delineated by the author. (1773) PART III. plate XXVII. Courtesy of the Mitchell Library, State Library of New South Wales.
11. Sydney Parkinson – Detail of Two Aborigines; one with markings on the chest and across the shoulders, resembling a crucifix. The other Aborigine is launching a spear from a spear-thrower with his right hand, and holding a shield with his left. Other sketches include two shields; a bark hut; four bark canoes, one containing an Aborigine paddling. Sketches probably made at Botany Bay, April 1770. © British Library Board Add. 9345, f. 14v.
12. LEFT. Photo of Australian bark shield from Botany Bay. (Before AD 1770). From Wikimedia Commons, the free media repository. Public Domain in UK. Also, PD-US. RIGHT. Detail of Five spears and a shield from New Zealand, Australia and New Guinea. © British Library Board ADD MS 23920, f. 35.
13. Photo of spears collected at Kamay (Botany Bay), 1770. © John MacDonald Cambridge University Museum of Archaeology and Anthropology currently on loan to Chau Chak Wing Museum, University of Sydney. Soon to be returned to the Gweagal people at Botany Bay.
14. Detail from Two Australian Aborigines and other drawings. © British Library Board Add. 9345, f. 14v.

CHAPTER 3

1. Daniel Solander. Detail from painting Omai (Mai), Sir Joseph Banks and Daniel Charles Solander. Painted 1775/1776 by William Parry. From Wikimedia Commons, the free media repository. Public Domain in UK. Also, PD-US.
2. The anonymous journal (A journal of a voyage round the world in his Majesty's ship *Endeavour*) / attributed to James Mario Matra (Magra). Journal entry page 113, (1st May 1770). Courtesy of National Library of Australia.
3. Captain John Gore (1780), by John Webber. From Wikimedia Commons, the free media repository. Public Domain in UK. Also, PD-US.
4. William Brougham Munkhouse (1768) by unknown author. From Wikimedia Commons, the free media repository. Public Domain in UK. Also, PD-US.
5. James Cook. A Sketch of Botany Bay. © British Library Board Add MS 7085, f.40. (All following charts in Chapter 3 are from this same source.)

CHAPTER 4

1. The artist of the chief mourner – Australian Aborigines in bark canoes © British Library Board ADD MS 15508, f. 10 (a) (no. 10).
2. James Cook. A Sketch of Botany Bay. © British Library Board Add MS 7085, f.40. (All following charts in Chapter 4 are from this same source.)
3. Captain John Gore (1780), by John Webber. From Wikimedia Commons, the free media repository. Public Domain in UK. Also, PD-US.
4. Herman Spöring. *Carcharhinus* sp. Pencil drawing. Catalogue of Natural History drawings commissioned by Joseph Banks on the Endeavour Voyage 1768–1771. Vol 3. 59.(1:55). Courtesy of Trustee of the Natural History Museum, London.
5. Herman Spöring. *Urolophus testaceus*, pencil sketch. Catalogue of the Natural History drawings commissioned by Joseph Banks on the *Endeavour* Voyage 1768–1771. Part 3: Zoology. Catalogue number 50. (1:46). Courtesy of Trustee of the Natural History Museum, London.
6. Herman Spöring. *Trygonorhina fasciata*, pencil sketch. Catalogue of the Natural History drawings commissioned by Joseph Banks on the *Endeavour* Voyage 1768–1771. Part 3: Zoology. Catalogue number 51. (1:47) Courtesy of Trustee of the Natural History Museum, London.
7. Herman Spöring. *Myliobatas australis*, pencil sketch. Catalogue of the Natural History drawings commissioned by Joseph Banks on the *Endeavour* Voyage 1768–1771. Part 3: Zoology. Catalogue number 52. (1:48) Courtesy of Trustee of the Natural History Museum, London.
8. Herman Spöring. *Aptychotrema banksia*, pencil sketch. Catalogue of the Natural History drawings commissioned by Joseph Banks on the *Endeavour* Voyage 1768–1771. Part 3: Zoology. Catalogue number 49. (1:45) Courtesy of Trustee of the Natural History Museum, London.

CHAPTER 5

1. Portion of A Chart of the East Coast of New Holland from Point Hickes to Smoaky Cape by Lieutenant J. Cook Commander of the *Endeavour*. (This and the following three charts in Chapter 5 are from this same source.) © British Library Board Add. 7085, f. 35.
2. James Cook. Cook's chart *Smoaky Cape to Cape Townsend by Lieutenant J. Cook Commander of the Endeavour*. (This and the following charts in Chapter 5 are from this same source.) © British Library Board Add 7085, f. 37.5.
3. Charles Praval. A View of Cape Byron and Mount Warning © British Library Board. Add Ms 7085, f. 36(c).

4. Charles Praval. Mount Warning © British Library Board. Add Ms 7085, f. 36(d).
  5. Charles Praval. A View of Sandy Cape © British Library Board. Add Ms 7085, f. 36(e).

CHAPTER 6
  1. Miniature portrait believed to be Isaac Smith. Courtesy of National Library of Australia. Bib ID 2104711.
  2. Portion of Smoaky Cape to Cape Townsend by Lieutenant J. Cook Commander of the *Endeavour*. (This and the following nine charts in Chapter 6 are from this same source.) © British Library Board Add 7085, f. 37.
  3. Charles Praval. A View of Cape Capricorn © British Library Board. Add Ms 7085, f. 36(f).
  4. Herman Spöring. *Portunus pelagicus*. Finished pencil drawing. Catalogue of the Natural History drawings commissioned by Joseph Banks on the *Endeavour* Voyage 1768–1771. Part 3: Zoology. Catalogue number 220. (3:7) Courtesy of Trustee of the Natural History Museum, London.
  5. Herman Spöring. *Portunus sanguinolentus*, pencil drawing. Catalogue of the Natural History drawings commissioned by Joseph Banks on the *Endeavour* Voyage 1768–1771. Part 3: Zoology. Catalogue number 219. (3:6) Courtesy of Trustee of the Natural History Museum, London.
  6. Cook's chart Cape Townsend to Cape Tribulation by Lieutenant J. Cook Commander of the *Endeavour*. (This and the following charts in Chapter 6 are from this same source.) © British Library Board Add 7085, f. 38.
  7. Portion of Cape Townsend to Cape Tribulation by Lieutenant J. Cook Commander of the *Endeavour*. © British Library Board Add 7085, f. 38.

CHAPTER 7
  1. Detail of Cape Townsend to Cape Tribulation by Lieutenant J. Cook Commander of the *Endeavour*. (This and the following charts in Chapter 7 are from this same source.) © British Library Board Add 7085, f. 38.
  2. Sydney Parkinson. North side of the entrance of Labyrinth Bay. New Holland. © British Library Board. ADD MS 9345, ff. 62v–63.
  3. Sydney Parkinson. Cumberland Isles, Islands at the South Entrance of the Streights. N. Holland. © British Library Board. ADD Ms 9345, ff. 64v-65.
  4. Sydney Parkinson. Two views of Pentecost Island. © British Library Board. ADD MS 9345, ff.63v-64(a).
  5. Charles Praval. A view of Pentecost Island at the Southern entrance of Whitsunday's Passage. © British Library Board. ADD MS 7085, f. 41(a).

CHAPTER 8
  1. Portion of Cape Townsend to Cape Tribulation by Lieutenant J. Cook Commander of the *Endeavour*. © British Library Board Add 7085, f. 38.
  2. Cook's chart Cape Tribulation to Endeavours Streights by Lieutenant J. Cook Commander of the *Endeavour*. (This and the following charts, and details of this chart in Chapter 8 are from this same source.) © British Library Board ADD MS 7085, f. 39.
  3. HMS Bark *Endeavour's* salvaged and restored small bower anchor, and one of six carriage guns hove overboard on 12th June 1770. Courtesy of James Cook Museum, Cooktown.
  4. Charles Praval. A view of the land about Endeavour River taken when the entrance bore WSW distance 1 Mile © British Library Board ADD MS 7085, f. 41(b).
  5. Richard Pickersgill. Portion of A Plan of the River on the East Coast of New Holland where his Maj: bark *Endeavour* repaired her Bottom After running on a reef of Rocks

where she lay 24 hours – June the 10th 1770. Courtesy of the Mitchell Library, State Library of New South Wales.
6. Sydney Parkinson. The *Endeavour* © British Library Board ADD MS 9345, f.50.

CHAPTER 9

1. Richard Pickersgill. A Plan of the River on the East Coast of New Holland where his Maj: bark *Endeavour* repaired her Bottom After running on a reef of Rocks where she lay 24 hours – June the 10th 1770. Courtesy of the Mitchell Library, State Library of New South Wales.
2. Sydney Parkinson. *The Endeavour* © British Library Board ADD MS 9345, f.57.
3. Sydney Parkinson. Detail from A view of the Endeavour River, on the coast of New Holland, where the ship was laid on shore, in order to repair the damage which she received on the rock. Engraving by Will Byrne from A journal of a voyage to the South Seas: in His Majesty's ship, the *Endeavour*: faithfully transcribed from the Papers of the last Sydney Parkinson, draughtsman to Joseph Banks, esq. on his late expedition with Dr. Solander around the world / embellished with views and designs delineated by the author. Hawkesworth (1773) III, pl. 19 (fp.557) Courtesy of the Mitchell Library, State Library of New South Wales.
4. TOP A View of Endeavour River, where the ship was laid ashore, in order to repair the damage, which she received on the rock. Engraving by Will Byrne, after a lost drawing by Parkinson. Hawksworth 1773. Curtesy of Wells' Cathedral Library.
   BOTTOM present-day Cooktown from similar perspective. © John MacDonald.
5. Sydney Parkinson. Detail from A view of the Endeavour River, on the coast of New Holland, where the ship was laid on shore, in order to repair the damage which she received on the rock. Engraving by Will Byrne from A journal of a voyage to the South Seas: in His Majesty's ship, the *Endeavour*: faithfully transcribed from the Papers of the last Sydney Parkinson, draughtsman to Joseph Banks, esq. on his late expedition with Dr. Solander around the world / embellished with views and designs delineated by the author. Hawkesworth (1773) III, pl. 19 (fp.557) Courtesy of the Mitchell Library, State Library of New South Wales.
6. Sydney Parkinson. The Pinnace © British Library Board ADD MS 9345, f. 22.
7. Sydney Parkinson. Nests of the White ant – Endeavour River © British Library Board ADD MS 23920, f. 37.
8. Richard Pickersgill. Detail from A Plan of the River on the East Coast of New Holland where his Maj: bark *Endeavour* repaired her Bottom After running on a reef of Rocks where she lay 24 hours – June the 10th 1770. Courtesy of the Mitchell Library, State Library of New South Wales.
9. James Cook. A Plan of the entrance of Endeavour River – New South Wales © British Library Board ADD MS 7085, f. 42.
10. Detail from Part of NEW HOLLAND. A draft of part of the shoals seen and sailed through by His Maj Bark the Endeavour in 1770 on the East Coast of New Holland by Richard Pickersgill. Courtesy of the Mitchell Library, State Library of New South Wales.
11. Catalogue of the Natural History drawings commissioned by Joseph Banks on the *Endeavour* Voyage 1768–1771. Part 3: Zoology. Courtesy of Trustee of the Natural History Museum, London.
    TOP LEFT – Pencil drawing of a Concertina fish *(Drepane punctata)* by Sydney Parkinson. Page 71. No 96. (2:21)
    TOP RIGHT – Pencil drawing of a Crested morwong (*Cheilodactylus (goniistius) Vestitus*) by Herman Spöring. Page 72. (2:23a)

BOTTOM LEFT – Pencil drawing of a Toad (*Lagocephalus spadiceus*) by Sydney Parkinson. Page 69. (1:65)

BOTTOM RIGHT – Pencil sketch of Epaulette shark (*Hemiscyllium ocellatum*) by Herman Spöring. Page 55. No 60. (1:56)

CHAPTER 10

1. Detail from Part of NEW HOLLAND. A draft of part of the shoals seen and sailed through by His Maj Bark the *Endeavour* in 1770 on the East Coast of New Holland by Richard Pickersgill. Courtesy of the Mitchell Library, State Library of New South Wales.
2. Sydney Parkinson. A view of the Endeavour River, on the coast of New Holland, where the ship was laid on shore, in order to repair the damage which she received on the rock. Engraving by Will Byrne from A journal of a voyage to the South Seas: in His Majesty's ship, the *Endeavour*: faithfully transcribed from the Papers of the last Sydney Parkinson, draughtsman to Joseph Banks, esq. on his late expedition with Dr. Solander around the world / embellished with views and designs delineated by the author. Hawkesworth (1773) III, pl. 19 (fp.557) Courtesy of the Mitchell Library, State Library of New South Wales.
3. Detail from Part of NEW HOLLAND. A draft of part of the shoals seen and sailed through by His Maj Bark the *Endeavour* in 1770 on the East Coast of New Holland by Richard Pickersgill. Courtesy of the Mitchell Library, State Library of New South Wales.
4. Ibid.
5. Richard Pickersgill. Detail from A Plan of the River on the East Coast of New Holland where his Maj: bark *Endeavour* repaired her Bottom After running on a reef of Rocks where she lay 24 hours – June the 10th 1770. Courtesy of the Mitchell Library, State Library of New South Wales.
6. Ibid.
7. Portion of Cook's chart Cape Tribulation to Endeavours Streights by Lieutenant J. Cook Commander of the *Endeavour*. © British Library Board ADD MS 7085, f. 39.
8. Mosquito – No copyright.
9. Portion of Cook's chart Cape Tribulation to Endeavours Streights by Lieutenant J. Cook Commander of the *Endeavour*. © British Library Board ADD MS 7085, f. 39.
10. Detail from Part of NEW HOLLAND. A draft of part of the shoals seen and sailed through by His Maj Bark the *Endeavour* in 1770 on the East Coast of New Holland by Richard Pickersgill. Courtesy of the Mitchell Library, State Library of New South Wales.
11. James Cook. *A Plan of the entrance of Endeavour River* – New South Wales © British Library Board ADD MS 7085, f. 42.
12. Sydney Parkinson. *Chelonia mydas*. Catalogue of the Natural History drawings commissioned by Joseph Banks on the *Endeavour* Voyage 1768–1771 Part 3: Zoology. Catalogue numbers 1.39 & 1.40 Courtesy of Trustee of the Natural History Museum, London.

CHAPTER 11

1. James Cook. A Plan of the entrance of Endeavour River – New South Wales © British Library Board ADD MS 7085, f. 42. (All following images in Chapter 11 are from this same source).
2. W.E. Roth. The Queensland Aborigines Vol. III, Plate IV.
3. Sydney Parkinson. The Longboat. Portion of Sketches of *Endeavour*'s boats © British Library Board ADD 9345, f. 21^v.

4. Charles Praval. A portrait of an Australian Aborigine © British Library Board ADD 15508, f.13, no. 15.
5. Ibid. (details)
6. A journal of a voyage to the South Seas: in His Majesty's ship, the *Endeavour*: faithfully transcribed from the papers of the late Sydney Parkinson, draughtsman to Joseph Banks, esq. on his late expedition with Dr. Solander around the world / embellished with views and designs delineated by the author. Page 152. Courtesy of the Mitchell Library, State Library of New South Wales.
7. Sydney Parkinson. Studies of Australian Aboriginal artefacts and other drawings. © British Library Board ADD 9345, f. 20.
8. Sydney Parkinson. The Yawl. Sketches of *Endeavour*'s boats © British Library Board ADD 9345, F.21.
9. Captain John Gore (1780), by John Webber. From Wikimedia Commons, the free media repository. Public Domain in UK. Also, PD-US.
10. Eastern wallaroo – *Macropus robustus*. Flickr – Ron Knight – sussexbirder.
11. Observations upon Animals made by the Naturalists of the *Endeavour* by G. B. SHARMAN, B.A., D.Sc. (1970) page 6. Courtesy of Mammalogy Team Australian Museum.
12. Charles Praval. Detail of A view of the land about Endeavour River taken when the entrance bore WSW distance 1 Mile © British Library Board ADD MS 7085, f. 41(b).
13. Observations upon Animals made by the Naturalists of the *Endeavour* by G. B. SHARMAN, B.A., D.Sc. (1970) page 6. Courtesy of Mammalogy Team Australian Museum.
14. H.M. Bark *Endeavour* by Ray Parkin page 347.

CHAPTER 12
1. Detail from Part of NEW HOLLAND. A draft of part of the shoals seen and sailed through by His Maj Bark the *Endeavour* in 1770 on the East Coast of New Holland by Richard Pickersgill. Courtesy of the Mitchell Library, State Library of New South Wales.
2. TOP Charles Praval. Detail of A view of the land about Endeavour River taken when the entrance bore WSW distance 1 Mile © British Library Board ADD MS 7085, f. 41(b). BOTTOM Present-day North Shore Beach and Mt Saunders © John MacDonald
3. James Cook. *A Plan of the entrance of Endeavour River – New South Wales* © British Library Board ADD MS 7085, f. 42.
4. Portion of Part of NEW HOLLAND. A draft of part of the shoals seen and sailed through by His Maj Bark the *Endeavour* in 1770 on the East Coast of New Holland by Richard Pickersgill. Courtesy of the Mitchell Library, State Library of New South Wales.

CHAPTER 13
1. James Cook. A Plan of the entrance of Endeavour River – New South Wales © British Library Board ADD MS 7085, f. 42. (All following plan images (and details of plan) in Chapter 13 are from this same source.)

CHAPTER 14
1. Portion of Part of NEW HOLLAND. A draft of part of the shoals seen and sailed through by His Maj Bark the *Endeavour* in 1770 on the East Coast of New Holland by Richard Pickersgill. Courtesy of the Mitchell Library, State Library of New South Wales.
2. Sydney Parkinson. Unfinished drawing *(Calyptorhynchus magnificus magnificus)* Banksian cockatoo – Catalogue of the Natural History drawings commissioned by Joseph Banks on the *Endeavour* Voyage 1768–1771 Part 3: Zoology. Catalogue number 10. (1:10) Courtesy of Trustee of the Natural History Museum, London.

3. Sydney Parkinson. Unfinished drawing *(Dasyurus hallucatus)* Quoll – Catalogue of the Natural History drawings commissioned by Joseph Banks on the *Endeavour* Voyage 1768–1771 Part 3: Zoology. Catalogue number 2. (1:2) Courtesy of Trustee of the Natural History Museum, London.
4. Grey Queensland Ring Tail Possum – *Pseudocheirus peregrinus* Ref. https://www.mammalage.com/contact/
5. Eastern Grey Kangaroo – *Macropus giganteas*. Atlas of Living Australia. Supplied by Leo Berzins.
6. Observations upon Animals made by the Naturalists of the *Endeavour* by G. B. SHARMAN, B.A., D.Sc. (1970) page 6. Courtesy of Mammalogy Team Australian Museum.
7. Sydney Parkinson. *Macropus sp.* 'Kanguru'. Animal springing. Catalogue of the Natural History drawings commissioned by Joseph Banks on the *Endeavour* Voyage 1768–1771 Part 3: Zoology. Catalogue number 4. (1:4) Page 35 Courtesy of Trustee of the Natural History Museum, London.
8. Sydney Parkinson. *Macropus sp.* 'Kanguru'. Animal crouched. Catalogue of the Natural History drawings commissioned by Joseph Banks on the *Endeavour* Voyage 1768–1771 Part 3: Zoology. Catalogue number 3. (1.3) Page 33. Courtesy of Trustee of the Natural History Museum, London.
9. Bulletin of the British Museum – Zoology Vol. 1 (1950–1953) Article 3, plate No 5.
10. Eastern Grey kangaroo – *Macropus gigantea*. Image Duncan McCaskill – creative commonsmRef. https://australian.museum/learn/animals/mammals/eastern-grey-kangaroo/#gallery-thumbnail
11. Observations upon Animals made by the Naturalists of the *Endeavour* by G.B. SHARMAN, B.A., D.Sc. (1970) page 6. Courtesy of Mammalogy Team Australian Museum.
12. James Cook. Portion of Cook's chart Cape Tribulation to Endeavours Streights by Lieutenant J. Cook Commander of the *Endeavour*. © British Library Board ADD MS 7085, f. 39.

CHAPTER 15

1. Portion of Cook's chart Cape Tribulation to Endeavours Streights by Lieutenant J. Cook Commander of the Endeavour. (This and the following charts in Chapter 15 are from this same source.) © British Library Board ADD MS 7085, f. 39.
2. Charles Praval. A View of the Land between Gores Mount and Cape Bedford taken from Turtle Reef © British Library Board ADD MS 7085, f. 41(c).
3. Charles Praval. A view of Cape Flattery bearing West 1 league © British Library Board ADD MS 7085, f. 41(d).
4. Detail from Part of NEW HOLLAND. A draft of part of the shoals seen and sailed through by His Maj Bark the *Endeavour* in 1770 on the East Coast of New Holland by Richard Pickersgill. Courtesy of the Mitchell Library, State Library of New South Wales.

CHAPTER 16

1. Detail from Part of NEW HOLLAND. A draft of part of the shoals seen and sailed through by His Maj Bark the *Endeavour* in 1770 on the East Coast of New Holland by Richard Pickersgill. Courtesy of the Mitchell Library, State Library of New South Wales. (This and all following plan images in Chapter 16 are from this same source.)

CHAPTER 17

1. Detail from Part of NEW HOLLAND. A draft of part of the shoals seen and sailed through by His Maj Bark the *Endeavour* in 1770 on the East Coast of New Holland by Richard Pickersgill. Courtesy of the Mitchell Library, State Library of New South Wales.
2. Charles Praval. A view of the Islands of Direction taken at the entrance of the Channel without the Reef © British Library Board ADD MS 7085, f. 41(e).
3. Portion of Cape Tribulation to Endeavours Streights by Lieutenant J. Cook Commander of the *Endeavour*. © British Library Board ADD MS 7085, f. 39.
4. Detail of Cape Tribulation to Endeavours Streights by Lieutenant J. Cook Commander of the *Endeavour*. © British Library Board ADD MS 7085, f. 39.
5. Portion of Cape Tribulation to Endeavours Streights by Lieutenant J. Cook Commander of the *Endeavour*. © British Library Board ADD MS 7085, f. 39.

CHAPTER 18

1. Portion of Cape Tribulation to Endeavours Streights by Lieutenant J. Cook Commander of the *Endeavour*. © British Library Board ADD MS 7085, f. 39. (This and the following charts in Chapter 18 are from this same source.)
2. A Chart of New South Wales, or the East Coast of New Holland. Discover'd and Explored by Lieutenant J. Cook, Commander of his Majesty's Bark *Endeavour*, in the Year MDCCLXX. Engraved by W. Whitchurch after Cook. Hawkesworth III, f. p.481. Courtesy of the Mitchell Library, State Library of New South Wales.
3. Extract from Cook's original journal – Journal of HMS *Endeavour*, 1768–1771 [manuscript] Page 753 fol. Bib ID 3525402 Courtesy of National Library of Australia.
4. Extracts from Joseph Banks – *Endeavour* journal, 15 August 1769–12 July 1771. Pages 291 & 292. Courtesy of the Mitchell Library, State Library of New South Wales.

EPILOGUE

1. This and the following seven maps (with minor alterations) are reproduced with permission from the Australian National University and National Library of Australia. https://webarchive.nla.gov.au/awa/20091011115149/http://southseas.nla.gov.au/journals/maps/05_atlantic.html
2. Batavia around 1780. Collectie Trope Museum De stad Batavia TM nr 3728-537. From Wikimedia Commons, the free media repository.

Acknowledgements

This is a work built on the labours of others.

First and foremost, we acknowledge the four journal keepers, the authors of this work – James Cook, Joseph Banks, Sydney Parkinson and James Mario Magra – who endured such hardship and danger and yet delivered such incredibly descriptive documents for our information and entertainment. How they did this under the conditions they endured simply beggars belief.

Closely following, we acknowledge those other mariners Richard Pickersgill, Robert Molineux, Daniel Solander and Herman Spöring, whose works we have rummaged for this publication.

Next come the great authorities like J. C. Beaglehole, Ray Parkin, and the Hakluyt Society on whom all Cook disciples depend so heavily.

Then the great museums and libraries; the Natural History Museum, Library and Archives (UK), British Library, National Library of Australia, National Museum of Australia, State Library of NSW (Mitchell Library), Queensland Museum, Museums Victoria, Greenwich and Australian Maritime museums, University of Otago (Hocken Collections) and Wells' Cathedral Library, UK (thanks Stuart), all who safeguard and share so generously the treasures the explorers brought back with them.

Our particular thanks to the Australian National University and National Library of Australia for their joint permission to reproduce their splendid Trove maps.

The outstanding Captain Cook Memorial Museum, Whitby, for all its insightful exhibitions (thanks Sophie Forgan).

So too those societies and associations and other bodies that keep the Cook story alive, and celebrate the explorers' achievements, like the Captain Cook Society (Australian and International) where those scholars John Robson, Cliff Thornton, Ian Boreham and Mal Nicolson ceaselessly work, and whose depth of knowledge has added so much to our understanding of the *Endeavour* story.

Our special thanks to John Robson (Captain Cook Society) for his advice, and for allowing us to plunder his monumental biographical work about those who sailed with Cook.

The *Endeavour* Replica, whose crew and skipper Chris Blake gave me a taste of the real thing.

Bob Hornsby, himself a replica builder, who fire-carved the Guugu Yimithirr canoe to recreate the other vessel that harboured alongside HM Bark *Endeavour* in Waalmbal Birri in 1770.

Eric Deeral, Senior Elder, sage and peacemaker, who laid the foundation stone of reconciliation in the Cooktown community. And Peter Scott, our Captain, whose leadership and example keeps the community on course; it's sails full set to catch that healing wind.

Erica Deeral, whose quiet water erodes the hardest prejudice.

Coleridge Bowen, strong man, inspirer, troubadour with song in his heart.

The Cooktown Re-enactment Association, especially Loretta Sullivan and Alberta Hornsby, that formidable and hard-working duo whose call for reconciliation from far north QLD carried right across the country to far away Canberra. What they achieved is here to stay.

Those current re-enactors who annually bring Cook and his companions alive and back to Endeavour River.

Our thanks to those many gallant re-enactors and their assistants, who were part of the golden years, when they broke the ice of a frozen history and brought us a true account of Cook's visit.

Thanks Lewis Jack, Dave Naylor, Fred Deeral, Robbie Harrigan, Mervin Yoren, William Bradley, Algon Naylor, Kegan Olbar, Damara Deeral, Dayle Kulka, Jake Stephens, Mary Cobus, Ginger Burns, Caroline Scotford Bowen, Jennifer Graff, Ray White, Carl Vogler, Jason Carol, Siezar Dewaal, Barry Bradley, Eric Batts, Chris Woltjen, Nelson Conboy, Phil Thompson, Ian White, Ty Leet, David Curtis, David Wilson, Richard Ball, Damon Ferdinand, Ethan Sieverding, Robert Scanlan, Evan Wong, Alex Dunn, Chris Voase, Sean Gillen, Paul Chang, Tony McKowen, Ben Wright, Greg Wagenmakers, Zak Wain, David Kamholtz, Chris Vela, Lauchlin Hook, Tyson and Amber Hang, Toby Gillen, Halfpenny, Ship's goat, Richard Lee, Serge Petelin, Darren Fietz, Rhonda Hill and Robin, Cathy Adams, Jill Bertwistle, Agata, Kimberly Sullivan, Vanessa Gillen, Allie Jenkin, Trevor Bambie, Helena Loncaric and the Pathway students, Stephen Gapps, and Peter Betino.

The Cooktown Council, Lee Robertson and Hopevale Council, Arts Queensland, Indigenous Coordination Centre Cairns, Australian government, Queensland Government, Diana (Cooktown Library), Troy Dennis (Hope Vale Arts and Crafts Centre), Pauline Voase (CDEP Hope Vale), Bev Grant, Jim Dolge, Barry Bowen, Tisha Gordon, Lawrence Deeral and the Hope Vale PCYC (Tenile), Fran Maddern, Peter Herman, Jim Symes, Billy Hyde Sound, Sam Neilly, Nico, Lyall and Pam (Cooktown Cruises), Charlie Martin, Billy Hodsoll (Treelopper Billy), Narelle Hine (Cooktown Local News), John and Nick (The Italian Restaurant), Alison Martin, Cooktown Police, Debra Taylor, Jullian

Blennerhasset (Cooktown State School), Karen Whippers, Roy McGuffle, Dianna Burns, Margaret Craker, Michel Chatenay, Jenny Vela, Des King, Sandy Lloyd, Judy Bennett and Willie Gordon (Guurrbi Tours), Penny Johnson (James Cook Museum), Helen Greaves (Computer Stuff), Bob Sullivan, Jacqui, Gary Ashworth, Nikki Gong, Carol Nunn, David Baker, Lynette Cockatoo, Lauren Erickson, Wenches, David, Tom and Robin (Apple UK), Mark (Apple Aust), One Ski Digital Media productions and Kate.

We thank our generous and talented friends Jason and Helen Daisley, who inspired and supported us and skippered us through those exciting Cook navigated seas.

Michael McCallion, who united our partnership, forever in our thoughts.

Antonia Macarthur and Ken Kelso, who set me on this adventure, and by whose example, encouragement and support kept me at the helm, my lasting thanks.

My father, who taught me the sea (Ronald Sommerville MacDonald RAAF S/Ldr 1916–2000).

My oldest mate Toot (coincidently the same name was used by the Tahitians when addressing Cook), and my son Harry, who lent their luminous voices to the website version of this project – https://captain-cook-continent-of-smoke.com – and our other son Charlie, who was always on hand to help with brilliant suggestions and solve numerous technical problems. Special and lasting thanks to Dennis Buck (Volunteer at the Captain Cook Memorial Museum, Whitby), George Fussey (Curator, Natural History Museum, Eton College) and Derek Smith (Seaman). These three heroes took on the marathon task of recording the voices of Cook, Banks and Parkinson for the *Endeavour*'s entire east coast journey for the website. Their contributions, dedication and patient endurance have brought to life the voyage as never before.

To those many others – Winfried Bruenken, Geni, Tess Egan, Simon (Museums Victoria), Donna (QM Images), Kylie Morrow (Atlas of Living Australia), Deborah Dickson-Smith (Diveplanit Travel), Schomynv, J. J. Harrison, John Tan, Ixitixel, Mike Prince, Aviceda, Eric Guinther, PHGCOM, Jeff Wright (Queensland Museum), Jinesh PS, Ben (Ben's Waterfowl), Gopala Krishna, Bernard Dupont, Robert Webster, Daniela, Nick Hobgood, Dominic Sherony, Julia Burgher, Andrew Mercer, Wildfeuer, Miya.m, Tatiana Gerus, Florence Trentin, fearlessRich, Dr Avishai Teicher, B.J.Hensen, Djambalawa, Marcus Stigwan, Mudasir Zainuddin, Strobilomyces, Mark Marathon, Stev Fitzgerald, Anne Hoggett and Nate Lawrence (Lizard Island Research Station), Toby Hudson, Andrew Green, Chaloklum, Ron Knight, Leo Berzins and Duncan McCaskill.

To Wikimedia Commons, the free media repository, Public Domain, and to those numerous public-spirited individuals who have allowed free access to their work on Wikimedia Commons, we salute you. Without you this work would never have seen the light of day.

To others, who by unintentional oversight find no mention here, whose work and expertise have informed this project, you know who you are, we beg your indulgence and thank you for your contributions.

And last, but not least, we wish to thank our publishers Pen and Sword Books, for their faith in taking us on, and for their flexibility and patience. Thanks Harriet, Charles, Jon and Gaynor.

I also want to thank my wife Clare who taught herself from scratch all the skills of modern computing and graphic art needed to make this project possible. The brightest star in my firmament, and my star for steering by.

> John and Clare MacDonald – The Cooktown Re-enactment Association